List of Symbols

$\boldsymbol{A}$	electromagnetic wave amplitude
$\boldsymbol{B}$	magnetic field
$\hat{\boldsymbol{d}}$	operator of the electric dipole moment
e	electron charge
$e_{\alpha\beta\gamma}$	Levi–Civita tensor, symbol
$\boldsymbol{E}$	electric field
$h = 2\pi\hbar$	Planck's constant
$\hat{R}^{il}$	spatial dispersion operator
$\hat{H}^{int}$	Hamiltonian of the light-matter interaction
$\hat{H}_0$	Hamiltonian of molecular electrons
$\boldsymbol{J}$	current density
I	light intensity
k, $\boldsymbol{k}$	wave number, wave vector
L	Lagrange function of the optical electrons, length
$\hat{\boldsymbol{m}}$	operator of the magnetic dipole moment
m	electron mass
n	complex refractive index
N_0	number density of the molecules
$\boldsymbol{P}$	electric field polarization response (electrical polarization)
$\hat{q}^{ij}$	operator of the quadrupole moment
$\boldsymbol{r}$	coordinate
r	degree of polarization for partially polarized light
$\boldsymbol{S} = \{S_1, S_2, S_3\}$	Stokes vector
$S_\alpha = \{S_0, S_1, S_2, S_3\}$	Stokes parameters
t	time
U	potential energy of the electrons
$\boldsymbol{v}$	electron velocity
$\mathscr{A}$	vector potential
δ_{ij}	Kronecker delta symbol
γ	damping parameter of the oscillator
$\Gamma^{(1)}_{ijk}$, Γ_{ijk}, $\Gamma^{(1,s)}_{ijk}$, $\Gamma^{(1,a)}_{ijk}$	nonlocality tensor, symmetric and anti-symmetric parts
$\Gamma^{(2)}_{ijkn}$, Γ_{ijkn}	second order nonlocal susceptibility
$\Gamma^{(3)}_{ijknl}$, Γ_{ijknl}	third order nonlocal susceptibility
ϵ_{ij}	dielectric tensor
η	ellipticity angle of the polarization ellipse
Θ	polarization azimuth of the polarization ellipse
λ	wavelength
$\boldsymbol{\rho}$	relative coordinate
$\sigma^{(i)}$	the Pauli matrices
τ	time delay between two events
$\chi^{(2)}_{ijk}$, χ_{ijk}	second order local susceptibility
$\chi^{(3)}_{ijkn}$, χ_{ijkn}	third order local susceptibility
ψ	medium response function
ω	angular frequency of electromagnetic oscillation
$\boldsymbol{\Omega} = \{\Omega_1, \Omega_2, \Omega_3\}$	self-action vector
ω_0	natural frequency of the oscillator
$\boldsymbol{\Omega}^L = \{\Omega^L_1, \Omega^L_2, \Omega^L_3\}$	self-action vector in linear case

Polarization of Light in Nonlinear Optics

Polarization of Light in Nonlinear Optics

Yu. P. Svirko and N.I. Zheludev
Department of Physics and Astronomy,
University of Southampton, UK

JOHN WILEY & SONS
Chichester · New York · Weinheim · Brisbane · Singapore · Toronto

Baffins Lane, Chichester,
West Sussex PO19 1UD, England

National 01243 779777
International (+44) 1243 779777
e-mail (for orders and customer service enquiries): cs-books@wiley.co.uk
Visit our Home Page on http://www.wiley.co.uk
or http://www.wiley.com

Other Wiley Editorial Offices

John Wiley & Sons, Inc., 605 Third Avenue,
New York, NY 10158-0012, USA

WILEY-VCH Verlag GmbH, Pappelallee 3,
D-69469 Weinheim, Germany

Jacaranda Wiley Ltd, 33 Park Road, Milton,
Queensland 4064, Australia

John Wiley & Sons (Asia) Pte Ltd, Clementi Loop #02-01,
Jin Xing Distripark, Singapore 129809

John Wiley & Sons (Canada) Ltd, 22 Worcester Road,
Rexdale, Ontario M9W, 1LI, Canada

Library of Congress Cataloging-in-Publication Data
Svirko, Yu P.
Polarization of light in nonlinear optics / Yu P. Svirko and N.I. Zheludev.
p. cm.
Includes bibliographical references and index.
ISBN 0-471-97640-7 (cloth : alk. paper)
1. Nonlinear optics. 2. Polarization (Light) I. Zheludev, N. I. II. Title.
QC446.2.S95 1998
535.5′2—dc21 97-48746
CIP

British Library Cataloguing in Publication Data
A catalogue record for this book is available from the British Library

ISBN 0 471 97640 7

Typeset in 10/12pt Times by Keytec Typesetting Ltd, Bridport, Dorset
Printed and bound in Great Britain by Biddles Ltd, Guildford, Surrey
This book is printed on acid-free paper responsibly manufactured from sustainable forestry, in which at least two trees are planted for each one used for paper production

To Elena, Ilya and Ivan

Contents

1 Introduction: Polarization of Light in Spectroscopy and Fundamental Physics

Nonlinearity in dissipative media shall be observable not only with respect to absorption. This absorption is linked to dispersion, therefore generally speaking, the speed of light in the medium shall also depend on the light's power. For the same reason, generally, a dependence on the light's power, i.e. a violation of the superposition principle, must be seen in other optical properties of the medium, such as birefringence, dichroism, polarization rotatory power, etc.

S.I. Wawilow, 1950

The observer should hold a blue filter and a polaroid before his eye and view a luminous filed . . . for a sufficient interval of time to allow the brushes seen at first completely to fade away. He should then suddenly remove the polaroid but allow the blue filter to remain in place. He will then see the brushes once again, but turned through a right angle. In other words the fovea then perceives with enhanced brightness that part of the incident light which was cut off by the polaroid . . .

Sir C.V. Raman, 1968

The building of crystal optics which erection began more than a century ago might only be completed if the spatial dispersion is taken into account

V.M. Agranovich and V.L. Ginzburg, 1979

In this chapter we will provide a short introduction to the study of linear and nonlinear polarization effects in this century.

The study of the polarization state of light offers unique spectroscopic opportunities. These arise from the astonishing accuracy and virtually unlimited dynamic range of polarization measurements and the amazing sensitivity of the light polarization state to the optical properties of a medium. This sensitivity is a result of polarization phenomena being essentially vectorial interference effects. Since the beginning of the twentieth century, polarization spectroscopy has been systematically used, and has produced exciting results, in materials science, chemistry, biology and physics. A number of excellent books have been written to describe polarization phenomena in linear optics including some recently published [1, 2]. Many other famous works such as [3–9] have payed a considerable attention to this subject.

Although nonlinear optics enjoyed comprehensive monographic coverage—we only mention a few very well known texts here [10–15], so far there has not been a single text devoted solely to polarization effects in nonlinear optics. Nevertheless, interest in polarization effects in nonlinear optics and spectroscopy is booming, a fact which is reflected, in particular, in the increasing number of research publications devoted to the subject every year. A new international conference, which concentrated on nonlinear polarization effects in lasers and spectroscopy (PELS), was launched in Toronto, Canada, in 1997, and a follow-up conference is scheduled to take place in Southampton, UK, in the year 2000 (for more information see: *http://www.phys.soton.ac.uk/PELS/INDEX.HTM*). The growing interest in nonlinear polarization effects and its exciting spectroscopic applications stimulated us to write this book, which presents a theoretical framework for describing the evolution of the polarization state of intense light in nonlinear media and discusses key experimental works in this field.

In this chapter we will give a short historical review of research in polarization phenomena in linear optics and those works in polarization nonlinear optics that are not covered in the rest of the book. We do not aim to give a comprehensive overview of this enormous and very diverse body of work. Rather, we intend to comment on some of the results which we believe were crucial for the development of the subject. In this chapter, where possible, we will try to avoid double references with the rest of the book.

Classical pre-laser polarimetry, or ellipsometry to use a term credited to Rothen [16], is a common spectroscopic technique that involves measuring the polarization state of light reflected from or transmitted through the sample under investigation. It is a technique that has been elaborated upon and refined to a very high level. From a very early stage it was recognized that one of the main fields where polarimetry could be applied was in the study of surfaces, pre-surface layers and interfaces. The ellipsometric concepts of reflection and transmission from bulk materials and both isotropic and anisotropic planar structures (including uniaxial and biaxial substrates and films, and multilayered film–substrate systems) were developed for these purposes. The reverse problem, in which ellipsometrically measured ratios of reflection or transmission coefficients are used to find the real and imaginary parts of the complex refractive coefficient and film thicknesses has also been investigated in detail. Modern inverse theories, based on numerical methods, account for the opportunities provided by multiple angle ellipsometry and spectroellipsometry and can deal with complex layered structures. Important early applications of ellipsometry included the study of the optical properties of materials in bulk, thin-film and multilayered phases. It was very successfully used in researching semiconductor interfaces such as the air–SiO_2–Si system. Other applications included the investigation of gas films condensed on to gold mirrors and the adsorption (including the dynamics of adsorption) of molecular and atomic species on to surfaces in contact with ambient gases or liquids. In these cases ellipsometry provided nondestructive and *in situ* measurements. It was used to investigate the chemisorption of oxygen on to cleaved silicon, other gases on to some metals and the oxidation of semiconductor and metal surfaces, particularly GaAs and other materials important to the electronics industry. In the study of the electrochemical electrode process, polarimetry provided an essentially nonperturbing method of analysis which could be used simultaneously with conventional electrochemical techniques. Interesting results have also been obtained in the study of discontinuous films and surfaces with rough boundaries. These concepts are all covered in Refs. [17], [18] and [19].

Spectropolarimetry, i.e. the study of polarization of light of different wavelengths interacting with the medium, has become a major tool in chemistry and life sciences. Organic chemistry deals mainly with bulk quantities of isotropic liquids, many of which show optical activity (i.e. the ability to rotate the polarization azimuth of a light wave). This rotatory power is due to the mirror image asymmetry (chirality) of molecules and is an intrinsic property of living material. In fact, most of the asymmetric molecules in living organisms are coiled in the manner of a left-handed spring and do not coinside with their mirror images. A natural scientist would recall Descartes' aphorism highlighting the extent of left–right asymmetry in nature: 'Any man who, upon looking down at his bare feet, doesn't laugh, has either no sense of symmetry or no sense of humor.' The study of polarization rotation due to the optical activity of living substances is a fundamental part of the ongoing inquiry into the origins of optical activity in nature [20–22]. On a more practical note, observations of natural and magnetic-field-induced optical rotatory dispersion and circular dichroism are invaluable in stereochemical studies, in particular in the study of steroids, ketones, amino acids, polypeptides and proteins, liquid crystals and organometallic compounds [23–26]. Magnetic circular dichroism and birefringence (the optical Faraday effect), induced by a static magnetic field via the Zeeman effect, are also widely used in the spectroscopy of biological and organic molecules. The study of natural optical activity and polarization rotation induced by magnetic fields now covers not only electronic but also vibrational transitions and Raman optical activity [27]. Another application of polarimetry in physical chemistry is polarized emission spectroscopy, a technique developed at the beginning of this century. It is based on the following idea: molecules/atoms radiatively decay from excited electronic states to ground states by fluorescence or phosphorescence via an intermediate level. The fluorescence polarization, as it depends on the orientation of the emission dipole, yields information about the orientation of molecules and carriers in solids. This technique has also become popular in solid state physics where it is known as hot luminescence polarimetry

[28, 29]. In medicine ellipsometry has become a unique research tool. Laser-based polarimetry offers significant advantages in the chiral analysis of organic materials using techniques based on high-performance liquid chromatography [30]. It is also applied to the study of immunological antigen–antibody reactions in thin films, the dynamics of blood coagulation and the interaction of blood with foreign surfaces. Scanning laser polarimetry has recently been employed for the detection of nerve fibre bundle defects [31].

Contrary to popular belief, the human eye is sensitive to the polarization state of light. In 1968 Sir C.V. Raman described an experiment which demonstrated this sensitivity at high intensity and in a limited spectral range, and wrote:

> One of the most remarkable of our visual facilities is the ability to recognize polarized light and to locate its plane of polarization ... though restricted to the blue-violet sector of the spectrum, the ability to detect polarization belongs to the same category of visual phenomena as the perception of light, form and colour and it stands in the closest relationship with these perceptions [32].

This sensitivity is, in fact, a nonlinear phenomenon and may be seen only in a very intense light. It is hardly noted by most people and such polarization sensitivity is, of course, insufficient for spectroscopic applications. Instead sophisticated apparatuses are used for the purpose of measuring light polarization accurately. The theory of polarimetry developed rapidly in conjunction with the advancement of polarization-sensitive instrumentation. In the 1960s and 1970s the null ellipsometer and then the automated null ellipsometer became routine and commercially available polarimetric instruments. More elaborate versions were developed including a motor-driven self-nulled ellipsometer [33] and a self-nulled ellipsometer with a Faraday cell [34] and electrooptic variable retarders [35]. The rotation analyser ellipsometer with analogue and digital signal detection was a very useful development (that described in Ref. [36] is a particularly good example). Polarimeters with piezobirefringent polarization modulators became popular [37] and digital processing of information allowed the use of low repetition rate lasers in precise polarization instrumentation [38].

Since the 1980s polarization measurements have been key elements in the study of fundamental symmetries and, in particular, in the study of symmetry breaking effects in physics (e.g. one can view polarization azimuth rotation in naturally optically active media as a manifestation of the lack of molecular mirror symmetry). In particular, optical polarimetry played an indispensable role in the study of parity violation in atoms. It was first pointed out by Yakov Zel'dovich that where parity is not conserved, rotation of the plane of polarization is possible in an isotropic substance that does not contain chiral molecules [39] (also see Refs. [40] and [41]). After the polarization effect was first predicted, a number of experimental investigations were launched with great enthusiasm at various laboratories. The observation of optical rotation, in experiments on bismuth at Novosibirsk, were reported in 1978 [42, 43]. Subsequently, optical rotation measurements were carried out on bismuth in Moscow [44], Oxford [45] and Seattle [46]; the last of these research groups also reported the effect in lead [47]. Coverage of later parity nonconservation rotation experiments in atoms can be found in publications [48] to [51] and the review articles [52, 53]. Indeed, optical experiments have demonstrated cases where the mirror symmetry of stable atoms is broken during absorption and emission of light. This is, however, a very small effect. As the light traverses a cell of atoms, even in very favourable conditions, its polarization azimuth rotates by only a very small amount, typically 10^{-8}–10^{-10} rad. These results support the theory of unification of electromagnetic and weak interactions. The interpretation of these experiments is based on the exchange of weak neutral Z_0 bosons between the electrons and the nucleus of an atom. The original experiments of this type complemented the discovery of weak currents in accelerator experiments. These polarization rotation experiments also convinced the scientific community that reliable spectroscopic polarization measurements with nanoradian sensitivities were possible.

Ellipsometric measurements were used to look for evidence of time-reversal symmetry breaking in high-temperature copper oxide superconductors. This broken reversibility is supposed to occur because of the alleged fractional statistics of the carriers [54], [55]. In this model known as the anyon model of superconductivity a spontaneous violation of the discrete symmetries of time reversal and two-dimensional reflection symmetry was expected due to the fractional statistics of carriers, while combined *PT* was thought to remain a valid symmetry. Broken symmetry could lead to an antisymmetric contribution to the dielectric tensor of the superconductor and may manifest itself as polarization rotation of light reflected from the medium. Experimental studies of this effect, however, produced conflicting results. Evidence of optical circular differential effects on a scale of 10^{-2}–10^{-3} rad was found in reflection measurements on a one-domain crystal of $Yba_2Cu_3O_{7-\delta}$ and transmission measurements on a single crystal of $Bi_2Sr_2CaCu_2O_8$ near the superconducting temperature [56]. Similar positive results were reported from materials such as $YBa_2Cu_3O_{7-\delta}$ films [57] (also see Refs. [58] and [59]). On the other hand, evidence against the optical manifestation of broken time-reversibility in high-T_c superconductors was reported in Refs. [60] and [61] from reciprocity experiments and reflectivity measurements on $Bi_2Sr_2CaCu_2O_8$ and Yba_2Cu_7. The question of the optical manifestation of broken time-reversibility in cuprate materials remains unsolved and is still an intriguing problem. The question of time-reversal symmetry of the medium Hamiltonian is not equivalent to that of time-reversal symmetry of the light–medium interaction Hamiltonian. It has been suggested [62] that the latter symmetry may be broken while the crystal Hamiltonian remains perfectly *P* and *T* symmetric. Such time nonreversal effects can only emerge in nonlocal light–matter interactions and also lead to a polarization effect. (see Appendices 4 and 5 of this book). As part of the investigation of symmetry violation, very sensitive polarization experiments have been developed to search for the electric dipole moments of atoms, molecules and neutrons. These experiments will play an important role in determining which theoretical model correctly describes the violation of the time-reversal invariance principle [63, 64]. An interesting experiment is planned to detect the magnetic-field-induced birefringence of a vacuum [65].

In 1950, i.e. 24 years after S.I. Wawilov and W.L. Lewshin reported the observation of the first nonlinear optical effect of departure from the Beer–Lambert absorption law with light of high intensity [66], Wawilow suggested that nonlinear polarization effects should be seen in birefringence, dichroism and polarization rotatory power (see the epigraph to this chapter) [67]. This was, of course, long before the invention of lasers and Wawilow believed that only the sort of conditions found inside stars could produce the huge light intensities that would make such effects possible. Nevertheless, it was his comments that first made nonlinear, light-intensity-dependent, polarization effects a matter of scientific discussion. Now we can ask a key question related to the subject of this book. What new nonlinear optics brought to the study of light polarization? A brief answer is: the intensity dependence of known polarization effects, polarization-sensitive interactions of two or more waves, polarization effects involving changes in the frequency of light and polarization instability. This book is devoted mainly to the first group of effects. We discuss the appearance of conventional polarization phenomena in intense light fields without frequency transformation and consider two wave nonlinear interactions leading to intensity-dependent changes in the polarization state of light. This will be discussed in Chapters 2 to 9. Here we will briefly review those works on the polarization effect in nonlinear optics that will not be covered in the following chapters.

With the invention of lasers, in particular pulsed q-switched and mode-locked lasers, polarization spectroscopy gained two new dimensions: it became possible to study nonlinear polarization effects using light waves of different frequencies and to achieve temporal resolutions that were impossible with conventional light sources. By manipulating the delay between laser pulses, one may study the dynamics of nonlinear polarization effects and measure transient optical nonlinearities (some references on transient polarization spectroscopy can be found in Chapter 7). It was only logical that newly discovered laser sources, such as mode-locked lasers, were employed to modify traditional ellipsometric techniques: reflective ellipsometry of semiconductors was greatly enhanced by the use of picosecond lasers (see Fig. 1.1). With picosecond time resolution it became possible to measure the transitional dynamics of the electron–hole plasma in semiconductors excited to a nonstationary state by a short light pulse [68]. By manipulating the frequencies of interacting waves one can study the complex dispersion characteristics of optical nonlinearities. One of the most powerful nonlinear spectroscopic techniques is four-wave-mixing spectroscopy where three waves, generally of different frequencies ω_1, ω_2 and ω_3, interact with the

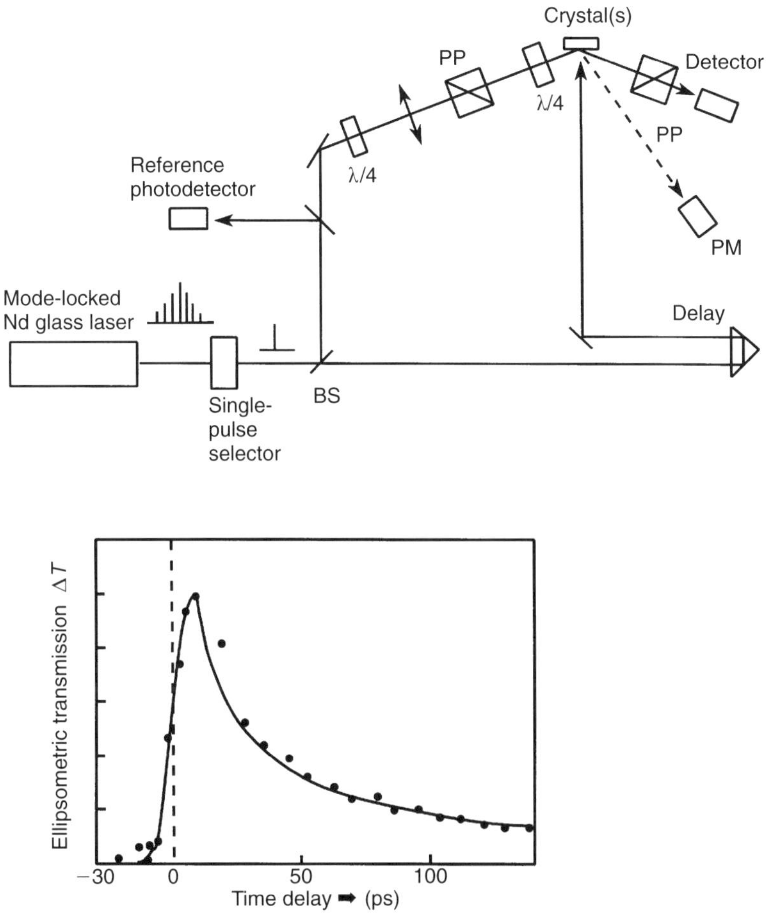

Fig. 1.1 Transient optically induced change δn in the refractive index of germanium n is measured by monitoring the polarization state of a picosecond probe light pulse reflected from the crystal interface. The ellipsometer transmission increases when a pump light pulse creates nonequilibrium free carriers. The ellipsometer transmission $\Delta T \propto \delta n/n$ depends on the density of free carriers. It was measured as a function of the delay between optical excitation and the probe pulse with picosecond accuracy. (Reproduced by permission of the American Physical Society from Ref. [68])

nonlinear medium to generate a fourth wave, at a frequency $\omega = |\omega_1 \pm \omega_2 \pm \omega_3|$. In particular, the scheme know as coherent anti-Stokes–Raman spectroscopy (CARS) has become very popular. In CARS the molecules are excited at the beat frequency $\Omega = \omega_1 - \omega_2$ between the two incident waves. Scattering of the wave at frequency ω_1 on the molecular excitation leads to the generation of a new coherent signal at the anti-Stokes frequency $\omega_a = 2\omega_1 - \omega_2$.

Thus, this is a process in which the Raman scattering is not a product of thermal fluctuations of the molecular oscillators, but is due to specifically prepared, coherent molecular oscillations [15, 69]. In the polarization version of CARS the polarization state of light scattered at frequency $\omega_a = 2\omega_1 - \omega_2$ is detected while $\omega_1 - \omega_2$ is scanned across the molecular resonance. The real advantage of the polarization measurement here is that the polarization state of the signal at ω_a does not depend on the intensity of the waves involved in the nonlinear process. As a result, the polarization spectra can be accurately measured with fluctuating pulse lasers. For example, Fig. 1.2 shows ellipsometric spectra for the anti-Stokes signal from bensole in the proximity of the strong 1178 cm^{-1} Raman line [70]. Polarization CARS spectra can be measured where strong broad fluorescence makes spontaneous Raman spectroscopy impossible [71]. By adjusting the polarization states of the interacting beams the nonresonant background signal can be appreciably suppressed [72]. Polarization CARS is now a well-established spectroscopic technique used in chemistry, biology and life sciences [73–76, 69]. Another form of polarization-sensitive four-wave-mixing spectroscopy is Raman-induced Kerr effect spectroscopy (RIKES). In RIKES there are two waves at frequencies ω_1 and ω_2 with the frequency $\omega_1 - \omega_2$ close to a combinational molecular resonance. If the probe wave is initially linearly polarized, it becomes elliptically polarized after passing through the medium. This ellipticity appears as a result of the birefringence induced by the presence of the waves at ω_1 and ω_2. Here the polarization spectra are measured when the frequency $\omega_1 - \omega_2$ is scanned across a Raman resonance of the medium. Thus RIKES is a form of four-wave-mixing process with partially degenerate frequencies: $\omega_2 = \omega_1 - \omega_1 + \omega_2$, (also see Refs. [69, 77]). This effect is relevant to the frequency degenerate pump–probe optical Kerr effect. Together with other frequency degenerate pump–probe polarization effects it will be analysed in Chapter 7.

Measurements of the polarization state of the second harmonic signal can be used to study surface magnetism and probe the chiral properties of surfaces and thin films. In this instance the polarization properties of light at twice the frequency of the incident wave are examined. It has been shown that surface magnetization can induce a number of nonlinear susceptibility elements, responsible for the second harmonic generation process, that would otherwise vanish [78]. These induced elements are shown to be detectable using second harmonic generation with appropriate polarization combinations so as to change the polarization state of the second harmonic signal. The influence of surface magnetization was first seen at an Fe(001) surface [79] . A much stronger effect was then observed in an Fe/Cr multilayer on an SiO_2 substrate where second harmonic rotation of 22° was detected [80]. Similar experiments were later performed using cobalt on Cu(001) and using Cu/Co/Cu(001) multilayers [81]. Polarization effects have been observed in second harmonic generation from chiral molecules absorbed at an air/water interface or into solid substrates and

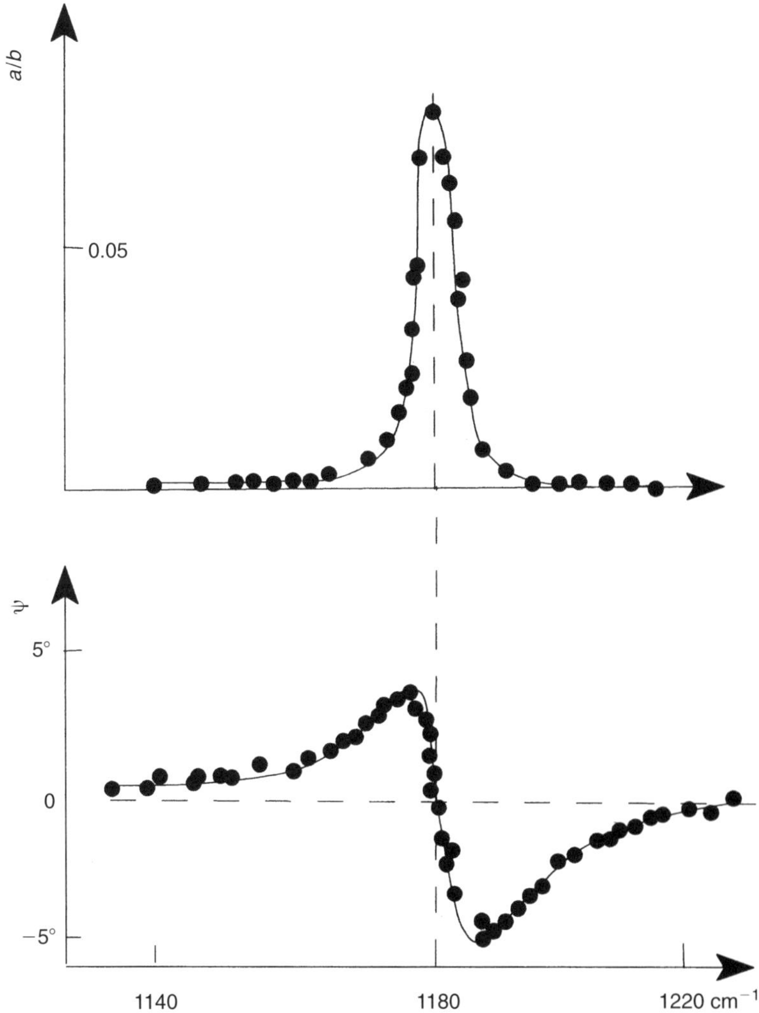

Fig. 1.2 Polarization state of the CARS signal in benzol when the frequency $(\omega_1 - \omega_2)/(2\pi c)$ is scanned across a Raman resonance of the medium. The upper curve presents the polarization ellipse axes ratio, while the lower curve shows the polarization azimuth. (Reproduced by permission of American Institute of Physics from Refs [70] and [69])

from Langmuir–Blodgett films [82–85]. This technique can be used to uniquely determine the relative complex values of the nonlinear coefficients and thus to completely characterize the nonlinear chiral response [86]. In certain circumstances it is possible to distinguish between the enantiomers of a chiral sample.

We are unable, in this book, to explore in detail a very exciting area of nonlinear optics, namely that of dynamical polarization phenomena and instability in lasers. The history and state of the art of this subject was recently covered in the excellent reviews [87–91]. Polarization instability in lasers was reported in a number of early works. The first paper published on the subject mentioned that the polarization of light emitted by a gas laser jumps, seemingly at random, between two orthogonal states (this gives rise to the expression 'randomly polarized lasers') [92]. This polarization flip was then observed using an He–Ne laser in the 1.15 μm [93] and 632 nm [94] lines of neon and later in the 3.39 μm emission line [95]. These lines are naturally linearly polarized. Emission switching between two circularly polarized modes at 1.52 μm was also seen using an He–Ne laser [96]. There are numerous works on laser polarization dynamics (for more references and more recent results see [99] to [105]).

Polarization instability in passive nonlinear optical devices (i.e. not lasers) is another fascinating topic which we will only touch on in Chapter 5. Initially, experimental and theoretical publications on polarization instability in atomic gases were the results of a study of two-level atoms with a degenerate Zeeman ground state [106]. If a gas of such atoms was placed in a Fabri–Perot cavity and pumped by intense, linearly polarized, resonant light, the polarization of the output radiation could assume either a linear polarizaton or right- or left-handed elliptical polarization, with the linear polarization being unstable. Such a system thus tends towards spontaneous polarization symmetry breaking at high intensities. An adiabatic reduction in light intensity then produces a switching of the polarization to a linear state. The first experiments of this sort were performed at the $^2S_{1/2} \rightarrow {}^2P_{1/2}$ transition in sodium vapour [107]. Self-oscillations of the polarization state in a similar system in a magnetic field were reported in Refs. [108] and [109]. In the works that followed, various aspects of the polarization instability in optical resonators (or with other forms of optical feedback such as a retro-reflecting mirror) were investigated. The polarization state was observed to behave in a complicated chaotic manner with 'windows' of periodic polarization oscillations [110–118].

Polarization instability and multistability effects were also studied in cubic and birefringent crystals [119–131] and an introduction into the origins of polarization instability in cubic crystals can be found in Chapter 5. Polarization instability has also been theoretically investigated in media with nonlinear gyrotropy (nonlinear optical activity, see Chapter 4) [131–133] and in systems showing strong two-photon absorptions [134] such as biexcitonic resonances in semiconductors [135–137]. The self-rotation of an intense light wave's polarization ellipse is the most common and comprehensively investigated the nonlinear polarization effect. A detailed account of this effect is given in Chapter 4. No polarization instability develops if a single propagating wave is considered, but a spatial polarization instability appears when two counter-propagating waves interact with nonlinear isotropic media [138, 139]. When the waves are orthogonally linearly or circularly polarized, the polarization of the waves does not change upon propagation. In other cases, the coupling between counter-propagating waves leads to periodic, multistable or chaotic spatial distributions of polarization in each beam [140–142]. If the medium nonlinearity is noninstantaneous, polarization instability in the time domain and even polarization chaos can be predicted [143, 144]. Further developments in this area have included the introduction of the 'polarization soliton' [145], 'polarization domain' [146] and spatial polarization in stability [103, 147, 148]. Some important works on polarization instability in birefringent media will be referred to in Chapter 5, but for more references see Refs. [149] to [156]. A review of the subject of polarization instability, covering early works can be found in Ref. [157].

In this book we will concentrate on the foundations of the polarization-sensitive nonlinear optics. We examine effects such as polarization ellipse self-rotation, nonlinear optical activity, the nonlinear anisotropic effect, induced gyrotropic dichroism. We will speak about the pump–probe effects known as the inverse Faraday effect, the optical Kerr effect, the pump–probe nonlinear anisotropic effect and pump–probe induced gyrotropic dichroism. We also analyse the appearance of these effects in the polarization state of light reflected from nonlinear media and discuss polarization effects in partially polarized light. We will develop a systematic approach to the phenomenological and microscopic theory of the optical nonlinearities involved in these polarization effects.

The authors of this book are grateful to the Leverhulme Trust and NATO for the financial support which allowed us to concentrate on the preparation on the manuscript in the UK. Y.P. Svirko is permanently with General Physics Institute, Russian Academy of Science, Moscow.

We are also deeply grateful to our colleagues, research associates and former research students who worked side by side with us on the experimental and theoretical aspects of the polarization of light and without whom this book would not have been possible, in particular to S.A Akhamnov, S.P. Apanasevich, P. Bennett, A.R. Bungay, F.V. Bunkin, D. Dhanjal, G. Fillipova, O. Garibian, Z.M. Kostov, A.I. Kovrigin, V.I. Kuznetsov, G.A. Liakhov, M. Kuwata-Gonokami, V.A. Makarov, A. Malinowski, L. Meisner, D. Paraschuk, A.D. Petrenko, S.V. Popov, I.S. Ruddock, S. Saltiel, I.S. Shatwell, R.S. Zadoyan, B.V. Zhdanov. We appreciate the valuable technical help of K. MacDonald with preparation of the manuscript.

REFERENCES

1. E. Collett, *Polarization of Light: Fundamentals and Applications*. Marcel Dekker, New York, 1993.
2. S. Huard, *Polarization of Light*, John Wiley & Sons, Chichester, 1997.
3. M. Born and E. Wolf, *Principles of Optics*, Pergamon Press, Oxford, 1970.
4. R.W. Ditchburn, *Light*. Academic Press, London, 1976.
5. R.D. Guenter, *Modern Optics*. John Wiley & Sons, New York, 1990.

6. S.G. Lipson, H. Lipson and D.S. Tannhauser, *Optical Physics*. Cambridge University Press, Cambridge, 1995.
7. S.A. Akhmanov and S. Yu. Nikitin, *Physical Optics*, Clarendon Press, Oxford, 1997.
8. V.M. Agranovich and V.I. Ginzburg, *Crystal Optics with Spatial Dispersion, and Excitons*. Springer-Verlag, Berlin, 1984.
9. L. Mandel and E. Wolf, *Optical Coherence and Quantum Optics*. Cambridge University Press, Cambridge, 1995.
10. S.A. Akhmanov and R.V. Khohlov, *Problems of Nonlinear Optics*, Gordon and Breach, New York, 1972.
11. N. Boembergen, *Nonlinear Optics*, Benjamin, New York, 1965.
12. P.N. Butcher and D. Cotter, *The Elements of Nonlinear Optics*. Cambridge University Press, Cambridge, 1990.
13. R.W. Boyd, *Nonlinear Optics*. Academic Press, San Diego, 1992.
14. A. Yariv and P. Yeh, *Optical Waves in Crystals*. John Wiley & Sons, New York, 1984.
15. Y.R. Shen, *The principles of Nonlinear Optics*. Wiley-Interscience, New York, 1984.
16. A. Rothen, *Rev. Sci. Instrum.*, **16**, 26 (1945).
17. R.M.A. Azzam and N.M. Bashara, *Ellipsometry and Polarized Light*, North-Holland, Amsterdam, 1977.
18. D.S. Klinger, J.W. Lewis and C.E. Randall, *Polarized Light in Optics and Spectroscopy*, Academic Press, Boston, 1990.
19. D.E. Aspnes, *Surface Science*, **101**, 84 (1980).
20. D.C. Walker (ed.), *Origins of Optical Activity in Nature*, Elsevier Scientific Publishing, Amsterdam, 1979.
21. V.I. Goldanskii, *Phys. Chem. A*, **101**, 3424 (1997).
22. S.F. Mason, *Molecular Optical Activity and the Chiral Discriminations*, Cambridge University Press, Cambridge, 1982.
23. F. Ciardelli and P. Salvadori (eds.), *Fundamental Aspects and Recent Developments in Optical Rotatory Dispersion and Circular Dichroism*, Heyden & Son Ltd, London, 1973.
24. C. Djerassi, *Optical Rotatory Dispersion*, McGraw-Hill, New York, 1960.
25. G. Snatzke (ed.), *Optical Rotatory Dispersion and Circular Dichroism in Organic Chemistry*, Heyden & Son Ltd, London, 1967.
26. V.I. Sokolov, *Chirality and Optical Activity in Organometallic Compounds*, Gordon and Breach Science Publishers, New York, 1990.
27. L.D. Barron, *Molecular Light Scattering and Optical Activity*, Cambridge University Press, Cambridge, 1982.
28. B.P. Zakharchenya, *J. Luminescence*, **24**, 669 (1981).
29. A.O. Anijalg, P.M. Saari, T.B. Tamm, K.E. Timpmann and A.M. Freiberg, *Kvantovaya Electronika*, **9**, 2449 (1982).
30. D.M. Goodall, *Trac-trends in Analyt. Chem.*, **12**, 177 (1993).
31. M. Mathieu and J. Funk, *Investigative Ophthalmology and Visual Sci.*, **37**, 5029 (1996).
32. C.V. Raman, *The Physiology of Vision*, The Indian Academy of Sciences, Bangalore, 1968.
33. H. Tkazaki, *Appl. Opt.*, **5**, 759 (1966).
34. H.J. Mathieu, D.E. McClure and R.H. Muller, *Rev. Sci. Instrum.*, **45**, 798 (1974).
35. T. Yamaguchi and H. Hasunuma, *Sci. Light (Japan)*, **16**, 64 (1967).
36. P.S. Hauge and F.H. Dill, *IBM J. Res. Devel.*, **17**, 472 (1973).
37. S.N. Jasperson, D.K. Burge and R.C. O'Handley, *Surf. Sci.*, **37**, 548 (1973).
38. B.V. Zhdanov, N.I. Zheludev, A.I. Kovrigin and D.V. Yakovlev, *Instruments and Experimental Techniques*, **24**, 1284 (1981).

39. Ya.B. Zel'dovich, *Soviet Phys. JETP*, **9**, 682 (1959).
40. V.N. Baier and I.B. Khriplovich, *Soviet Phys. JETP*, **12**, 959 (1961).
41. M.A. Bouchiat and C. Bouchiat, *Phys. Lett. B*, **48**, 111 (1974).
42. L.M. Barkov and M.S. Zolotarev, *JETP Lett.*, **28**, 503 (1978).
43. L.M. Barkov and M.S. Zolotarev, *Phys. Lett. B*, **85**, 308 (1979).
44. G.N. Birich, Yu.V. Bogdanov, S.I. Kanorsky, I.I. Sobelman, V.N. Sorokin, I.I. Struk and E.A. Yukov, *Soviet Phys. JETP*, **60**, 442 (1984).
45. J. Taylor and P.G.H. Sandras, DPhil Thesis, Oxford, 1984.
46. J.H. Hollister, G.R. Apperson, L.L. Lewis, T.P. Emmons, T.G. Vold and E.N. Fortson, *Phys. Rev. Lett.*, **46**, 643 (1981).
47. T.P. Emmons, J.M. Reeves and E.N. Fortson, *Phys. Rev. Lett.*, **51**, 2089 (1983).
48. N.H. Edwards, S.J. Phipp, P.E.G. Baird and S. Nakayama, *Phys. Rev. Lett.*, **74**, 2654 (1995); *Journal de Physic B*, 1861, **29** (1996).
49. M.J.D. Macpherson, K.P. Zetie, R.B. Warrington, D.N. Stacey and J.P. Hoare, *Phys. Rev. Lett.*, **67**, 2784 (1991).
50. P.A. Vetter, D.M. Meekhof, P.K. Majumder, S.K. Lamoreaux and E.N. Fortson, *Phys. Rev. Lett.*, **74**, 2658 (1995); *Phys. Rev. A*, **52**, 1895 (1995).
51. R.B. Warrington, C.D. Thompson and D.N. Stacey, *Europhys. Lett.*, **24**, 641 (1993).
52. E.D. Commins, *Physica Scripta*, **6**, 468 (1987); **46**, 92 (1992); *Nature* (*Lond*), **326**, 203 (1988).
53. M.A. Bouchiat and C. Bouchiat, *C.R. Acad. Sci. Paris*, **322**, serie IIB, 381 (1996); *Reports on Progress in Physics*, **66**, 1351 (1997).
54. X.G. Wen and A. Zee, *Phys. Rev. Lett.*, **62**, 2873 (1989); *Phys. Rev. B*, **43**, 5595 (1991).
55. B.I. Halperin, J.March-Russell and F. Wolczek, *Phys. Rev. B*, **40**, 8726 (1989).
56. H.J. Weber, D. Weitbrecht, D. Brach, A.L. Shelankov, H. Keiter, W. Weber, T. Wolf, J. Geerk, G. Linker, G. Roth, P.C. Splittgerberhunnekes and G. Guntherodt, *Solid State Commun.*, **76**, 511 (1990).
57. K.B. Lyons, J. Kwo, J.F. Dillon, G.P. Espinosa, M. Mcglashanpowell, A.P. Ramirez and L. Schneemeyer, *Phys. Rev. Lett.*, **64**, 2949 (1990); K.B. Lyons, J.F. Dillon, E.S. Hellman, E.H. Hartford and M. McGlashanpowell, *Phys. Rev. B*, **43**, 11408 (1991).
58. K.B. Lyons, J.F. Dillon Jr, S.J. Duclos, C.B. Eom, H.L. Kao, J. Kwo, J.M. Phillips and M.P. Siegal, *Phys. Rev. B*, **47**, 8195 (1993).
59. K.B. Lyons, J.F. Dillon Jr, E.S. Hellman, E.H. Hartford and M. Mcglashanpowell, *Phys. Rev. B*, **43**, 11408 (1991).
60. T.W. Lawrence, A. Szoke and R.B. Laughin, *Phys. Rev. Lett.*, **69**, 1439 (1992).
61. S. Spielman, J.S. Dodge, L.W. Lombardo, C.B. Eom, M.M. Fejer, T.H. Geballe and A. Kapitulnik, *Phys. Rev. Lett.*, **68**, 3472 (1992).
62. Yu.P. Svirko and N.I. Zheludev, *Opt. Lett.*, **20**, 1809 (1995).
63. L.R. Hunter, *Science*, **252**, 73 (1991).
64. S.A. Murthy, D. Krause Jr, Z.L. Li and L.R. Hunter, *Phys. Rev. Lett.*, **63**, 965 (1989).
65. D. Bakalov, G. Cantatore, G. Carugno, S. Carusotto, P. Favaron, F. Dellavalle, I. Gabrielli, U. Gastaldi, E. Iacopini, P. Micossi, E. Milotti, R. Onofrio, R. Pengo, F. Perrone, G. Petrucci, E. Polacco, C. Rizzo, G. Ruoso, E. Zavattini, G. Zavattini, *Nucl. Phys. B*, **35**, 180 (1994).
66. S.I. Wawilow and W.L. Lewshin, *Z. Phys.*, **35**, 932 (1926).
67. S.I. Wawilow, *Microstuctura Sveta (Microstructure of Light)*, USSR Academy of Sciences Publishing, Moscow, 1950, p. 73.
68. D.H. Auston and C.V. Shank, *Phys. Rev. Lett.*, **32**, 1120 (1974).
69. S.A. Akhmanov and N.I. Koroteev, *Methods of Nonlinear Optics in Spectroscopy of Light Scuttering*, Nauka, Moscow, 1981.

70. S.A. Akhmanov, A.F. Bunkin, N.I. Koroteev and S.G. Ivanov, *JETP Lett.*, **25**, 444 (1977).
71. W.P. Deboeij, G.W. Lucassen, C. Otto and J. Greve, *J. Raman Spectrosc.*, **24**, 383 (1993).
72. T. Scholten, G.W. Lucassen, E. Koelewijn, F.F.M. Demul and J. Greve, *J. Raman Spectrosc.*, **20**, 503 (1989); G.W. Lucassen, W.P. Deboeij and J. Greve, *J. Raman Spectrosc.* **24**, 383 (1993).
73. A.Y. Chikishev, N.I. Koroteev, C. Otto and J. Greve, *J.Raman Spectrosc.*, **27**, 893 (1996).
74. A. Voroshilov, C. Otto and J. Greve, *Appl. Spectrosc.*, **50**, 78 (1996); *J. Chem. Phys.*, **106**, 2589 (1997).
75. F.W. Schneider, R. Brakel and H. Spiegel, *Berichte Der Bunsen Gessellschaft Fur Physikalische Chemie*, **93**, 304 (1989).
76. T. Kamisuki, S. Takeuchi, N. Akamatsu, Y. Adachi, S. Maeda and C. Hirose, *J. Chem. Phys.*, **88**, 4592 (1988).
77. D. Henman, R.W. Hellwarth, M.D. Levinson and G. Martin, *Phys. Rev. Lett.* **36**, 189 (1976).
78. R. Pan, H.D. Wei and Y.R. Shen, *Phys. Rev. B*, **39**, 1229 (1989).
79. J. Reif, J.C. Zink, C.-M. Schneider and J.Kirschner, *Phys. Rev. Lett.*, **67**, 2878 (1991).
80. B. Koopmans, M.G. Koerkamp and T. Rasing, *Phys. Rev. Lett.*, **74**, 3692 (1995).
81. H.A. Wierenga, W. Dejong, M.W.J. Prins, Th. Rasing, R. Vollmer, A. Kirilyuk, H. Schwabe and J.H. Kirschner, *Phys. Rev. Lett.*, **74**, 1462 (1995); H.A. Wierenga, M.W.J. Prins and T. Rasing, *Physica B*, **204**, 281 (1995).
82. T. Petralli-Mallow, T.M. Wong, J.D. Byers, H.I. Yee and J.M. Hicks, *J. Phys. Chem.*, **97**, 1383 (1993).
83. M.J. Crawford, S. Haslam, J.M. Probert, Y.A. Gruzdkov and J.G. Frey, *Chem. Phys. Lett.*, **230**, 260 (1994).
84. J.J. Maki, T. Verbiest, M. Kauranen, S. Van Elshocht and A. Persoos, *J. Chem. Phys.*, **105**, 767 (1996).
85. J.D. Byers, H.I. Yee and J.M. Hicks, *J. Chem. Phys.*, **101**, 6233 (1994).
86. V. Volkov, Yu.P. Svirko, V.F. Kamalov, L. Song and M.A. El-Sayed, *Biophys. J.*, **73**, (1997).
87. N.B. Abraham and P. Mandel, *Dynamical Polarization Phenomena in Lasers*, Notes of Lecture Presented at First International Conference and School on *Polarization Effects in Lasers and Spectroscopy*, Toronto, 1997.
88. G.M. Stephan and A.D. May, *Quantum Semiclass. Opt.* **10**, 19 (1998).
89. R. Vilaseca, C. Serrat, A. Kul'minskii, G.J. de Valcárcel, E. Roländ and R. Corbalán, *Quantum Semiclass. Opt.*, **10**, 37 (1998).
90. R. Müller and A. Klehr, *Quantum Semiclass. Opt.*, **10**, R37 (1998).
91. C.L. Tang, *Quantum Semiclass. Opt.*, **10**, R51 (1998).
92. H. De Lang, *Philips Res. Rep.* **19**, 429 (1964).
93. D. Culshaw and J. Kannelaud, *Phys. Rev.*, **141**, 237 (1966).
94. E.Yu. Andreyva, K.D. Teryokhin and S.A. Fridrikov, *Opt. Spectrosk.*, **27**, 809 (1969).
95. R.L. Fork, W.J. Tomilson and L.J. Heilos, *Appl. Phys. Lett.*, **8**, 162 (1966).
96. A. Le Floch, G. Ropars, J.M. Lenormand and R. Le Naour, *Phys. Rev. Lett.*, **52**, 918 (1984).
97. N.B. Abraham, M.D. Matlin and R.S. Gioggia, *Phys. Rev. A*, **53** 3514 (1996).
98. J. Martingegalado, M.S. Miguel and N.B. Abraham, *Optics Letters*, **21**, 351 (1996).
99. M. San Miguel, Q. Feng and J.V. Moloney, *Phys. Rev. A*, **52**, 1728 (1995).
100. J. Martin-Regalago, F. Prati, M. San Miguel and N.B. Abraham, *IEE J. Quant. Electr.*, **33**, 765 (1997).

101. G. Stephan and D. Hugon, *Phys. Rev. Lett.*, **55**, 703 (1985); A.D. May and G. Stephan, *J. Opt. Soc. Am. B*, **6**, 2355 (1989).
102. A.P. Voitovich, L.P. Svirina and V.N. Severikov, *Opt. Commun.*, **111**, 370 (1994).
103. L.A. Melnikov, A.I. Konukhov and M.V. Ryabinina, *Proc. SPIE*, **3177**, 49 (1997).
104. M.P. Van Exter, R.F.M. Hendriks and J.P. Woerdman, *Phys. Rev. A*, **57**, 2080 (1998).
105. A.P. Voitovich and V.N. Serikov, *Lasery s Anisotropnimi Resonatorami*, Nauka i Technika, Minsk, 1988.
106. M. Kitano, T. Yabuzaki and T. Ogawa, *Phys. Rev. Lett.*, **46**, 926 (1981).
107. S. Cecchi, G. Giusfredi, E. Petriella and P. Salieri, *Phys. Rev. Lett.*, **49**, 1928 (1982).
108. M. Kitano, T. Yabuzaki and T. Ogawa, *Phys. Rev. A*, **24**, 3156 (1981).
109. J. Mlynek, F. Mitchke, R. Deserno and W. Lange, *Phys. Rev. A*, **29**, 1297 (1984).
110. H.J. Carmichael, C.M. Savage and D.F. Walls, *Phys. Rev. Lett.*, **50**, 163 (1983).
111. E. Ciacobino, *Opt. Commun.*, **56**, 249 (1985).
112. G.G. Adonts, G.P. Dzhotyan and E.G. Kanetsyan, *Soviet J. Quantum Electron*, **14**, 777 (1984).
113. M.W. Hamilton, R.J. Ballagh and W.J. Sandle, *Z. Phys. B*, **49**, 263 (1982).
114. I.P. Areshev and V.K. Subashiev, *Zhurnal Tekhnicheskoi Fiziki*, **55**, 799 (1985).
115. I.P. Aresev, N.N. Rosanov, T.A. Murina and V.K. Subashiev, *Opt. Commun.*, **47**, 414 (1983).
116. T. Yabuzaki, T. Okamoto, M. Kitano and T. Ogawa, *Phys. Rev. A*, **29**, 1964 (1984).
117. R.J. Ballagh and V. Jain, *Phil. Trans. R. Soc. London A*, **313**, 445 (1984).
118. C. Parriger, P. Hannaford, W.J. Sandle and R.J. Ballagh, *Phys. Rev. A*, **31**, 4043 (1985).
119. M.I. Dykman and G.G. Tarasov, *Soviet Phys. Solid State*, **24**, 1361 (1982); *Optics Comm.*, **49**, 79 (1984).
120. J. Yumoto and K. Otsuka, *Phys. Rev. Lett.*, **54**, 1806 (1985).
121. K.L. Sala, *Phys. Rev. A*, **29**, 1944 (1984).
122. G. Gregori and S. Wabnitz, *Phys. Rev. Lett.*, **56**, 600 (1986).
123. M.V. Tratnik and J.E. Sipe, *Phys. Rev. A*, **35**, 2965 (1987).
124. M.V. Tratnik and J.E. Sipe, *Phys. Rev. A*, **35**, 2976 (1987).
125. M.V. Tratnik and J.E. Sipe, *Phys. Rev. A*, **36**, 4817 (1988).
126. M.I. Dykman, *Soviet Phys. JETP*, **64**, 927 (1986).
127. J. Yumoto and K. Otsuka, *Phys. Rev. A*, **34**, 4445 (1986).
128. V.A. Makarov, S.M. Pershin, A.A. Podshivalov, R.S. Zadoyan and N.I. Zheludev, *Opt. Lett.*, **8**, 557 (1983).
129. N.I. Zheludev, R.S. Zadoyan, A.I. Kovrigin, V.A. Makarov, S.M. Pershin and A.A. Podshivalov, *Soviet J. Quantum Electron*, **13**, 843 (1983).
130. M.I. Dykman and G.G. Tarasov, *Soviet Phys. JETP*, **59**, 1180 (1984).
131. S.A. Akhmanov, N.I. Zheludev and Yu.P. Svirko, *Bull. Acad. Sci. USSR Phys. Ser. Physics*, **46**, 53 (1982).
132. V.A. Makarov, A.V. Matveeva, and M.M. Stol'nits, *Bull. Acad. Sci. USSR Phys. Ser. Physics*, **50**, 155 (1986).
133. V.A. Makarov and A.V. Matveeva, *Soviet J. Quantum Electron*, **18**, 89 (1988).
134. H.H. Ritze and A. Bandilla, *Phys. Lett. A*, **78**, 447 (1980).
135. K. Cho and T. Itoh, *Solid State Commun.*, **52**, 287 (1984).
136. M. Inoue, *Phys. Rev. B*, **33**, 1317 (1986).
137. H.H. Kranz and H. Hang, *Phys. Rev. B*, **34**, 2554 (1986).
138. A.E. Kaplan, *Opt. Lett.*, **8**, 560 (1983).
139. A.E. Kaplan and C.T. Law, *IEEE J. Quantum Electron.*, **QE-21**, 1529 (1985).
140. S. Wabnitz and G. Gregory, *Opt. Commun.*, **59**, 72 (1986).
141. A.A. Maier, *Soviet J. Quantum Electron.*, **14**, 157 (1984).

142. B. Daino, G. Gregori and S. Wabnitz, *J. Appl. Phys.*, **58**, 4512 (1985).
143. A.L. Gaeta, R.W. Boyd, J.R. Achkerhalt and R.W. Milonni, *Phys. Rev. Lett.*, **58**, 2432 (1987); *J. Opt. Soc. Amer.*, **13**, 9 (1986).
144. D.J. Gauthier, M.C. Mclcuit and R.W. Boyd, *Phys. Rev. Lett.*, **61**, 1827 (1988).
145. M.V. Tratnik and J.E. Sipe, *Phys. Rev. Lett.*, **58**, 1104 (1987).
146. V.E. Zakharov and A.V. Mikhailov, *JETP Lett.*, **45**, 349 (1987).
147. R. Holzner and S. Dangel, *Quantum Semiclass. Opt.*, **10**, R1 (1998).
148. W. Lange, A. Aumann, T. Ackemann and E. Büthe, *Quantum Semiclass. Opt.*, **10**, R23 (1998).
149. H.G. Winful, *Appl. Phys. Lett.*, **47**, 213 (1985).
150. H.G. Winful, *Opt. Lett.*, **11**, 33 (1986).
151. M. Roman, S. Kielich and W. Gadomski, *Phys. Rev. A*, **34**, 351 (1986).
152. F. Matera and S. Wabnitz, *Opt. Lett.*, **11**, 467 (1986).
153. A. Vatarescu, *Electron. Lett.*, **23**, 379 (1987).
154. S. Trillo, S. Wabnitz, W.C. Banyai, N. Finlayson, C.T. Seaton, G.I. Stegeman and R.H. Stolen, *Appl. Phys. Lett.*, **53**, 837 (1988).
155. K.J. Blow, N.J. Doran and D. Wood, *Opt. Lett.*, **12**, 202 (1987).
156. S. Trillo and S. Wabnitz, *Phys. Rev. A*, **36**, 3881 (1987).
157. N.I. Zheludev, *Sov. Phys. Usp.*, **32**, 357 (1989).

FURTHER READING

E.R. Pike, in L. Brawn, A. Pais and B. Pippard (eds.), *Twentieth Century Optics*, AIP, New York, and IOP, Bristol, 1995.

2 Constitutive Equations In Nonlinear Optics

In this chapter we will begin with Maxwell's equations and introduce the constitutive equations for the linear and nonlinear optical response of a medium. The ideas of causality and nonlocality will also be introduced. We will present the constitutive equations using the concept of local and nonlocal optical susceptibilities. We will show how some general features of optical response can be found by analysing the structure of the constitutive equation and the symmetry of the optical susceptibility tensors.

2.1 MAXWELL'S EQUATIONS, CAUSALITY AND NONLOCALITY

Maxwell's equations describe the basic laws for the electric field **E** and magnetic field **B** of a light wave interacting with a medium [1, 2]:

$$\begin{aligned}
\operatorname{curl}\mathbf{E} &= -\frac{1}{c}\frac{\partial \mathbf{B}}{\partial t} \\
\operatorname{div}\mathbf{B} &= 0 \\
\operatorname{curl}\mathbf{B} &= \frac{1}{c}\frac{\partial \mathbf{E}}{\partial t} + \frac{4\pi}{c}\mathbf{J} \\
\operatorname{div}\mathbf{E} &= 4\pi\sigma
\end{aligned} \tag{2.1}$$

Here **J** and σ are, respectively, the electric current density and electric charge density in the medium. If there are no external currents flowing through the medium and no uncompensated charges (we will be analysing this case later), Maxwell's equations may be rewritten as follows:

$$\begin{aligned}
\operatorname{curl}\mathbf{E} &= -\frac{1}{c}\frac{\partial \mathbf{B}}{\partial t} \\
\operatorname{div}\mathbf{B} &= 0 \\
\operatorname{curl}\mathbf{B} &= \frac{1}{c}\frac{\partial \mathbf{D}}{\partial t} \\
\operatorname{div}\mathbf{D} &= 0
\end{aligned} \tag{2.2}$$

Here we have introduced the electric displacement vector $\mathbf{D} = \mathbf{E} + 4\pi\mathbf{P}$ and the vector of polarization response:

$$\mathbf{P}(\mathbf{r},\ t) = \int_{-\infty}^{t} \mathbf{J}(\mathbf{r},\ t')\,\mathrm{d}t' \tag{2.3}$$

The number of variables in Maxwell's equations is greater than the number of equations, and therefore these equations alone are not sufficient for describing light propagation. We need an additional relationship between the optical response of the medium (i.e. the vectors of polarization response **P** or electric displacement **D**) and the wave fields. This additional relationship accounts for the material properties of the medium and could, for example, be of the form $\mathbf{P} = \mathbf{P}(\mathbf{E},\ \mathbf{B})$.

It follows from the first of Maxwell's equations (2.2) that **E** and **B** are related:

$$\mathbf{B}(\mathbf{r},\ t) = \mathbf{B}(\mathbf{r},\ t = -\infty) - c\int_{-\infty}^{t} \mathrm{d}t' \,\mathrm{curl}\,\mathbf{E}(\mathbf{r},\ t') \tag{2.4}$$

The first term on the right-hand side of formula (2.4) is a static magnetic field that is not relevant to the oscillating fields of the electromagnetic wave. We will not be considering magnetized media, and so we can drop this term. Thus, on resorting to equation (2.4) we can always exclude **B** from the material relation $\mathbf{P} = \mathbf{P}(\mathbf{E},\ \mathbf{B})$, reducing it to the following functional form [1]:

$$\mathbf{P} = \mathbf{P}(\mathbf{E}) \tag{2.5}$$

Specific forms of this relation are known as the constitutive equations of the medium. In equation (2.5), **P** depends on the coordinate and, of course, on time. Equation (2.5) must obey the causality principle: ***P*** at any given moment is determined by the current and preceding values of the fields of the light wave, but not by future excitations. This retrospective relationship between **P** and **E** leads to the dependence of the optical response on the frequency of the light wave. Similarly, the optical response is not necessarily local; i.e. the response could depend on excitations not only at the point of observation but also in some vicinity around it at earlier moments. Such nonlocality leads to the dependence of the optical response on the wave vector of the light wave. This is known as spatial dispersion.

As mentioned above, equation (2.5) is not suitable for the treatment of the optical effects in magnetized media or in the presence of a static magnetic field. For such tasks, a more complex approach involving two constitutive equations is necessary [3]. Of course, as long as we do not consider static magnetization, this treatment is equivalent to that based on equation (2.5) (see Ref. [4]).

In equation (2.5) the relationship between **P** and **E** could be nonlinear; i.e. an increase in the magnitude of **E** will lead to a disproportionate change of **P**. Such nonlinear constitutive equations are often used for the description of nonlinear optical effects, which will be a matter of primary concern in this book. Usually (but not always) the nonlinear optical response can be expressed as a power series in the electric field strength:

$$\mathbf{P}(\mathbf{r},\,t) = \mathbf{P}^{(1)}(\mathbf{r},\,t) + \mathbf{P}^{(2)}(\mathbf{r},\,t) + \mathbf{P}^{(3)}(\mathbf{r},\,t) + \cdots \tag{2.6}$$

Here the upper indices stand for the order of the term with respect to the electric field strength. The presentation of the polarization response as a series, equation (2.6), implies some limits to the strength of the electric field: higher order terms must be much smaller than the preceding lower order terms. The first term on the right-hand side of equation (2.6) describes the linear optical response and increases proportionally with **E**.

2.2 LINEAR CONSTITUTIVE EQUATIONS

We will begin by considering the constitutive equation (2.6) when all the terms, apart from $\mathbf{P}^{(1)}(\mathbf{r},\,t)$, are ignored. Such a curtailed constitutive equation is known as a linear constitutive equation. Since **E** and **P** are vector quantities, we need to have three relations of this type, one for each projection $P_x^{(1)}$, $P_y^{(1)}$, $P_z^{(1)}$, in the laboratory Cartesian frame. The most general way to account for the nonlocality and causality of linear response is an integral of the form:

$$P_i^{(1)}(\mathbf{r},\,t) = \sum_{j=x,y,z} \int_V \mathrm{d}\mathbf{r}_1 \int_{-\infty}^{t} \mathrm{d}t_1 \psi_{ij}^{(1)}(\mathbf{r}-\mathbf{r}_1;\, t-t_1) E_j(\mathbf{r}_1,\, t_1) \tag{2.7}$$

Here the subscripts enumerate the Cartesian axes x, y, z. In equation (2.7) $\psi_{ij}^{(1)}(\mathbf{r}-\mathbf{r}_1,\, t-t_1)$ will be referred to as the response function of the medium. This function describes the optical response at point $\mathbf{r}$ and time t resulting from the electric field at point $\mathbf{r}_1$ at time t_1 (see Fig. 2.1). Optical properties of the medium are crucially dependent on how the function $\psi_{ij}^{(1)}(\mathbf{r}-\mathbf{r}_1,\, t-t_1)$ depends on $\rho = \mathbf{r} - \mathbf{r}_1$ and $\tau = t - t_1$. In molecular media, the function $\psi_{ij}^{(1)}(\rho,\,\tau)$ would normally have a maximum at $\rho = 0$ and diminish for $|\rho| \geqslant a$, where a is the typical size of a molecule. If the response in a particular location only depends on excitation at this location, i.e. the function $\psi_{ij}^{(1)}(\rho,\,\tau)$ is only non-zero only at $\rho = 0$, the medium is said to have a local optical response. For example, we can think of the highly localized response of an unmoving atom whose diameter is much smaller than the wavelength λ of the light wave propagating in the medium. Even if the response cannot be seen as truly local, in optics $\lambda \gg a$, usually, and integral (2.7) may be expanded over the small parameter a/λ. This parameter is known as the spatial dispersion parameter. The spatial dispersion effects in linear optics are comprehensively documented in Ref. [5].

In this book we will be dealing only with monochromatic or nearly monochromatic light waves. The electric field of an electromagnetic wave of frequency ω

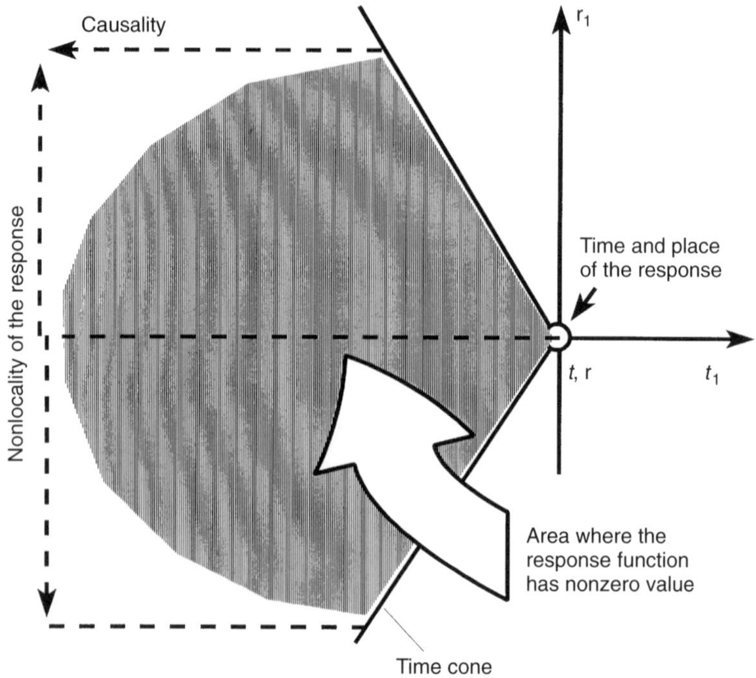

Fig. 2.1 Optical response at a particular point depends on the history of excitation in a certain area surrounding the point

and the corresponding linear optical response may be written in terms of their complex spectral amplitudes $\mathbf{E}(\omega, \mathbf{r})$ and $\mathbf{P}^{(1)}(\omega, \mathbf{r})$ as follows:

$$\begin{aligned} \mathbf{E}(\mathbf{r}, t) &= \mathbf{E}(\omega, \mathbf{r})\,\mathrm{e}^{-\mathrm{i}\omega t} + [\mathbf{E}(\omega, \mathbf{r})]^{*}\mathrm{e}^{\mathrm{i}\omega t} \\ \mathbf{P}^{(1)}(\mathbf{r}, t) &= \mathbf{P}^{(1)}(\omega, \mathbf{r})\,\mathrm{e}^{-\mathrm{i}\omega t} + [\mathbf{P}^{(1)}(\omega, \mathbf{r})]^{*}\mathrm{e}^{\mathrm{i}\omega t} \end{aligned} \tag{2.8}$$

By substituting equations (2.8) into (2.7), we rewrite equation (2.7) in the spectral representation:

$$P_i^{(1)}(\omega, \mathbf{r}) = \sum_{j=x,y,z} \int_V \mathrm{d}\boldsymbol{\rho} \int_0^{\infty} \mathrm{d}\tau \psi_{ij}^{(1)}(\boldsymbol{\rho}, \tau)\,\mathrm{e}^{\mathrm{i}\omega\tau} E_j(\omega, \mathbf{r} - \boldsymbol{\rho}) \tag{2.9}$$

Since the response function $\psi_{ij}^{(1)}(\boldsymbol{\rho}, \tau)$ diminishes at $|\boldsymbol{\rho}| \gg \lambda$, the spectral component of the wave's electric field which spatially oscillates with period λ can be expanded about $\mathbf{r}$ in the Taylor series:

$$\mathbf{E}(\omega, \mathbf{r}-\rho) = \mathbf{E}(\omega, \mathbf{r}) + \frac{\partial \mathbf{E}(\omega, \mathbf{r})}{\partial \mathbf{r}}(-\rho) + \frac{1}{2}\sum_{i=x,y,z}\sum_{j=x,y,z}\frac{\partial^2 \mathbf{E}(\omega, \mathbf{r})}{\partial r_i \partial r_i}\rho_i\rho_j + \ldots \tag{2.10}$$

If one kept only the first term on the right-hand side of equation (2.10), this would be called a local approximation. If the first and second terms were kept, it would be known as a first-order spatial dispersion approximation. Some phenomena in the optics of crystals, plasma and gases are due to the third term in equation (2.10). These are second-order spatial dispersion effects. In this book, however, we will be dealing only with optical effects which can be described in the local and first-order approximations. Therefore, only the first two terms on the right-hand side of equation (2.10) will be retained for further analysis.

After substituting expansion (2.10) into equation (2.9) we arrive at the following form of the linear constitutive equation:

$$P_i^{(1)}(\omega, \mathbf{r}) = \frac{1}{4\pi}\sum_{j=x,y,z}[\epsilon_{ij}(\omega) - \delta_{ij}]E_j(\omega, \mathbf{r}) + \frac{1}{4\pi}\sum_{j=x,y,z}\sum_{n=x,y,z}\Gamma^{(1)}_{ijn}(\omega)\frac{\partial E_j(\omega, \mathbf{r})}{\partial r_n} \tag{2.11}$$

Here δ_{ij} is the Kronecker delta symbol, $\epsilon_{ij}(\omega)$ is a material tensor known as the dielectric tensor:

$$\epsilon_{ij}(\omega) = \delta_{ij} + 4\pi\int_V \mathrm{d}\rho\int_0^\infty \mathrm{d}\tau\psi^{(1)}_{ij}(\rho, \tau)\,\mathrm{e}^{\mathrm{i}\omega\tau} \tag{2.12}$$

and $\Gamma^{(1)}_{ijn}(\omega)$ is a material tensor called the nonlocality tensor:

$$\Gamma^{(1)}_{ijn}(\omega) = -4\pi\int_V \mathrm{d}\rho\left[\rho_n\int_0^\infty \mathrm{d}\tau\psi^{(1)}_{ij}(\rho, \tau)\,\mathrm{e}^{\mathrm{i}\omega\tau}\right] \tag{2.13}$$

Further on in this book we will use Einstein's tensor notation where the symbol of a summation (i.e. Σ) is sometimes dropped, but the summation is kept presumed over the repeated coordinate indices. In this notation constitutive equation (2.11) becomes

$$P_i^{(1)}(\omega, \mathbf{r}) = \frac{1}{4\pi}\{[\epsilon_{ij}(\omega) - \delta_{ij}]E_j(\omega, \mathbf{r}) + \Gamma^{(1)}_{ijn}(\omega)\nabla_n E_j(\omega, \mathbf{r})\} \tag{2.14}$$

or, in the time domain,

$$P_i^{(1)}(\mathbf{r}, t) = \frac{1}{4\pi}\{[\epsilon_{ij}(\omega) - \delta_{ij}]E_j(\mathbf{r}, t) + \Gamma^{(1)}_{ijn}(\omega)\nabla_n E_j(\mathbf{r}, t)\}\,\mathrm{e}^{-\mathrm{i}\omega t} + \text{c.c.} \tag{2.15}$$

In equations (2.14) and (2.15) we have used the gradient operator symbol $\nabla_n = \partial/\partial r_n$. For an electromagnetic wave propagating with wave vector $\mathbf{k}$, the spectral amplitude $\mathbf{E}(\omega, \mathbf{r})$ harmonically depends on the coordinate $\mathbf{r}$, i.e.

$\mathbf{E}(\omega, \mathbf{r}) = \mathbf{A} \exp\{\mathrm{i}\mathbf{k}\cdot\mathbf{r}\}$. Therefore, constitutive equation (2.14) may be presented as follows:

$$P_i^{(1)}(\omega, \mathbf{r}) = P_i^{(1)}(\omega, \mathbf{k})\, \mathrm{e}^{\mathrm{i}(\mathbf{k}\cdot\mathbf{r})} = \frac{1}{4\pi}[\epsilon_{ij}(\omega) - \delta_{ij} + \mathrm{i}k_n\Gamma_{ijn}^{(1)}(\omega)]A_j\, \mathrm{e}^{\mathrm{i}(\mathbf{k}\cdot\mathbf{r})} \quad (2.16)$$

This constitutive equation will be widely used in this book to describe linear optical effects.

2.3 NONLINEAR CONSTITUTIVE EQUATIONS

In a very similar manner we can derive the constitutive equations for describing nonlinear optical effects following Ref. [6] and Ref. [7]. We will find such equations of the second and third order in the strength of the electric field of the wave. Starting with the principle of causality and accounting for nonlocality, we present equation (2.6) as follows:

$$\begin{aligned}
P_i(\mathbf{r}, t) &= P_i^{(1)}(\mathbf{r}, t) + P_i^{(2)}(\mathbf{r}, t) + P_i^{(3)}(\mathbf{r}, t) \\
&= \int_V \mathrm{d}\rho \int_0^\infty \mathrm{d}\tau \psi_{ij}^{(1)}(\rho, \tau) E_j(\mathbf{r} - \rho, t - \tau) \\
&\quad + \int_V \mathrm{d}\rho \int_V \mathrm{d}\rho_1 \int_0^\infty \mathrm{d}\tau \int_0^\infty \mathrm{d}\tau_1 \psi_{ijk}^{(2)}(\rho, \tau; \rho_1, \tau_1) \\
&\quad \times E_j(\mathbf{r} - \rho, t - \tau) E_k(\mathbf{r} - \rho_1, t - \tau_1) \\
&\quad + \int_V \mathrm{d}\rho \int_V \mathrm{d}\rho_1 \int_V \mathrm{d}\rho_2 \int_0^\infty \mathrm{d}\tau \int_0^\infty \mathrm{d}\tau_1 \int_0^\infty \mathrm{d}\tau_2 \psi_{ijkl}^{(3)}(\rho, \tau; \rho_1, \tau_1; \rho_2, \tau_2) \\
&\quad \times E_j(\mathbf{r} - \rho, t - \tau) E_k(\mathbf{r} - \rho_1, t - \tau_1) E_l(\mathbf{r} - \rho_2, t - \tau_2)
\end{aligned} \quad (2.17)$$

Here $\psi_{ijk}^{(2)}(\rho, \tau; \rho_1, \tau_1)$ and $\psi_{ijkl}^{(3)}(\rho, \tau; \rho_1, \tau_1; \rho_2, \tau_2)$ are higher order response functions of the medium. For example, $\psi_{ijk}^{(2)}(\rho, \tau; \rho_1, \tau_1)$ describes the medium reaction at point $\mathbf{r}$ and time t resulting from excitations at point $\mathbf{r} - \rho$ at time $t - \tau$ and at point $\mathbf{r} - \rho_1$ at time $t - \tau_1$. It has been shown [6] that the second-order response function $\psi_{ijk}^{(2)}(\rho, \tau; \rho_1, \tau_1)$ is invariant under the simultaneous interchange of the spatial–temporal arguments and the corresponding Cartesian indices: $\psi_{ijk}^{(2)}(\rho, \tau; \rho_1, \tau_1) = \psi_{ikj}^{(2)}(\rho_1, \tau_1; \rho, \tau)$. This property may be generalized to the higher order nonlinear response functions, and is known as intrinsic permutation symmetry. In particular, the third-order response function obeys the rule $\psi_{ijkl}^{(3)}(\rho, \tau; \rho_1, \tau_1; \rho_2, \tau_2) = \psi_{ijlk}^{(3)}(\rho, \tau; \rho_2, \tau_2; \rho_1, \tau_1) = \psi_{ikjl}^{(3)}(\rho_1, \tau_1; \rho, \tau; \rho_2, \tau_2) = \psi_{ilkj}^{(3)}(\rho_2, \tau_2; \rho_1, \tau_1; \rho, \tau)$.

Let us assume that the light field is a set of monochromatic waves, the frequencies of which are enumerated by the subscript s:

$$\mathbf{E}(\mathbf{r},\, t) = \sum_s \{\mathbf{E}(\omega_s,\, \mathbf{r}) \exp(-\mathrm{i}\omega_s t) + [\mathbf{E}(\omega_s,\, \mathbf{r})]^* \exp(\mathrm{i}\omega_s t)\} \qquad (2.18)$$

This form ensures that the electric field of the light wave $\mathbf{E}(\mathbf{r},\, t)$ is a real quantity. In this presentation the spectrum has both 'positive' and 'negative' spectral components. Instead of explicitly writing down the terms containing the 'positive' and 'negative' components, we could agree that the summation always takes place over 'positive' and 'negative' frequencies while the corresponding spectral components obey the following relation:

$$[\mathbf{E}(\omega_s,\, \mathbf{r})]^* = \mathbf{E}(-\omega_s,\, \mathbf{r}) \qquad (2.19)$$

If such agreement is followed, we can drop the second term on the right-hand side of equation (2.18):

$$\mathbf{E}(\mathbf{r},\, t) = \sum_s \mathbf{E}(\omega_s,\, \mathbf{r}) \exp(-\mathrm{i}\omega_s t) \qquad (2.20)$$

Presentation (2.20) will become important later, when specific forms of the constitutive equations serving specific optical effects will be derived.

Resorting (2.20), the second-order optical response, quadratic in the electric field strength, is given by

$$\begin{aligned} P_i^{(2)}(\mathbf{r},\, t) = \sum_{a,b} [&\chi_{ijk}^{(2)}(\omega_a + \omega_b,\, \omega_a,\, \omega_b) E_j(\omega_a,\, \mathbf{r}) E_k(\omega_b,\, \mathbf{r}) \\ &+ \Gamma_{ijkn}^{(2)}(\omega_a + \omega_b,\, \omega_a,\, \omega_b) E_k(\omega_b,\, \mathbf{r}) \nabla_n E_j(\omega_a,\, \mathbf{r})]\, \mathrm{e}^{-\mathrm{i}(\omega_a+\omega_b)t} \end{aligned} \qquad (2.21)$$

Here, again, we have presumed a summation over repeated indices $j,\, k,\, n = x,\, y,\, z$ and introduced two material tensors, the second-order local susceptibility

$$\chi_{ijk}^{(2)}(\omega_a + \omega_b,\, \omega_a,\, \omega_b) = \int_V \mathrm{d}\rho_1 \int_V \mathrm{d}\rho_2 \int_0^\infty \mathrm{d}\tau_1 \int_0^\infty \mathrm{d}\tau_2 \psi_{ijk}^{(2)}(\rho_1,\, \tau_1;\, \rho_2,\, \tau_2)\, \mathrm{e}^{\mathrm{i}\omega_a\tau_1 + \mathrm{i}\omega_b\tau_2} \qquad (2.22)$$

and the second-order nonlocal susceptibility

$$\begin{aligned} &\Gamma_{ijkn}^{(2)}(\omega_a + \omega_b,\, \omega_a,\, \omega_b) = \\ &\quad -2\int \mathrm{d}\rho_1 \rho_{1n} \int \mathrm{d}\rho_2 \int_0^\infty \mathrm{d}\tau_1 \int_0^\infty \mathrm{d}\tau_2 \psi_{ijk}^{(2)}(\rho_1,\, \tau_1;\, \rho_2,\, \tau_2)\, \mathrm{e}^{\mathrm{i}\omega_a\tau_1 + \mathrm{i}\omega_b\tau_2} \end{aligned} \qquad (2.23)$$

The third-order optical response, cubic in the electric field strength, is given by

$$P_i^{(3)}(\mathbf{r}, t) = \sum_{a,b,c}[\chi_{ijkl}^{(3)}(\omega_a + \omega_b + \omega_c, \omega_a, \omega_b, \omega_c)E_j(\omega_a, \mathbf{r})E_k(\omega_b, \mathbf{r})E_l(\omega_c, \mathbf{r})$$

$$+ \Gamma_{ijkln}^{(3)}(\omega_a + \omega_b + \omega_c, \omega_a, \omega_b, \omega_c)E_k(\omega_b, \mathbf{r})$$

$$\times E_l(\omega_c, \mathbf{r})\nabla_n E_j(\omega_a, \mathbf{r})]\, \mathrm{e}^{-\mathrm{i}(\omega_a+\omega_b+\omega_c)t} \tag{2.24}$$

Here, once more, a summation is presumed over repeated indices $j, k, l, n = x, y, z$. The third-order local and nonlocal susceptibility tensors are given by

$$\chi_{ijkl}^{(3)}(\omega_a + \omega_b + \omega_c, \omega_a, \omega_b, \omega_c) = \int_V \mathrm{d}\rho_1 \int_V \mathrm{d}\rho_2 \int_V \mathrm{d}\rho_3 \int_0^\infty \mathrm{d}\tau_1 \int_0^\infty \mathrm{d}\tau_2 \int_0^\infty \mathrm{d}\tau_3$$

$$\times\, \psi_{ijkl}^{(3)}(\rho_1, \tau_1; \rho_2, \tau_2; \rho_3, \tau_3)\, \mathrm{e}^{\mathrm{i}\omega_a\tau_1+\mathrm{i}\omega_b\tau_2+\mathrm{i}\omega_c\tau_3} \tag{2.25}$$

and

$$\Gamma_{ijkln}^{(3)}(\omega_a + \omega_b + \omega_c, \omega_a, \omega_b, \omega_c) = -3\int \mathrm{d}\rho_1\rho_{1n} \int \mathrm{d}\rho_2 \int \mathrm{d}\rho_3 \int_0^\infty \mathrm{d}\tau_1 \int_0^\infty \mathrm{d}\tau_2 \int_0^\infty \mathrm{d}\tau_3$$

$$\times\, \psi_{ijkl}^{(3)}(\rho_1, \tau_1; \rho_2, \tau_2; \rho_3, \tau_3)\, \mathrm{e}^{\mathrm{i}\omega_a\tau_1+\mathrm{i}\omega_b\tau_2+\mathrm{i}\omega_c\tau_3} \tag{2.26}$$

The forms of the constitutive equations introduced above dictate some restraints on the susceptibility tensors. Beyond that, severe restrictions on the susceptibility tensors can be imposed by the symmetry of the medium itself. It is important to know these symmetries in order to correctly apply the constitutive equations. These restrictions often result in relations between the susceptibility tensor components that can dramatically simplify the analysis of optical effects. Therefore, prior to calculations, it is desirable to identify the additional symmetries of the susceptibilities.

2.4 SYMMETRIES OF THE SUSCEPTIBILITY TENSORS

Since the electric field and the displacement vectors in Maxwell's equations are physically measurable quantities, they must be purely real. This has already been accounted for in equation (2.20). Moreover, the left-hand sides of equations (2.15), (2.21) and (2.24) must be real. Therefore the following symmetry, known as the reality condition, is required for the dielectric and nonlocality tensors:

$$[\epsilon_{ij}(\omega)]^* = \epsilon_{ij}(-\omega) \tag{2.27}$$

$$[\Gamma_{ijn}^{(1)}(\omega)]^* = \Gamma_{ijn}^{(1)}(-\omega) \tag{2.28}$$

Similar relations exist for the second-hand and third-order tensors of local and nonlocal nonlinear response:

$$[\chi^{(2)}_{ijk}(\omega_a + \omega_b, \omega_a, \omega_b)]^* = \chi^{(2)}_{ijk}(-\omega_a - \omega_b, -\omega_a, -\omega_b) \tag{2.29}$$

$$[\Gamma^{(2)}_{ijkn}(\omega_a + \omega_b, \omega_a, \omega_b)]^* = \Gamma^{(2)}_{ijkn}(-\omega_a - \omega_b, -\omega_a, -\omega_b) \tag{2.30}$$

$$[\chi^{(3)}_{ijkl}(\omega_a + \omega_b + \omega_c, \omega_a, \omega_b, \omega_c)]^* = \chi^{(3)}_{ijkl}(-\omega_a - \omega_b - \omega_c, -\omega_a, -\omega_b, -\omega_c) \tag{2.31}$$

$$[\Gamma^{(3)}_{ijkln}(\omega_a + \omega_b + \omega_c, \omega_a, \omega_b, \omega_c)]^* = \Gamma^{(3)}_{ijkln}(-\omega_a - \omega_b - \omega_c, -\omega_a, -\omega_b, -\omega_c) \tag{2.32}$$

Therefore the reality condition may be formulated as follows: simultaneous reversal of the signs of all frequency arguments is equivalent to complex conjugation of the susceptibility.

The constitutive equations (2.21) and (2.24) assume a summation over different frequencies. The local contribution to the second-order optical response, which is described by $\chi^{(2)}_{ijk}(\omega_a + \omega_b, \omega_a, \omega_b)$, must clearly be independent of the order in which spectral components $\mathbf{E}(\omega_a, \mathbf{r})$ and $\mathbf{E}(\omega_b, \mathbf{r})$ appear in expression (2.21). Consequently, a simultaneous permutation of the electric fields and corresponding frequencies in the argument of the second-order susceptibility should not alter the optical response. Therefore material equation (2.21) imposes the following symmetry restrain on the susceptibility:

$$\chi^{(2)}_{ijk}(\omega_a + \omega_b, \omega_a, \omega_b) = \chi^{(2)}_{ikj}(\omega_a + \omega_b, \omega_b, \omega_a) \tag{2.33}$$

The local second-order susceptibility does not change under a simultaneous permutation of the Cartesian indices j and k and corresponding frequency arguments ω_a and ω_b.

Similarly, we can derive for the third-order cubic optical susceptibility:

$$\begin{aligned}\chi^{(3)}_{ijkl}(\omega_a + \omega_b + \omega_c, \omega_a, \omega_b, \omega_c) &= \chi^{(3)}_{ikjl}(\omega_a + \omega_b + \omega_c, \omega_b, \omega_a, \omega_c)\\ &= \chi^{(3)}_{ilkj}(\omega_a + \omega_b + \omega_c, \omega_c, \omega_b, \omega_a)\\ &= \chi^{(3)}_{ijlk}(\omega_a + \omega_b + \omega_c, \omega_a, \omega_c, \omega_b)\end{aligned} \tag{2.34}$$

The symmetries given by equations (2.33) and (2.34) are known as the frequency permutation symmetries of the susceptibility tensors. No symmetry of this sort is appropriate for the dielectric tensor because it depends on only one frequency.

If nonlocality is taken into account, the optical response depends not only on the spectral components of the fields but also on their gradients. The spectral components of the electric fields under the gradient operator are distinguishable from other fields and consequently are not permutable with the other spectral components of the field. Therefore, the second-order nonlocal susceptibility tensor $\Gamma^{(2)}_{ijkn}(\omega_a + \omega_b, \omega_a, \omega_b)$ does not have any frequency permutation symmetry.

The nonlocal contribution to the optical response, which is given by $\Gamma^{(3)}_{ijkln}(\omega_a+\omega_b+\omega_c, \omega_a, \omega_b, \omega_c)$, is independent only of the order in which the various spectral components of the electric field, enumerated by the subscripts b and c, appear in the expression (2.24):

$$\Gamma^{(3)}_{ijkln}(\omega_a+\omega_b+\omega_c, \omega_a, \omega_b, \omega_c) = \Gamma^{(3)}_{ijlkn}(\omega_a+\omega_b+\omega_c, \omega_a, \omega_c, \omega_b) \quad (2.35)$$

In later chapters we will consider absorption-less media in which the overall loss of the energy of the propagating waves is negligible. In this case optical susceptibilities obey some important additional frequency permutation relations which may be deduced from energy conservation considerations. For linear susceptibilities the additional permutation relations are

$$\begin{aligned} \epsilon_{ij}(\omega) &= \epsilon_{ji}(-\omega) = [\epsilon_{ji}(\omega)]^* \\ \Gamma^{(1)}_{ijn}(\omega) &= \Gamma^{(1)}_{jin}(-\omega) = -[\Gamma^{(1)}_{jin}(\omega)]^* \end{aligned} \quad (2.36)$$

For the second-order susceptibilities the frequency permutation relations for absorption-less media are

$$\begin{aligned} \chi^{(2)}_{ijk}(\omega_a+\omega_b, \omega_a, \omega_b) &= \chi^{(2)}_{jki}(-\omega_a, \omega_b, -\omega_a-\omega_b) \\ &= \chi^{(2)}_{kij}(-\omega_b, -\omega_a-\omega_b, \omega_a) \\ \Gamma^{(2)}_{ijkn}(\omega_a+\omega_b, \omega_a, \omega_b) + \Gamma^{(2)}_{jkin}(-\omega_a, \omega_b, -\omega_a-\omega_b) & \\ + \Gamma^{(2)}_{kijn}(-\omega_b, -\omega_a-\omega_b, \omega_a) &= 0 \end{aligned} \quad (2.37)$$

The absence of absorption imposes the following restraints on the cubic susceptibilities:

$$\begin{aligned} \chi^{(3)}_{ijkl}(\omega_a+\omega_b+\omega_c, \omega_a, \omega_b, \omega_c) &= \chi^{(3)}_{jkli}(-\omega_a, \omega_b, \omega_c, -\omega_a-\omega_b-\omega_c) \\ &= \chi^{(3)}_{klij}(-\omega_b, \omega_c, -\omega_a-\omega_b-\omega_c, \omega_a) \\ &= \chi^{(3)}_{lijk}(-\omega_c, -\omega_a-\omega_b-\omega_c, \omega_a, \omega_b) \\ \Gamma^{(3)}_{ijkln}(\omega_a+\omega_b+\omega_c, \omega_a, \omega_b, \omega_c) + \Gamma^{(3)}_{jklin}(-\omega_a, \omega_b, \omega_c, -\omega_a-\omega_b-\omega_c) & \\ + \Gamma^{(3)}_{klijn}(-\omega_b, \omega_c, -\omega_a-\omega_b-\omega_c, \omega_a) & \\ + \Gamma^{(3)}_{lijkn}(-\omega_c, -\omega_a-\omega_b-\omega_c, \omega_a, \omega_b) &= 0 \end{aligned} \quad (2.38)$$

A detailed derivation of the above frequency permutation relations is given in Ref. [7].

The crystal symmetry of an optical medium places the most powerful restrictions on the symmetry of the optical response. The presence or absence of an inversion centre, planes of symmetry and rotation axes in the crystalline structure are very important. Generally, the lower the symmetry of a crystal, the wider the range of optical and nonlinear optical effects that can be observed in it. In media of high symmetry many optical effects are forbidden. For example, birefringence is possible in most crystals, but is forbidden in highly symmetric, isotropic media, such as liquids. The conditions that a medium imposes on the observability of optical effects may be formulated as requirements on the optical susceptibility tensors. For instance, in isotropic media all diagonal elements of the dielectric tensor are equal, while in the crystalline case all of the diagonal elements may be different. Usually the symmetry of a medium forces some elements of the susceptibility tensors to be zero and sets up relationships between others. The higher the symmetry of a medium, the smaller the number of nonzero elements and the more relationships there are between elements. The nonzero components of the material tensors and linear relations between them may be found in various sources [7–9]. A specialized computer program is supplied with Ref. [7] to calculate the symmetries of high-rank susceptibility tensors. A new CD encyclopedia of the symmetries of high-rank material tensors will be published shortly [10]. In Appendices 1 and 2 we list the tensors of the local and nonlocal optical susceptibilities of order three for crystals of different symmetry classes. They will be widely used in this book.

The presentation of a susceptibility tensor depends on the coordinate system in which it is written. The crystal form (cell) is connected with the so-called crystallographic system of coordinates, which is constructed on the basis of vectors coinciding with the edges of the unit cell of the crystal (see, for example, Refs. [7] to [9]). This coordinate system is not always convenient because its basis vectors are not orthonormal in all crystal symmetries. It is therefore more popular to use Cartesian coordinate frames with pairwise orthogonal axes which are known as the natural coordinate systems. For crystals of all symmetries there is a certain widely accepted way of directing the Cartesian axes X, Y, Z of the natural system. The prescriptions for the natural coordinate system installation may be found in various sources [7, 8]. In fact, they are simple and logical (see Table 2.1). For example, the Z axis in the hexagonal and trigonal systems is chosen along the six- and threefold symmetry axes respectively, the main distinctive symmetry elements of the structures. In cubic crystals all three Cartesian axes X, Y, Z are chosen along the edges of the cubic unit cell and so on.

For practical reasons, however, it is often necessary to describe nonlinear optical effects in a Cartesian coordinate system which is different from the natural one. Such a coordinate system may be imposed by the laboratory environment; for example, one may want to have one axis of the coordinate frame along the laser beam propagation direction and the other perpendicular to

Table 2.1 Rules for choosing natural coordinates X, Y, Z^a (after Ref. 8)

System	Class	X	Y	Z
Triclinic	$\bar{1}$, 1	belongs to (010)	$\perp$(010)	[001]
Monoclinic (first set)	2 (2$\parallel$$Y$) m ($m\perp Y$) $2/m$ (2$\parallel$$Y$)	$\perp$(100)	[010]	[001]
Monoclinic (second set)	2 (2$\parallel$$Z$) m ($m\perp Z$) $2/m$ (2$\parallel$$Z$)	[100]	$\perp$(010)	[001]
Orthorhombic Tetragonal Cubic	222, $mm2$, mmm, 4, $\bar{4}$, $4/m$, 422, $4mm$, $\bar{4}2m$, $4/mmm$, 23, $m3$, 432, $\bar{4}3m$, $m3m$	[100]	[010]	[001]
Trigonal Hexagonal	3, $\bar{3}$, 32, $3m$, $\bar{3}m$, 6, $\bar{6}$, $6/m$, 622, $6mm$, $\bar{6}2m$ $6/mmm$	[2$\bar{1}\bar{1}$0]	[01$\bar{1}$0]	[0001]

[a]For trigonal and hexagonal systems directions for X, Y and Z axes are quoted using Miller-Bravais notation (see, for example, F.C. Phillips *An Introduction to Crystallography*, Longman, London, 1977, p. 84). For all other systems, Millerian symbols are used (as quoted, p. 42).

the optical table. If necessary, the susceptibility tensor T may recalculated in the laboratory Cartesian system as follows:

$$T'_{i_1 i_2 \ldots i_R} = c_{i_1 j_1} \ldots c_{i_R j_R} T_{j_1 j_2 \ldots j_R} \tag{2.39}$$

Here T' is the tensor in the laboratory system of coordinates and c_{ij} is the matrix of transformation from the natural to the laboratory system. The elements of this transformation matrix are in fact the cosines of the angles between the axes of the natural frame XYZ and the new coordinate frame $X'Y'Z'$ (see Fig. 2.2):

$$\begin{aligned} c_{11} &= \cos(\angle XOX') \\ c_{12} &= \cos(\angle XOY') \\ &\ldots \\ &\ldots \\ c_{33} &= \cos(\angle ZOZ') \end{aligned} \tag{2.40}$$

The elements of the transformation matrix may be expressed in a standard fashion using only three angles, known as the Euler angles (see, for example, Ref. [11]). The Euler angles unambiguously define the rotation from the natural coordinate system to the laboratory one.

We now consider nonlinear optical properties of media of the most widespread symmetry, i.e. the symmetry of an isotropic medium. Obviously here the

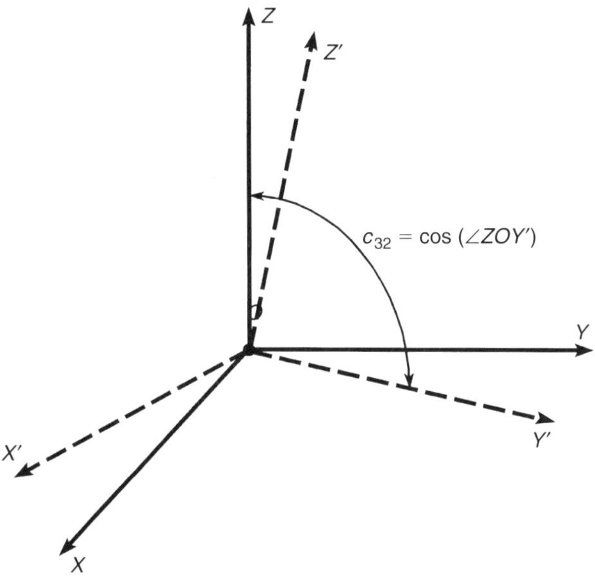

Fig. 2.2 Mutual orientation of two Cartesian coordinate frames is determined by the angles between their axes. The cosines of these angles give the transformation matrix c_{ij}

Cartesian system may be chosen arbitrarily since all directions are equivalent. In isotropic media the susceptibility tensors may be presented via two elementary tensors, a totally symmetric tensor of rank two which is known as the Kronecker δ symbol:

$$\begin{aligned} &\delta_{xx} = \delta_{yy} = \delta_{xx} = 1 \\ &\delta_{ij} = 0 \qquad \text{if } i \neq j \end{aligned} \tag{2.41}$$

and totally antisymmetric third-rank tensor known as the Levi–Civita symbol e_{ijk}:

$$\begin{aligned} &e_{xyz} = e_{zxy} = e_{yzx} = 1 \\ &e_{xzy} = e_{zyx} = e_{yxz} = -1 \\ &e_{ijk} = 0 \qquad \text{if } i = j \text{ or } i = k \text{ or } k = j \end{aligned} \tag{2.42}$$

In terms of these two elementary tensors the first-, second- and third-order susceptibility tensors for an isotropic medium are as follows:

$$\begin{aligned} \epsilon_{ij}(\omega) &= \epsilon\delta_{ij} \\ \Gamma^{(1)}_{ijn}(\omega) &= \Gamma^{(1)} e_{ijn} \end{aligned} \tag{2.43}$$

$$\begin{aligned} \chi^{(2)}_{ijk}(\omega_a+\omega_b,\omega_a,\omega_b) &= \chi^{(2)} e_{ijk} \\ \Gamma^{(2)}_{ijkn}(\omega_a+\omega_b,\omega_{a,}\omega_b) &= \Gamma^{(2)}_1\delta_{ij}\delta_{kn} + \Gamma^{(2)}_2\delta_{ik}\delta_{jn} + \Gamma^{(2)}_3\delta_{in}\delta_{jk} \end{aligned} \tag{2.44}$$

$$\chi^{(3)}_{ijkl}(\omega_a+\omega_b+\omega_c,\omega_a,\omega_b,\omega_c) = \chi^{(3)}_1\delta_{ij}\delta_{kl} + \chi^{(3)}_2\delta_{ik}\delta_{jl} + \chi^{(3)}_3\delta_{il}\delta_{jk}$$

$$\begin{aligned} \Gamma^{(3)}_{ijkln}(\omega_a+\omega_b+\omega_c,\omega_a,\omega_b,\omega_c) = {} & \Gamma^{(3)}_1(\delta_{il}e_{jkn} + \delta_{jl}e_{ikn} + \delta_{ik}e_{jln} + \delta_{jk}e_{iln}) \\ & + \Gamma^{(3)}_2\delta_{ij}e_{kln} + \Gamma^{(3)}_3(\delta_{jn}e_{kli} + \delta_{in}e_{klj}) \\ & + \Gamma^{(3)}_4\delta_{kl}e_{ijn} + \Gamma^{(3)}_5(\delta_{ln}e_{ijk} + \delta_{kn}e_{ijl}) \\ & + \Gamma^{(3)}_6(\delta_{jn}e_{kli} - \delta_{im}e_{klj}) \end{aligned} \tag{2.45}$$

Here ϵ, $\Gamma^{(1)}$, $\chi^{(2)}$, $\Gamma^{(2)}_{1-3}$, $\chi^{(3)}_{1-3}$ and $\Gamma^{(3)}_{1-6}$ are scalar quantities.

2.5 CONSTITUTIVE EQUATIONS FOR SPECIFIC OPTICAL EFFECTS

Constitutive equations (2.21) and (2.24) are the most general forms of relations between optical excitation and response written in terms of optical susceptibilities. We will now specify them for important cases of light propagation in quadratic and cubic media.

Let us first consider the interaction of an electromagnetic wave of frequency ω with a medium having quadratic nonlinearity. In general the optical response of the medium is given by the constitutive equation (2.21). Accordingly to (2.19), the electric field associated with the light wave at the frequency ω has two complex-conjugated spectral components, $\mathbf{E}(\omega,\mathbf{r}) = [\mathbf{E}(-\omega,\mathbf{r})]^*$. This implies that the summation in equation (2.21) is taken over $\omega_a = \pm\omega$ and $\omega_b = \pm\omega$. As a result the sum in equation (2.21) will have four terms and the polarization response $\mathbf{P}^{(2)}(\mathbf{r}, t)$ will have spectral components at frequencies $\pm 2\omega$ and zero frequency:

$$
\begin{aligned}
\mathbf{P}^{(2)}(\mathbf{r}, t) = {} & \chi^{(2)}_{ijk}(2\omega, \omega, \omega)E_j(\omega, \mathbf{r})E_k(\omega, \mathbf{r})\mathrm{e}^{-2i\omega t} \\
& + \Gamma^{(2)}_{ijkn}(2\omega, \omega, \omega)\nabla_n E_j(\omega, \mathbf{r})E_k(\omega, \mathbf{r})\mathrm{e}^{-2i\omega t} \\
& + \chi^{(2)}_{ijk}(0, \omega, -\omega)E_j(\omega, \mathbf{r})E^*_k(\omega, \mathbf{r}) \\
& + \Gamma^{(2)}_{ijkn}(0, \omega, -\omega)\nabla_n E_j(\omega, \mathbf{r})E^*_k(\omega, \mathbf{r}) + \text{c.c.}
\end{aligned} \tag{2.46}
$$

Therefore excitation at frequency ω generates a polarization response oscillating at the doubled frequency and a static polarization response of zero frequency. Now from equation (2.46) we can write down the constitutive equation for the second harmonic generation process:

$$
\begin{aligned}
\mathbf{P}^{(2)}(2\omega, \mathbf{r}) = {} & \chi^{(2)}_{ijk}(2\omega, \omega, \omega)E_j(\omega, \mathbf{r})E_k(\omega, \mathbf{r}) \\
& + \Gamma^{(2)}_{ijkn}(2\omega, \omega, \omega)E_k(\omega, \mathbf{r})\nabla_n E_j(\omega, \mathbf{r})
\end{aligned} \tag{2.47}
$$

It has a local (first) term and a nonlocal (second) term that contain spatial derivatives of the electric field. Since the last two frequencies in the arguments of the local susceptibility are equal, the frequency permutation symmetry (2.33) requires $\chi^{(2)}_{ijk}(2\omega, \omega, \omega)$ to be symmetrical with respect to permutations of the second and third Cartesian indices:

$$
\chi^{(2)}_{ijk}(2\omega, \omega, \omega) = \chi^{(2)}_{ikj}(2\omega, \omega, \omega) \tag{2.48}
$$

Constitutive equation (2.47) describes the optical response of the quadratic medium at the double frequency when the incident electric field has an arbitrary dependence on coordinates. In particular, this equation is suitable for describing the interaction of two light waves at frequency ω propagating in different directions (noncollinear second harmonic generation).

In the simplest case of the single-beam second harmonic generation process, equation (2.47) can be simplified. In this case the spectral components at $+\omega$ and $-\omega$ are $\mathbf{A}\exp(i\mathbf{k}\cdot\mathbf{r})$ and $\mathbf{A}^*\exp(-i\mathbf{k}\cdot\mathbf{r})$ respectively, where $\mathbf{A}$ is the wave amplitude and $\mathbf{k}$ is the wave vector. Therefore, the constitutive equation (2.47) takes the following form:

$$
P^{(2)}_i(2\omega, \mathbf{r}) = [\chi^{(2)}_{ijk}(2\omega, \omega, \omega) + ik_n\Gamma^{(2)}_{ijkn}(2\omega, \omega, \omega)]A_jA_k\,\mathrm{e}^{2i(\mathbf{k}\cdot\mathbf{r})} \tag{2.49}
$$

This constitutive equation clearly shows that a sole light wave with frequency ω and wave vector $\mathbf{k}$ generates an optical response of frequency 2ω which propagates with wave vector $2\mathbf{k}$. Note that the form of the constitutive equation(2.49) implies that only part of the second-order nonlocal susceptibility tensor which is symmetrical with respect to the permutation of the second and third Cartesian indexes,

$$\Gamma^{s}_{ijkn}(2\omega, \omega, \omega) = \Gamma^{s}_{ikjn}(2\omega, \omega, \omega) = \tfrac{1}{2}[(\Gamma^{(2)}_{ijkn}(2\omega, \omega, \omega) + \Gamma^{(2)}_{ikjn}(2\omega, \omega, \omega)] \tag{2.50}$$

will contribute to the polarization response at the frequency 2ω. This is because the polarization response in equation (2.49) depends on $A_j A_k$, which is symmetrical with respect to permutation of the indices j and k. Therefore, after the summation over those 'dummy' Cartesian indices only the part of the tensor $\Gamma^{(2)}_{ijkn}(2\omega, \omega, \omega)$ that is symmetrical with respect to permutation of j and k will survive (see Ref. [7] for details). Correspondingly, the constitutive equation for the single-beam second harmonic generation process may be written in terms of the spectral component of the polarization response [7]:

$$P^{(2)}_{i}(2\omega, 2\mathbf{k}) = [\chi^{(2)}_{ijk}(2\omega, \omega, \omega) + 2ik_n\Gamma^{s}_{ijkn}(2\omega, \omega, \omega)]A_j A_k \tag{2.51}$$

In equation (2.51) the tensors of local quadratic susceptibility $\chi^{(2)}_{ijk}(2\omega, \omega, \omega)$ and nonlocal quadratic susceptibility $\Gamma^{s}_{ijkn}(2\omega, \omega, \omega)$ are symmetrical with respect to the permutation of the second and third Cartesian indices.

If an electromagnetic wave of frequency ω interacts with a medium having cubic nonlinearity, $\mathbf{P}^{(3)}(\mathbf{r}, t)$ is given by the constitutive equation (2.24) where $\omega_a = \pm\omega$, $\omega_b = \pm\omega$ and $\omega_c = \pm\omega$. The sum in equation (2.24) has eight terms and the polarization response has spectral components at frequencies $\pm 3\omega$ and $\pm\omega$:

$$\begin{aligned} P^{(3)}_{i}(\mathbf{r}, t) = {} & [\chi^{(3)}_{ijkl}(3\omega, \omega, \omega, \omega)E_j(\omega, \mathbf{r})E_k(\omega, \mathbf{r})E_l(\omega, \mathbf{r}) \\ & + \Gamma^{(3)}_{ijkln}(3\omega, \omega, \omega, \omega)E_k(\omega, \mathbf{r})E_l(\omega, \mathbf{r})\nabla_n E_j(\omega, \mathbf{r})]\mathrm{e}^{-3\mathrm{i}\omega t} \\ & + [3\chi^{(3)}_{ijke}(\omega, \omega, \omega, -\omega)E_j(\omega, \mathbf{r})E^{*}_{i}(\omega, \mathbf{r})E_k(\omega, \mathbf{r})E^{*}_{e}(\omega, \mathbf{r}) \\ & + 2\Gamma^{(3)}_{ijkln}(\omega, \omega, \omega, -\omega)E_k(\omega, \mathbf{r})E^{*}_{l}(\omega, \mathbf{r})\nabla_n E_j(\omega, \mathbf{r}) \\ & + \Gamma^{(3)}_{ijkln}(\omega, -\omega, \omega, \omega)E_k(\omega, \mathbf{r})E_i(\omega, \mathbf{r})\nabla_n E^{*}_{j}(\omega, \mathbf{r})]\mathrm{e}^{-\mathrm{i}\omega t} + \text{c.c.} \end{aligned} \tag{2.52}$$

Clearly, the light wave at frequency ω generates an optical response in the medium oscillating at the frequency of the third optical harmonic (the first two terms in equation (2.52)) and at the frequency of the incident wave (the last terms in equation (2.52)). From equation (2.52) we can write down the equation for the spectral component of the optical response at frequency ω:

$$P_i^{(3)}(\omega, \mathbf{r}) = 3\chi_{ijkl}^{(3)}(\omega, \omega, \omega, -\omega)E_j(\omega, \mathbf{r})E_k(\omega, \mathbf{r})E_l^*(\omega, \mathbf{r})$$
$$+ 2\Gamma_{ijkln}^{(3)}(\omega, \omega, \omega, -\omega)E_k(\omega, \mathbf{r})E_l^*(\omega, \mathbf{r})\nabla_n E_j(\omega, \mathbf{r})$$
$$+ \Gamma_{ijkln}^{(3)}(\omega, -\omega, \omega, \omega)E_k(\omega, \mathbf{r})E_l(\omega, \mathbf{r})\nabla_n E_j^*(\omega, \mathbf{r}) \quad (2.53)$$

In the derivation of equations (2.53) and (2.52) we have used the frequency permutation relations (2.34) and (2.35).

Since the second and third frequencies in the arguments of the local cubic susceptibility are equal, the frequency permutation symmetry (2.34) requires $\chi_{ijkl}^{(3)}(\omega, \omega, \omega, -\omega)$ to be symmetrical with respect to permutations of the second and third Cartesian indices:

$$\chi_{ijkl}^{(3)}(\omega, \omega, \omega, -\omega) = \chi_{ikjl}^{(3)}(\omega, \omega, \omega, -\omega) \quad (2.54)$$

Constitutive equation (2.53) describes the optical response of the cubic medium at the incident frequency when the incident electric field has an arbitrary coordinate dependence. In particular, this equation is suitable when there are several light waves of the same frequency propagating in different directions.

In the simplest case of a single light wave, the constitutive equation (2.53) describes the process of light wave self-action and may be significantly simplified. Indeed, by substituting $\mathbf{E}(\omega, \mathbf{r}) = \mathbf{A}\exp(\mathrm{i}\mathbf{k}\cdot\mathbf{r})$ into equation (2.53) we can rewrite it as follows:

$$P_i^{(3)}(\omega, \mathbf{r}) = [3\chi_{ijkl}^{(3)}(\omega, \omega, \omega, -\omega) + 2ik_n\Gamma_{ijkln}^{(3)}(\omega, \omega, \omega, -\omega)$$
$$- ik_n\Gamma_{ilkjn}^{(3)}(\omega, -\omega, \omega, \omega)]A_jA_kA_l^*\mathrm{e}^{\mathrm{i}(\mathbf{k}\cdot\mathbf{r})} \quad (2.55)$$

To arrive at equation (2.55) we have replaced the dummy index j with l and vice versa in the last term on the right-hand side of equation (2.53). The optical response given by equation (2.55) depends on A_jA_k, which is symmetrical with respect to permutation of the indices j and k. Therefore only that part of the tensor $\Gamma_{ijkln}^{(3)}(\omega, \omega, \omega, -\omega)$ which is symmetrical with respect to permutation of j and k will contribute to the second term on the right-hand side in equation (2.55). Consequently, the constitutive equation (2.55) may be rewritten in terms of the spectral component of the optical response as follows [7]:

$$P_i^{(3)}(\omega, \mathbf{k}) = (\chi_{ijkl} + ik_n\Gamma_{ijkln})A_jA_kA_l^* \quad (2.56)$$

Here we have dropped frequency arguments and simplified the notation by introducing the following tensors, which are symmetrical with respect to the permutation of the second and third indices:

$$\chi_{ijkl} = \chi_{ikjl} = 3\chi_{ijkl}^{(3)}(\omega, \omega, \omega, -\omega) \quad (2.57)$$

and

$$\Gamma_{ijkln} = \Gamma_{ikjln} = \Gamma^{(3)}_{ijkln}(\omega, \omega, \omega, -\omega) + \Gamma^{(3)}_{ikjln}(\omega, \omega, \omega, -\omega) - \Gamma^{(3)}_{ilkjn}(\omega, \omega, \omega, -\omega) \tag{2.58}$$

The constitutive equation (2.56) will be widely used in this book.

2.6 UNITS FOR OPTICAL SUSCEPTIBILITIES

The nonlinear constitutive equations in this book are written in esu units of physical quantities. Generally, they are of the form:

$$\mathbf{P}^{(n)} = \chi^{(n)}\mathbf{E}^n + i\mathbf{k}\Gamma^{(n)}\mathbf{E}^n \tag{2.59}$$

where $n > 1$. The quantities entering the equations have the following dimensions:

$\mathbf{P}^{(n)}$, $\mathbf{E}$	stat volt cm^{-1} or cm$^{-1/2}$ g$^{1/2}$ s^{-1}
$\chi^{(n)}$	stat volt^{1-n} cm^{n-1}
$\Gamma^{(n)}$	stat volt^{1-n} cm^{n}
$\mathbf{k}$	cm^{-1}

The same nonlinear constitutive equation may be presented in the SI system of units:

$$\mathbf{P}^{(n)} = \epsilon_0(\chi^{(n,\mathrm{SI})}\mathbf{E}^n + i\mathbf{k}\Gamma^{(n,\mathrm{SI})}\mathbf{E}^n) \tag{2.60}$$

Here the permittivity of a vacuum $\epsilon_0 = 8.854 \times 10^{-12}\mathrm{F/m}$. In SI units the quantities entering the constitutive equation have the following dimensions:

$\mathbf{P}^{(n)}$	Cm^{-2}
$\mathbf{E}$	volt m^{-1}
$\chi^{(n,\mathrm{SI})}$	volt^{1-n} m^{n-1}
$\Gamma^{(n,\mathrm{SI})}$	volt^{1-n} m^{n}
$\mathbf{k}$	m^{-1}

The following relationships hold between the values of the nonlinear susceptibilities in SI and esu units:

$$\frac{\chi^{(n,\mathrm{SI})}}{\chi^{(n)}} = 4\pi(3 \times 10^4)^{1-n} \tag{2.61}$$

and

$$\frac{\Gamma^{(n,\mathrm{SI})}}{\Gamma^{(n)}} = 400\pi(3 \times 10^4)^{1-n} \tag{2.62}$$

In this book the linear constitutive equations used are written in esu units (see equation (2.14), for example):

$$P_i^{(1)} = \frac{1}{4\pi}\{[\epsilon_{ij} - \delta_{ij}]E_j + \Gamma_{ijn}^{(1)}\nabla_n E_j\}$$

In SI units the same equation will be presented as

$$P_i^{(1)} = \epsilon_0(\chi_{ij}^{(1,SI)}E_j + \Gamma_{ijn}^{(1,SI)}\nabla_n E_j)$$

The following relations between coefficients of the equations hold:

$$\chi_{ij}^{(1,SI)} = \epsilon_{ij} - \delta_{ij}$$

$$\Gamma_{ijn}^{(1,SI)} = 100\Gamma_{ijn}^{(1)}$$

Some sources use a form of the constitutive equation that is slightly different from equation (2.59). It operates not with the polarization response vector **P**, but the electric displacement vector $\mathbf{D} = \mathbf{E} + 4\pi\mathbf{P}$. In terms of the electric displacement, the constitutive equation in the esu units is

$$\mathbf{D}^{(n)} = \epsilon^{(n)}\mathbf{E}^n + i\mathbf{k}G^{(n)}\mathbf{E}^n \tag{2.63}$$

In esu units **D** has the same dimension as **P**. The following relations exist between the nonlinear susceptibilities $\chi^{(n)}$ and $\Gamma^{(n)}$ and the nonlinear permitivities $\epsilon^{(n)}$ and $G^{(n)}(n > 1)$:

$$\begin{aligned} 4\pi\chi^{(n)} &= \epsilon^{(n)} \\ 4\pi\Gamma^{(n)} &= G^{(n)} \end{aligned} \tag{2.64}$$

Constitutive equation (2.63) may be presented in SI units:

$$\mathbf{D}^{(n)} = \epsilon_0(\epsilon^{(n)}\mathbf{E}^n + i\mathbf{k}G^{(n)}\mathbf{E}^n) \tag{2.65}$$

The same relationship as in equations (2.61) and (2.62) holds between the values of the nonlinear permitivities in SI and esu units ($h > 1$):

$$\frac{\epsilon^{(n,\mathrm{SI})}}{\epsilon^{(n)}} = 4\pi(3 \times 10^4)^{1-n} \tag{2.66}$$

and

$$\frac{G^{(n,\mathrm{SI})}}{G^{(n)}} = 400\pi(3 \times 10^4)^{1-n} \tag{2.67}$$

In this presentation the linear constitutive equation for the electric displacement in esu units is

$$D_i^{(1)} = \epsilon_{ij}E_j + \Gamma_{ijn}^{(1)}\nabla_n E_j$$

while in SI units it will be

$$D_i^{(1)} = \epsilon_0(\epsilon_{ij}E_j + G_{ijn}^{(1,SI)}\nabla_n E_j)$$

where

$$G_{ijn}^{(1,SI)} = 100\ \Gamma_{ijn}^{(1)}$$

REFERENCES

1. L.D. Landau, E.M. Lifshitz and L.P. Pitaevski, *Electrodynamics of Continuous Media*, Pergamon Press, Oxford, 1984.
2. J. Van Kranendonk and J.E. Sipe, *Prog. Optics*, **15**, 245 (1977).
3. E.J. Post, *Formal Structure of Electromagnetic*, North-Holland, Amsterdam, 1962.
4. A.R. Bungay, Yu.P Svirko and N.I. Zheludev, Phys. Rev. B, **47**, 11730 (1993).
5. V.M. Agranovich and V.L. Ginsburg, *Spatial Dispersion in Crystal Optics and Theory of Excitons*. Interscience Publishers, New York, 1966.
6. P.N. Butcher and D. Cotter, *The Elements of Nonlinear Optics*, Cambridge University Press, Cambridge, 1990.
7. S.V. Popov, Yu.P. Svirko and N.I. Zheludev, *Susceptibility Tensors for Nonlinear Optics*, IOP, Bristol and Philadelphia, 1995.
8. Yu.I. Sirotin and M.P. Shaskolskaya, *Foundation of Physics of Crystals*, Nauka, Moscow, 1979 (in Russian).
9. B.K. Vainshtein, *Modern Crystallography. I. Symmetry of Crystals*. Springer-Verlag, Heidelberg, 1981.
10. Yu. P. Svirko, S.V. Popov and N.I. Zheludev, *CD Encyclopedia of Material Tensors*, John Wiley & Sons, Chichester (in preparation).
11. G.A. Korn and T.M. Korn, *Mathematical Handbook*, McGraw-Hill, New York, 1968, Ch. 14.

FURTHER READING

S.A. Akhmanov and R.V. Khohlov, *Problems of Nonlinear Optics*, Gordon and Breach, New York, 1972.

S.A. Akhmanov and S.Yu. Nikitin, *Physical Optics*, Clarendon Press, Oxford, 1997.

R.W. Boyd, *Nonlinear Optics*, Academic Press, San Diego, 1992.

S.R. De Groot, *The Maxwell Equations*, North-Holland, Amsterdam, 1969.

Y.R. Shen, *The Principles of Nonlinear Optics*, Wiley-Interscience, New York, 1984.

3 Polarization of Light and the Evolution Equation for the Stokes Parameters in Linear Optics

In this chapter we will provide the essential foundation for the future consideration of nonlinear polarization effects. We will introduce the Stokes parameters and will relate the Stokes formalism to the traditional measures of light polarization, such as polarization azimuth and ellipticity. The evolution equations for the Stokes parameters in a linear optical medium will be derived. The basic polarization effects of linear and circular birefringence and dichroism will be considered and illustrated by presenting them on the Poincaré sphere. General solutions to the Stokes parameter evolution equations will be derived and the criteria for the eigenpolarizations of the medium will be found. Finally, the equation for the evolution of the degree of polarization of partially polarized light will be derived.

3.1 POLARIZATION OF LIGHT

By their very nature, the spatial–temporal fields of electromagnetic waves are vectors. To account for this we shall introduce the idea of light polarization. Since a number of specialized books have been devoted to the description of light polarization (see, for example, Refs. [1] to [4]), we will only briefly review this subject. We will define the polarization for a plane transverse electromagnetic wave where the electric and magnetic field vectors are perpendicular to the direction of propagation. A generalization of the idea of light polarization for divergent light beams has been developed in Ref. [5]. Let us set up a right-handed Cartesian coordinate frame with the Z axis pointing along the direction of wave propagation. We will consider a wave forming a narrow spectral band around a central frequency ω:

$$E_{x,y}(z,\, t) = A_{x,y}(z,\, t)\,\mathrm{e}^{-\mathrm{i}\omega t+\mathrm{i}kz} + \mathrm{c.c.} \tag{3.1}$$

Here $\mathbf{A}$ is the wave amplitude and k is the wave number (modulus of the wave vector $\mathbf{k}$). Such a wave is called a plane wave because its wavefront is a plane: an observer facing the optical source sees the vector $\mathbf{E}$ oscillating in the XY plane, which is perpendicular to the direction of propagation.

Representation (3.1) implies that anizotropy along the direction of propagation is small, the wave remains predominantly transverse and the core complex refractive index, $n = ck/\omega$, may be introduced. The medium anisotropy results in different evolution rates for $A_x(z,\, t)$ and $A_y(z,\, t)$ with z. This difference will eventually lead to a change in the light polarization state. For a wave with a narrow spectral band, the amplitude $A_{x,y}(z, t)$ changes insignificantly over the period of the wave: $|\partial A_{x,y}(z, t)/\partial t| \ll \omega |A_{x,y}(z, t)|$.

If E_x and E_y oscillate in phase, that is $A_x(z, t) = \alpha A_y(z, t)$, with α being a real constant, formula (3.1) represents the electric field of a linearly polarized plane wave. The wave is called linearly polarized because the observer sees the oscillating vector tracing out a straight line in the *XY* plane. The plane containing the wave vector **k** and the vector **E** is called the plane of polarization. If $A_y(z, t) = 0$, the observer sees the vector **E** oscillating along *X*. Such a wave is referred to as linearly polarized along the *X* direction. If $A_x(z, t) = 0$, the wave is referred to as linearly polarized along the *Y* direction.

If there is a phase shift between E_x and E_y, that is $A_x(\mathrm{z, t}) = \alpha A_y(z, t)$, with α being a complex number, the wave is said to have elliptical polarization: an observer facing the wave source sees the end of the vector **E** tracing out an ellipse. Circular polarization is a specific case of elliptical polarization where $\alpha = \pm i$. It is called circular because an observer sees the end of the vector **E** tracing out a circle. Circularly polarized waves may be left- or right-handed. If $A_x(z, t) = iA_y(z, t)$, it is said that the wave is a left-handed circularly polarized (LCP) wave: for the observer looking in the direction from which the light is coming, the end of the electric field vector would appear to trace out a circle in a counterclockwise direction. If $A_x(z, t) = -iA_y(z, t)$, it is said that the wave is a right-handed circularly polarized (RCP) wave: the end of the electric field vector would appear to trace out a circle in a clockwise direction. Right-handed elliptically polarized waves (REP) and left-handed elliptically polarized (LEP) waves are defined analogously.

The polarization state of an elliptically polarized wave may be completely specified by two parameters, the polarization azimuth and the ellipticity angle (or just ellipticity for the sake of shortening) (see Fig. 3.1). The polarization azimuth Θ is the angle measured between the *X* direction of the right-handed Cartesian frame and the major (longest) axis of the polarization ellipse. Θ may take values between 0 and π.

The ellipticity angle η is defined by enclosing the polarization ellipse in a rectangular 'box' as shown in Fig. 3.1; η is the angle between the main diagonal of this box and its longest side. The angle is given a positive sign for right elliptically polarized light and a negative sign for left elliptically polarized light. Therefore the polarization angle varies from 0 to $\pi/4$ for right elliptically polarized waves and from 0 to $\pm\pi/4$ for left elliptically polarized waves. Right and left circularly polarized waves correspond to $\eta = -\pi/4$ while $\eta = 0$ corresponds to a linearly polarized wave.

Definitions of the wave polarization state based on the polarization azimuth and ellipticity angle are intuitively clear and simple. However, because of the very high frequency of light waves, an experiment in which the trace of the end of the electric field vector is followed is impossible. Another limitation is that the polarization ellipse concept is only applicable when totally polarized light is

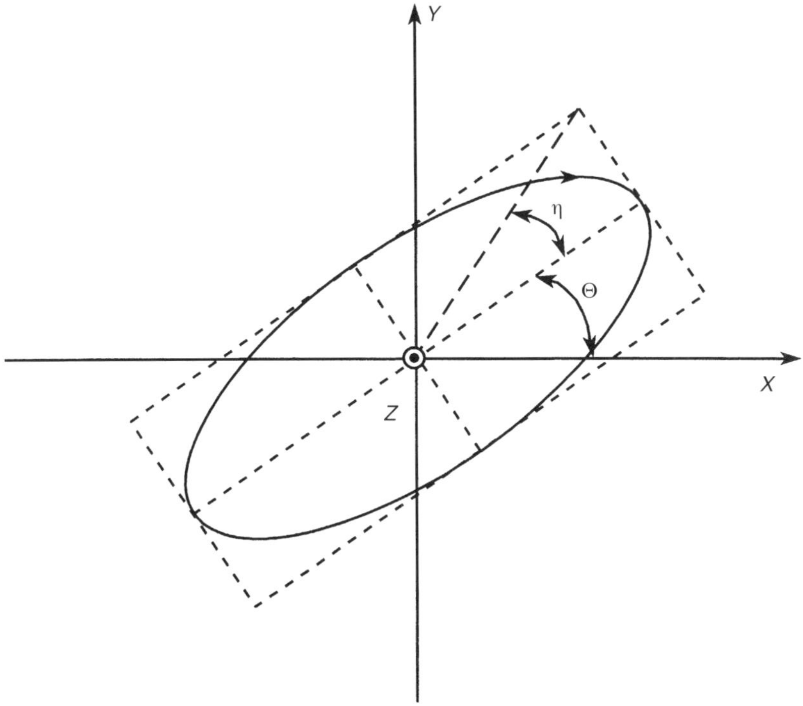

Fig. 3.1 Definition of the polarization azimuth Θ and the ellipticity angle η. The elliptical trace of the end of the electric field vector as seen by an observer facing the light source is presented. The sketch corresponds to a right-handed elliptically polarized wave

being considered. It is not suitable for partially polarized light. Therefore we need to find a way of relating the parameters Θ and η to quantities, such as intensity, that can be measured directly in an optical experiment. In 1852 George Gabriel Stokes found that the polarization of light could be represented by a set of four observables (measurable parameters) now known as the Stokes parameters [6]. In what follows we will define the Stokes parameters and show how they are related to the parameters of polarization ellipse, Θ and η.

We start by introducing Wolf's coherency matrix:

$$g_{ij} = \begin{pmatrix} \langle A_x(z, t)A_x^*(z, t)\rangle & \langle A_x(z, t)A_y^*(z, t)\rangle \\ \langle A_y(z, t)A_x^*(z, t)\rangle & \langle A_y(z, t)A_y^*(z, t)\rangle \end{pmatrix} e^{-2\mathrm{Im}\{k\}z} \tag{3.2}$$

where $\langle . .\rangle$ denotes a time average over the photodetector integration time. The

Stokes parameters, $S_\alpha = \{S_0; S_1; S_2; S_3\}$, of the electromagnetic wave are defined as follows:

$$S_\alpha = \sigma_{ij}^{(\alpha)} g_{ji} \tag{3.3}$$

Here a summation is presumed over the indices i and j, and $\sigma_{ij}^{(\alpha)}$ are the Pauli spin matrices of quantum mechanics [7]:

$$\begin{aligned} \sigma^{(0)} &= \begin{pmatrix} 1 & 0 \\ 0 & 1 \end{pmatrix} \qquad \sigma^{(1)} = \begin{pmatrix} 1 & 0 \\ 0 & -1 \end{pmatrix} \\ \sigma^{(2)} &= \begin{pmatrix} 0 & 1 \\ 1 & 0 \end{pmatrix} \qquad \sigma^{(3)} = \begin{pmatrix} 0 & -\mathrm{i} \\ \mathrm{i} & 0 \end{pmatrix} \end{aligned} \tag{3.4}$$

The Pauli matrices have the following properties that will be used throughout this book:

$$\sigma_{ij}^{(0)}\sigma_{ji}^{(0)} = \sigma_{ij}^{(1)}\sigma_{ji}^{(1)} = \sigma_{ij}^{(2)}\sigma_{ji}^{(2)} = \sigma_{ij}^{(3)}\sigma_{ji}^{(3)} = 2 \tag{3.5}$$

$$\sigma_{ik}^{(\alpha)}\sigma_{kj}^{(\beta)} = \sigma_{ij}^{(0)}\delta_{\alpha\beta} + ie_{\alpha\beta\gamma}\sigma_{ij}^{(\gamma)} \tag{3.6}$$

In equation (3.6) α, β and $\gamma = 1$, 2 and 3, and $e_{\alpha\beta\gamma}$ is a totally antisymmetric Levi–Civita tensor ($e_{123} = e_{231} = e_{321} = -e_{213} = -e_{321} = -e_{231} = 1$ and other elements $e_{\alpha\beta\gamma}$ are zero).

Therefore the Stokes parameters are certain time-averaged quadratic forms of the Cartesian components of the light wave amplitude:

$$\begin{aligned} S_0 &= (\langle A_x A_x^* \rangle + \langle A_y A_y^* \rangle)\, \mathrm{e}^{-2\mathrm{Im}\{k\}z} \\ S_1 &= (\langle A_x A_x^* \rangle - \langle A_y A_y^* \rangle)\, \mathrm{e}^{-2\mathrm{Im}\{k\}z} \\ S_2 &= (\langle A_x A_y^* \rangle + \langle A_y A_x^* \rangle)\, \mathrm{e}^{-2\mathrm{Im}\{k\}z} \\ S_3 &= i(\langle A_x A_y^* \rangle - \langle A_y A_x^* \rangle)\, \mathrm{e}^{-2\mathrm{Im}\{k\}z} \end{aligned} \tag{3.7}$$

These four Stokes parameters will be often written as a string: $\{S_0, S_1, S_2, S_3\}$ or $S_0\{1, S_1/S_0, S_2/S_0, S_3/S_0\}$. For a wave propagating in a vacuum, the Stokes parameters may be easily measured with simple optical instruments and a photodetector. S_0 gives the light wave intensity, $I = cS_0/2\pi$, where c is the speed of light and may be measured directly by a photodetector. A photodetector equipped with a linear analyser allows the measurement of S_1 and S_2. S_1 is proportional to the intensity difference as measured with the analyser oriented along the X and Y axes. Similarly, S_2 is given by the intensity difference as measured with the analyser oriented along the bisector of the X and Y axes and the perpendicular direction. The Stokes parameter S_3 may be measured by a photodetector equipped with circular analysers. It is proportional to the intensity difference as measured with right and left circular analysers (see Refs [1] to [3]).

The introduction of the Stokes parameters allows a straightforward analytical definition of different polarization states of light, including partially polarized light. Unpolarized and partially polarized light come from sources such as the Sun, stars, light bulbs and so on. In unpolarized (or naturally polarized) light $A_x(z, t)$ and $A_y(z, t)$ fluctuate independently in time, giving a random distribution of the resulting **E** vector in the *XY* plane. If $A_x(z, t)$ and $A_y(z, t)$ are random they do not correlate: the time averages of the *XY* mixed forms, such as $\langle A_y(z, t)A_x^*(z, t)$, in equations (3.7) become zero and $\langle A_x(z, t)A_x^*(z, t) = \langle A_y(z, t)A_y^*(z, t)\rangle$. Therefore in unpolarized light only the parameter S_0, which is proportional to the intensity, is nonzero:

$$S_\alpha = S_0\{1; 0; 0; 0\} \tag{3.8}$$

Light is called totally polarized if the Stokes parameters satisfy the relation:

$$S_0 = \sqrt{S_1^2 + S_2^2 + S_3^2} \tag{3.9}$$

For an elliptically polarized wave where $A_x(z, t) = \alpha A_y(z, t)$ condition (3.9) is automatically satisfied. Therefore, an elliptically polarized wave is always totally polarized. As specific cases of elliptical polarization, circular and linearly polarized waves are also totally polarized.

Partially polarized light is a mixture of naturally polarized light and totally polarized light. All four Stokes parameters are independent, however:

$$S_0 > \sqrt{S_1^2 + S_2^2 + S_3^2} \tag{3.10}$$

The degree of polarization r for partially polarized light is defined as follows:

$$r = \frac{\sqrt{S_1^2 + S_2^2 + S_3^3}}{S_0} \tag{3.11}$$

For partially polarized light $0 < r < 1$, totally polarized light has $r = 1$, while natural (unpolarized) light has $r = 0$.

If the Stokes parameters are known, the polarization azimuth Θ and the ellipticity angle η may be returned as follows:

$$\tan 2\Theta = \frac{S_2}{S_1} \tag{3.12}$$

and

$$\sin 2\eta = \frac{S_3}{S_0} \tag{3.13}$$

If the light intensity, and therefore S_0, the degree of polarization r, the polarization azimuth Θ and the ellipticity angle η are known, using equations (3.11) to (3.13) one may return the Stokes parameters S_{1-3}:

$$S_1 = rS_0 \cos(2\eta)\cos(2\Theta)$$
$$S_2 = rS_0 \cos(2\eta)\sin(2\Theta) \tag{3.14}$$
$$S_3 = rS_0 \sin(2\eta)$$

Some examples are given below of different light polarization states as they are described in the formalism of the Stokes parameters:

Right-handed circular polarization:	$S_\alpha = S_0\{1;\ 0;\ 0;\ 1\}$
Left-handed circular polarization:	$S_\alpha = S_0\{1;\ 0;\ 0;\ -1\}$
Light linearly polarized along X:	$S_\alpha = S_0\{1;\ 1;\ 0;\ 0\}$
Light linearly polarized along Y:	$S_\alpha = S_0\{1;\ -1;\ 0;\ 0\}$
Light linearly polarized at 45° to X:	$S_\alpha = S_0\{1;\ 0;\ 1;\ 0\}$
Light linearly polarized at −45° to X:	$S_\alpha = S_0\{1;\ 0;\ -1;\ 0\}$

Numerous useful properties of the Stokes polarization parameters may be found in Refs. [1] and [2].

A particular polarization state may be represented as a point in a three-dimensional Stokes coordinate space by introducing the so-called Stokes vector **S** with Cartesian coordinates S_1, S_2 and S_3. The point lies on a sphere of radius $(S_1^2 + S_2^2 + S_3^2)^{1/2}$. This sphere in Stokes space is known as the Poincaré sphere (see Fig. 3.2). The sphere was first introduced in 1892 by Henri Poincaré who discovered that different polarization states may be represented on a sphere in exactly the same manner as the positions of stars in the stereoscopic projection of the sky [9].

If light is totally polarized, the radius of the Poincaré sphere is equal to S_0. The 'north' and 'south' poles of the sphere correspond to right and left circular polarizations. Points on the 'equator' represent linearly polarized light with the 'longitude' being twice the polarization azimuth Θ. The 'latitude' of a point will give twice the angle of ellipticity η of the wave. Antipodes, i.e. points at opposite ends of a diameter of the sphere, are known as orthogonal polarizations. For example, a pair of linear polarizations on opposite sides of the equator are orthogonal. Left- and right-handed circular polarizations, on opposite poles, are also orthogonal polarizations. If light is partially polarized the ratio between the Poincaré sphere radius and the Stokes parameter S_0 gives the degree of polarization $r = (S_1^2 + S_2^2 + S_3^2)^{1/2}/S_0$.

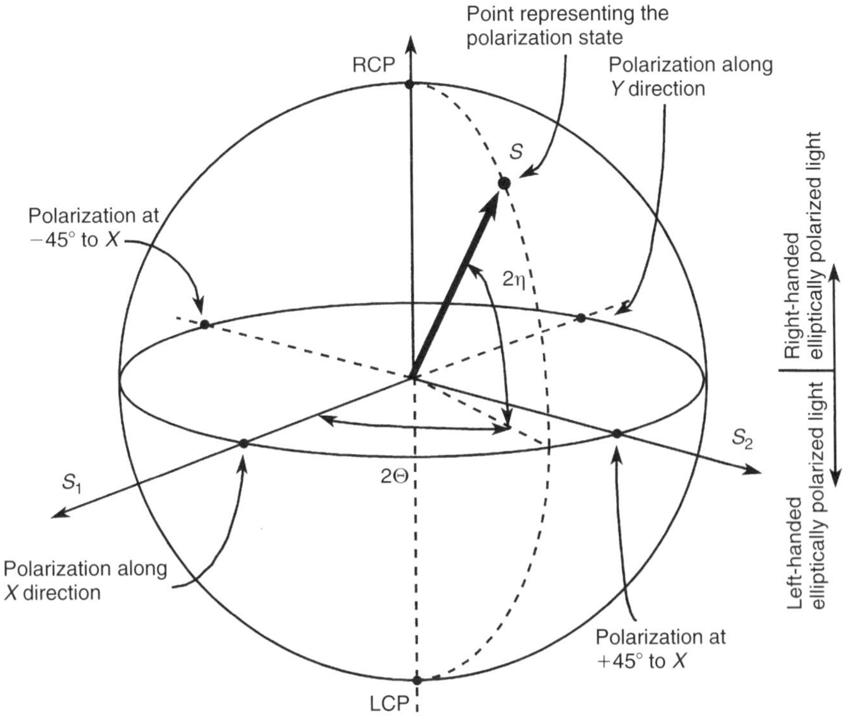

Fig. 3.2 Representation of the polarization state of light wave on the Poincaré sphere

3.2 THE EVOLUTION EQUATIONS FOR THE STOKES PARAMETERS

Maxwell's equations (2.2) combined with the constitutive equation (2.16) provide the starting point for the analysis of light propagation in linear media. For a plane, narrow-band electromagnetic wave this analysis may be considerably simplified by reducing the set of Maxwell's equations to an equation for the magnitude of the wave. From here one can calculate the electric field as a function of the propagation coordinate and then determine the polarization state of the wave. Alternatively, one may derive the evolution equations for the Stokes parameters of the wave, thus avoiding the intermediate stage of calculating the fields. Initially the evolution equations for the Stokes vector were derived for linear media [9, 10]. The formalism is easily extended to dissipation-free nonlinear media [11–13] and even to dissipative nonlinear media with nonlocal optical response [14] and partially polarized light [15].

We will now proceed to derive the evolution equations for the Stokes parameters and illustrate their application to polarization effects in linear optics. By applying the curl operator to both sides of the first Maxwell equation (2.2), taking the time derivative of the third Maxwell equation and then subtracting the results, we exclude the magnetic field **B** and arrive at the wave equation:

$$\text{curl curl } \mathbf{E} + \frac{1}{c^2}\frac{\partial^2 \mathbf{D}}{\partial t^2} = 0 \tag{3.15}$$

We will solve equation (3.15) using the 'slow envelope' approach: it will be presumed that the amplitude $\mathbf{A}(z, t)$ of the electric field of the wave in equation (3.1) is a 'slow' function of the coordinate of propagation z:

$$|\mathbf{A}| \gg \left|\frac{1}{k}\frac{\mathrm{d}\mathbf{A}}{\mathrm{d}z}\right| \gg \left|\frac{1}{k^2}\frac{\mathrm{d}^2\mathbf{A}}{\mathrm{d}z^2}\right| \tag{3.16}$$

Substituting equation (2.16) into (3.15), using $\mathbf{D} = \mathbf{E} + 4\pi \boldsymbol{P}$, and resorting equation (3.16) leads to

$$2ik\frac{\mathrm{d}A_i}{\mathrm{d}z} + \frac{\omega^2}{c^2}(\epsilon_{ij} + ik\Gamma^{(1)}_{ijz} - n^2\delta_{ij})A_j = 0 \tag{3.17}$$

or, in a shorter form,

$$\frac{\mathrm{d}A_i}{\mathrm{d}z} = \frac{\mathrm{i}\omega}{2c}u_{ij}A_j \tag{3.18}$$

where we have introduced

$$u_{ij} = \frac{1}{n}(\epsilon_{ij} + ik\Gamma^{(1)}_{ijz} - n^2\delta_{ij}) \tag{3.19}$$

and indices i and j may be either x or y.

We aim to derive the evolution equations for the Stokes parameters. To do that we expand u_{ij} over the Pauli matrices:

$$u_{ij} = \tfrac{1}{2}\Omega^{\mathrm{L}}_{\alpha}\sigma^{(\alpha)}_{ij} \tag{3.20}$$

Here a summation is presumed over $\alpha = 0, \ldots, 3$ and the expansion coefficients are

$$\Omega^{\mathrm{L}}_{\alpha} = \sigma^{(\alpha)}_{ji}u_{ij} \tag{3.21}$$

Now when equation (3.20) is substituted into equation (3.18), the later may be rearranged:

$$\frac{\mathrm{d}A_i}{\mathrm{d}z} = \frac{\mathrm{i}\omega}{4c}\Omega^{\mathrm{L}}_{\alpha}\sigma^{(\alpha)}_{ik}A_k \tag{3.22}$$

or, for the complex conjugated amplitude,

$$\frac{\mathrm{d}A^*_j}{\mathrm{d}z} = -\frac{\mathrm{i}\omega}{4c}\Omega^{\mathrm{L}*}_{\alpha}\sigma^{(\alpha)}_{kj}A^*_k \tag{3.23}$$

where $\alpha = 0, \ldots, 3$ and we have used the fact that $\sigma^{(\alpha)}_{ik} = [\sigma^{(\alpha)}_{ki}]^*$. Again, in equation (3.22) and (3.23) summations are presumed over repeated indices α and k. Now we multiply equation (3.22) by $A^*_j \sigma^{(\mu)}_{ji}$ and equation (3.23) by $A_i \sigma^{(\mu)}_{ji}$ and add the results:

$$\frac{\mathrm{d}}{\mathrm{d}z}(S_\mu\, \mathrm{e}^{2\mathrm{Im}\{k\}z}) = \frac{\mathrm{i}\omega}{4c}(\Omega^{\mathrm{L}}_\alpha A^*_j \sigma^{(\mu)}_{ji} \sigma^{(\alpha)}_{ik} A_k - \Omega^{\mathrm{L}*}_\alpha A^*_k \sigma^{(\alpha)}_{kj} \sigma^{(\mu)}_{ji} A_i) \tag{3.24}$$

Here μ and $\alpha = 0, \ldots, 3$.

Equation (3.24) may be conveniently presented as a set of two equations: a scalar equation for the Stokes parameter S_0 and a vectorial equation for the Stokes vector $\mathbf{S}$. Indeed, for $\mu = 0$, equation (3.24) becomes

$$\frac{\mathrm{d}}{\mathrm{d}z}(S_0\, \mathrm{e}^{2\mathrm{Im}\{k\}z}) = \frac{\mathrm{i}\omega}{4c}(\Omega^{\mathrm{L}}_\alpha A^*_j \sigma^{(\alpha)}_{jk} A_k - \Omega^{\mathrm{L}*}_\alpha A^*_k \sigma^{(\alpha)}_{ki} A_i) \tag{3.25}$$

In equation (3.25) summations are presumed over i, j and k. Using

$$A^*_j \sigma^{(\alpha)}_{jk} A_k = A^*_k \sigma^{(\alpha)}_{ki} A_i = S_\alpha\, \mathrm{e}^{2\mathrm{Im}\{k\}z} \tag{3.26}$$

equation (3.25) can be rewritten for the Stokes parameter S_0:

$$\frac{\mathrm{d}S_0}{\mathrm{d}z} = -2\,\mathrm{Im}\,\{k\}S_0 - \frac{\omega}{2c}\,\mathrm{Im}\,\{\Omega^{\mathrm{L}}_\alpha S_\alpha\} \tag{3.27}$$

For $\mu = 1$, 2 or 3, equation (3.24) can be rearranged using identities (3.6) as follows:

$$\frac{\mathrm{d}}{\mathrm{d}z}(S_\mu\, \mathrm{e}^{2\mathrm{Im}\{k\}z}) = \frac{\mathrm{i}\omega}{4c}[\Omega^{\mathrm{L}}_0 A^*_j \sigma^{(\mu)}_{jk} A^*_k + \Omega^{\mathrm{L}}_\alpha A_j(\sigma^{(0)}_{jk}\delta_{\alpha\mu} + ie_{\mu\alpha\gamma}\sigma^{(\gamma)}_{jk})A_k - \text{c.c.}] \tag{3.28}$$

Here α, μ and $\gamma = 1$, 2 and 3, and summations are presumed over α and γ. Now resorting definition (3.3) we arrive at the evolution equations for the components of the Stokes vector $\mathbf{S}$:

$$\frac{\mathrm{d}S_\mu}{\mathrm{d}z} = -2\,\mathrm{Im}\,\{k\}S_\mu - \frac{\omega}{2c}[\mathrm{Im}\,\{\Omega^{\mathrm{L}}_0 S_\mu + \Omega^{\mathrm{L}}_\mu S_0\} + e_{\mu\alpha\gamma}\,\mathrm{Re}\,\{\Omega^{\mathrm{L}}_\alpha S_\gamma\}] \tag{3.29}$$

The evolution equations for the Stokes vector $\mathbf{S}$ and Stokes parameter S_0 may be written in the compact vector form:

$$\begin{aligned} \frac{\mathrm{d}\mathbf{S}}{\mathrm{d}z} &= -\frac{\omega}{2c}\,\mathrm{Im}\,\{(4n + \Omega^{\mathrm{L}}_0)\mathbf{S} + \mathbf{\Omega}^{\mathrm{L}} S_0 + i[\mathbf{\Omega}^{\mathrm{L}} \times \mathbf{S}]\} \\ \frac{\mathrm{d}S_0}{\mathrm{d}z} &= -\frac{\omega}{2c}\,\mathrm{Im}\,\{(4n + \Omega^{\mathrm{L}}_0)S_0 + (\mathbf{\Omega}^{\mathrm{L}}\cdot\mathbf{S})\} \end{aligned} \tag{3.30}$$

where vector $\mathbf{\Omega}^{\mathrm{L}}$ with projections Ω_1^{L}, Ω_2^{L} and Ω_3^{L} (see formula (3.21)) is defined in the same coordinate space as $\mathbf{S}$. From the second equation in (3.30) we can conclude that the evolution of light intensity, which is proportional to S_0, depends on $\mathrm{Im}\,\{4n + \Omega_0^{\mathrm{L}}\}$ and $\mathrm{Im}\,\{\mathbf{\Omega}^{\mathrm{L}}\}$. Therefore a dissipation-less medium will correspond to $\mathrm{Im}\,\{n\} = 0$, $\mathrm{Im}\,\{\mathbf{\Omega}^{\mathrm{L}}\} = 0$ and $\mathrm{Im}\,\{\Omega_0^{\mathrm{L}}\} = 0$.

The set of evolution equations (3.30) is a very powerful tool in the study of light propagation effects. These equations may be applied to media of arbitrary symmetry. However, the scope of their use is limited to media with low anisotropy along the direction of propagation. In fact, equations (3.30) will work well as long as $\Delta n \ll n$, where Δn is the refractive index difference for the differently polarized eigenwaves of the medium.

3.3 ELEMENTARY POLARIZATION EFFECTS

In this section we will show how the evolution equations (3.30) may be used for describing basic optical propagation effects such as birefringence and optical activity.

Depending on the symmetry, media (crystals) are characterized by isotropic or anisotropic dielectric constants. In isotropic media the dielectric tensor ϵ_{ij} may be presented in a diagonal form where all components, apart from those on the main diagonal, are zero. Moreover, in isotropic media all diagonal elements are equal and, in general, are complex numbers: $\epsilon_{ij} = n^2\delta_{ij}$ (where n is the complex refractive index of the medium). In crystals of high symmetry (cubic classes) the dielectric tensor remains isotropic; however, in the majority of crystal symmetry classes it is anisotropic, i.e. the tensor has unequal diagonal components. Such anisotropy leads to the effect of double refraction. Furthermore, anisotropic crystals may be divided into uniaxial and biaxial crystals. In uniaxial crystals one of the three diagonal elements of the dielectric tensor is different from the other two which are equal. In biaxial crystals all three components are different. Anisotropy of crystals leads to the effect of double refraction, discovered by Erasmus Bartholin and analysed by Newton and Huygens at the end of the seventeenth century [16]. Anisotropy of the dielectric tensor leads to the effect of double refraction and some interesting and useful polarization effects. Such polarization effects appear in the zero-order spatial dispersion approximation, i.e. without accounting for nonlocality. For a description of these polarization effects the second term in the constitutive equation (2.16) may be dropped. (In some crystals, however, nonlocality is important as it leads to the effect of optical activity which coexists with double refraction. This will be considered later in this chapter). Here, using a simple example, we will illustrate how the evolution equations for the Stokes parameters may be used for the description of polarization phenomena due to the anisotropy of the dielectric tensor.

We will examine the evolution of light polarization in a crystalline medium with local optical response ($\Gamma^{(1)}_{ijn} = 0$) and anisotropic dielectric tensor:

$$\epsilon^{\mathrm{N}} = \begin{pmatrix} \epsilon_{11} & 0 & 0 \\ 0 & \epsilon_{22} & 0 \\ 0 & 0 & \epsilon_{33} \end{pmatrix} \tag{3.31}$$

presented in the natural Cartesian coordinate frame. The Stokes vector evolution equation (3.30) is written in the laboratory coordinate frame which, in general, is different from the natural coordinate frame of the crystal. Therefore the dielectric tensor (3.31) must be recalculated for the laboratory frame using formula (2.39).

We will examine a simple case where the light wave propagates along the Y^{N} axis of the natural frame. Therefore the Z axis of the laboratory frame is directed along Y^{N} (see Fig. 3.3). Let the X axis of the laboratory frame be directed at an angle $-\phi$ to X^{N}. This arrangement will allow us to describe important cases

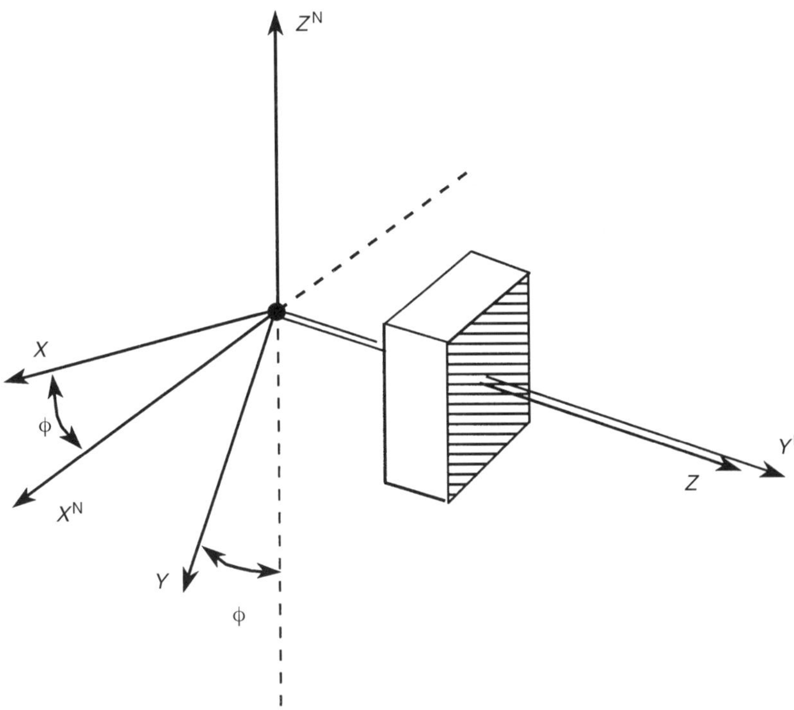

Fig. 3.3 Installation of the natural and laboratory coordinate frames for the description of polarization effects in birefringent crystals

such as those of an optical retarder and a dichroic linear polarizer. Here the matrix of transformation c_{ij} from the natural to the laboratory Cartesian frame is (see equation (2.40)):

$$c_{ij} = \begin{pmatrix} \cos\phi & 0 & \sin\phi \\ \sin\phi & 0 & -\cos\phi \\ 0 & 1 & 0 \end{pmatrix} \tag{3.32}$$

In the laboratory frame the dielectric tensor will become

$$\epsilon_{ij} = c_{ii'}c_{jj'}\epsilon^{\mathrm{N}}_{i'j'} = \begin{pmatrix} \epsilon_{xx} & \epsilon_{xy} & \epsilon_{xz} \\ \epsilon_{yx} & \epsilon_{yy} & \epsilon_{yz} \\ \epsilon_{zx} & \epsilon_{zy} & \epsilon_{zz} \end{pmatrix}$$

$$= \begin{pmatrix} [\cos^2\phi\,\epsilon_{11} + \sin^2\phi\,\epsilon_{33}] & [\sin\phi\cos\phi\,(\epsilon_{11} - \epsilon_{33})] & 0 \\ [\sin\phi\cos\phi\,(\epsilon_{11} - \epsilon_{33})] & [\sin^2\phi\,\epsilon_{11} + \cos^2\phi\,\epsilon_{33}] & 0 \\ 0 & 0 & \epsilon_{22} \end{pmatrix} \tag{3.33}$$

Now, substituting equation (3.33) into (3.19) we get

$$\begin{aligned} u_{xx} &= \frac{\epsilon_{xx} - n^2}{n} \\ u_{yy} &= \frac{\epsilon_{yy} - n^2}{n} \\ u_{xy} &= u_{yx} = \frac{2\epsilon_{xy}}{n} \end{aligned} \tag{3.34}$$

and using definition (3.21) we calculate Ω^{L}_0 and the components of the vector $\mathbf{\Omega}^{\mathrm{L}}$:

$$\begin{aligned} \Omega^{\mathrm{L}}_1 &= u_{xx} - u_{yy} = \frac{\epsilon_{11} - \epsilon_{33}}{n}\cos 2\phi \\ \Omega^{\mathrm{L}}_2 &= 2u_{xy} = \frac{\epsilon_{11} - \epsilon_{33}}{n}\sin 2\phi \\ \Omega^{\mathrm{L}}_3 &= 0 \\ \Omega^{\mathrm{L}}_0 &= u_{xx} = u_{yy} = \frac{\epsilon_{11} + \epsilon_{33} - 2n^2}{n} \end{aligned} \tag{3.35}$$

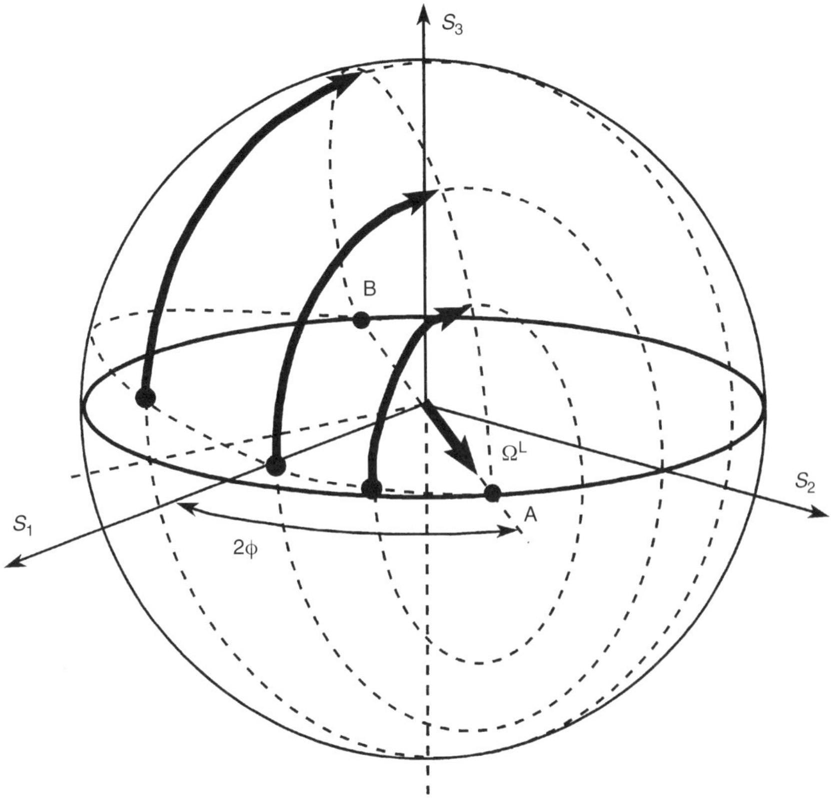

Fig. 3.4 The Poincaré sphere presentation of linear birefringence. Different initial ellipticities and polarization azimuths are shown. The vector $\mathbf{\Omega}^{\mathrm{L}}$ lies in the S_1S_2 plane and makes angle 2ϕ with the S_1 axis. When a light wave propagates through the medium, the point on the Poincaré sphere which represents the polarization state traces out a circular orbit perpendicular to the S_1S_2 plane. If the initial polarization state is represented by either point A or point B on the equator (orthogonal linear polarizations), the polarization state does not change. Therefore linear polarizations at ϕ and at $\phi + \pi/2$ to the X axis are the medium's eigenpolarizations

The wave propagation may be easily visualized in absorption-less birefringent media, where $\mathbf{\Omega}^{\mathrm{L}}$ is real and lies in the equatorial plane of the Poincaré sphere (Fig. 3.4). The evolution equation (3.30) will then take the following form:

$$\frac{\mathrm{d}\mathbf{S}}{\mathrm{d}z} = -\frac{\omega}{2c}[\mathbf{\Omega}^{\mathrm{L}} \times \mathbf{S}] \tag{3.36}$$

which describes a precession of the Stokes vector around vector $\mathbf{\Omega}^{\mathrm{L}}$. From any initial position the end of the Stokes vector of a propagating wave will trace out a vertical circle on the Poincaré sphere, perpendicular to the vector $\mathbf{\Omega}^{\mathrm{L}}$. Since the longitude of a point on the Poincaré sphere gives the polarization azimuth and the latitude gives the ellipticity, both of these parameters will oscillate during propagation. The polarization azimuth will go from one elliptical polarization to the opposite handed elliptical polarization via a linearly polarized state and so on.

Let us now analyse the general case, without disregarding dissipation. With $\mathbf{\Omega}^{\mathrm{L}}$ and Ω_0^{L} given by equations (3.35), the set of equations (3.30) may be rewritten as follows:

$$\begin{aligned}
\frac{\mathrm{d}S_1}{\mathrm{d}z} &= -\Delta S_1 - 2\delta'' \cos 2\phi\, S_0 - 2\delta' \sin 2\phi\, S_3 \\
\frac{\mathrm{d}S_2}{\mathrm{d}z} &= -\Delta S_2 - 2\delta'' \sin 2\phi\, S_0 + 2\delta' \cos 2\phi\, S_3 \\
\frac{\mathrm{d}S_3}{\mathrm{d}z} &= -\Delta S_3 - 2\delta' \cos 2\phi\, S_2 + 2\delta' \sin 2\phi\, S_1 \\
\frac{\mathrm{d}S_0}{\mathrm{d}z} &= -\Delta S_0 - 2\delta'' \cos 2\phi\, S_1 - 2\delta'' \sin 2\phi\, S_2
\end{aligned} \tag{3.37}$$

where we have introduced

$$\begin{aligned}
\Delta &= \frac{\omega}{2c}\,\mathrm{Im}\left\{2n + \frac{\epsilon_{11}+\epsilon_{33}}{n}\right\} \\
\delta &= \delta' + i\delta'' = \frac{\omega}{4c}\frac{\epsilon_{11}-\epsilon_{33}}{n}
\end{aligned} \tag{3.38}$$

The solutions to the set of equations (3.37) are given by

$$\begin{aligned}
S_0(L) &= [S_{00}\cosh(2\delta''L) - (S_{01}\cos 2\phi + S_{02}\sin 2\phi)\sinh(2\delta''L)]e^{-\Delta L} \\
S_1(L) &= [S_{01}(\cos^2 2\phi \cosh 2\delta''L + \sin^2 2\phi \cos 2\delta' L] + \\
&\quad S_{20}(\cosh 2\delta''L - \cos 2\delta' L)\sin 2\phi \cos 2\phi - \\
&\quad S_{00}\cos 2\phi \sinh 2\delta''L - S_{30}\sin 2\phi \sin 2\delta' L]e^{-\Delta L} \\
S_2(L) &= [S_{20}(\sin^2 2\phi \cosh 2\delta''L + \cos^2 2\phi \cos 2\delta' L) + \\
&\quad S_{10}(\cosh 2\delta''L - \cos 2\delta' L)\sin 2\phi \cos 2\phi - \\
&\quad S_{00}\sin 2\phi \sinh 2\delta''L + S_{30}\cos 2\phi \sin 2\delta' L]e^{-\Delta L} \\
S_3(L) &= [S_{03}\cos(2\delta' L) + (S_{01}\sin 2\phi - S_{02}\cos 2\phi)\sin(2\delta' L)]e^{-\Delta L}
\end{aligned} \tag{3.39}$$

From here we obtain the polarization azimuth and ellipticity at $z = L$:

$$\Theta(L) = \phi + \tfrac{1}{2}\tan^{-1}\left[\frac{\sin 2\eta_0 \sin(2\delta' L) + \sin(2\Theta_0 - 2\phi)\cos 2\eta_0 \cos 2\delta' L}{\cos(2\Theta_0 - 2\phi)\cos 2\eta_0 \cosh 2\delta'' L - \sinh 2\delta'' L}\right]$$

$$\eta(L) = \tfrac{1}{2}\sin^{-1}\left[\frac{\sin 2\eta_0 \cos 2\delta' L - \sin(2\Theta_0 - 2\phi)\cos 2\eta_0 \sin(2\delta' L)}{\cosh(2\delta'' L) - \cos(2\Theta_0 - 2\phi)\cos 2\eta_0 \sinh(2\delta'' L)}\right] \quad (3.40)$$

where $\Theta_0 = \Theta(0)$ and $\eta_0 = \eta(0)$ are the polarization azimuth and ellipticity of the wave at $z = 0$. From equations (3.40) it is clear that if $\eta_0 = 0$ and $\Theta_0 = \phi$ or $\Theta_0 = \phi + \pi/2$, the polarization does not change, that is $\Theta_0(L) \equiv \Theta_0$ and $\eta(L) \equiv 0$. Therefore linear polarizations with the electric field vector at ϕ and $\phi + \pi/2$ to the X axis are eigenpolarizations.

Let us consider an important case of low birefringence or short crystal length. Then, for a linearly polarized incident wave the polarization changes are small, that is $\Theta(L) - \Theta(0) \ll 1$ and $\eta(L) \ll 1$. Under these circumstances formulae (3.40) simplify to

$$\begin{pmatrix} \Theta(L) - \Theta(0) \\ \eta(L) \end{pmatrix} = \begin{pmatrix} \delta'' \\ \delta' \end{pmatrix} \sin[2\Theta(0) - 2\phi]$$

$$= \frac{\omega L}{4c} \times \begin{pmatrix} \mathrm{Im} \\ -\mathrm{Re} \end{pmatrix} \left\{ \frac{\epsilon_{11} - \epsilon_{33}}{n} \right\} \sin[2\Theta(0) - 2\phi] \quad (3.41)$$

Here the rotation of the polarization azimuth $\Theta(L) - \Theta(0)$ is proportional to $\mathrm{Im}\{(\epsilon_{11} - \epsilon_{33})/n\}$ and is due to the differential absorption of the two orthogonal linear eigenwaves: the polarization azimuth will tend towards the eigenpolarization with the lowest absorption coefficient. Propagation will yield ellipticity proportional to $\mathrm{Re}\{(\epsilon_{11} - \epsilon_{33})/n\}$ which is related to the eigenwave's differential refractive index.

Differential absorption and refraction of linearly polarized light is widely used in optical devices. Two important examples are quarter-wave plates (or $\lambda/4$-wave plates) and dichroic polarizers. A quarter-wave plate is a slab of birefringent low-absorption crystal ($\mathrm{Im}\{\epsilon_{ij}\} = 0$; $\mathrm{Im}\{n\} = 0$), often used to convert circularly into linearly polarized light and vice versa. The principle of operation of the quarter-wave plate may be easily explained on the Poincaré sphere (Fig 3.4). Let us start with right circularly polarized light, represented by the north pole of the sphere. If this light propagates along a birefringent direction, the point representing the polarization state will move towards the equator on a circular orbit perpendicular to vector $\mathbf{\Omega}$. According to formulae (3.40) it will reach the equator and the outgoing light will become linearly polarized, when $L = (\pi/4\delta') = \lambda n/2(\epsilon_{11} - \epsilon_{33})$. If, however, we start with linear polarization, after passing through a crystal slab of thickness L, the light wave will become left circularly polarized. For example, consider making a quarter-wave plate from the birefringent, crystalline material potassium dihydrogen phosphate (known by the abbreviation KDP). At $\lambda = 532$ nm it has $\epsilon_{11} = 2.287$ and $\epsilon_{33} = 2.162$ [17]; therefore a crystal slab of thickness $\lambda n/2(\epsilon_{11} - \epsilon_{33}) = 3.22\ \mu$m will make a quarter-wave plate at this wavelength. Such a slab of birefringent material is called a quarter-wave plate because it introduces a quarter-wave phase delay between the two eigenpolarizations of the crystal. Indeed, the difference in refractive indices for the eigenpolarizations $\Delta n \simeq (\epsilon_{11} - \epsilon_{33})/2n$ and therefore $\Delta nL \simeq \lambda/4$.

Equations (3.40) can also be very effective for describing absorptive birefringent media. As an elementary example we will take a dichroic polarizer based on strong linear dichroism, i.e. differential absorption of two eigenpolarizations. Polaroid film is the best-known example of this type of polarizer. In Polaroid film absorption is strongly anisotropic, that is $\delta''L \gg 1$ (see formulae 3.38). If a linearly polarized wave passes through such a polarizer, according to formulae (3.41) ellipticity tends to zero since $\eta(L) \propto \mathrm{e}^{-2\delta''L} \Rightarrow 0$. Similarly, the polarization azimuth tends to the eigenpolarization since $\Theta(L) - \phi \propto \mathrm{e}^{-2\delta''L} \Rightarrow 0$. Therefore, irrespective of the incident polarization state, light passing through the dichroic polarizer becomes linearly polarized with an azimuth at ϕ to the X axis.

It has been known since 1811, when Dominique Arago discovered the phenomenon, that in some media the plane of polarization rotates as light propagates through it. This effect was first seen in quartz crystals and was given the name of optical activity. In 1815 Biot found that the rotation was proportional to the pass length and, in the case of liquids, to the concentration of the optically active component in the solution. Because this rotation takes place naturally, it was called natural rotation. Referring to the constitutive equation (2.16), the optical activity effect is the simplest and most obvious manifestation of the optical response nonlocality. It is associated with the nonlocality tensor $\Gamma^{(1)}_{ijn}$ and is a first-order spatial dispersion effect. Optical activity is a well-documented phenomenon, a detailed description of which may be found in numerous books [18–22]. Here, using the simple example of optical activity in isotropic media, we will show how the evolution equation for the Stokes parameters may be used to describe polarization effects due to optical nonlocality.

Starting with constitutive equation (2.49):

$$P_i(\omega, \mathbf{k}) = \frac{1}{4\pi}(\epsilon_{ij} - \delta_{ij} + ik_n\Gamma^{(1)}_{ijn})A_j \tag{3.42}$$

we will consider optical activity in isotropic media due to the nonzero nonlocality tensor $\Gamma^{(1)}_{ijn}$. In isotropic media the dielectric tensor is diagonal, $\epsilon_{ij} = n^2\delta_{ij}$ (n is the complex refractive index). Isotropic media can belong to one of two symmetry point groups, $\infty\infty m$ (isotropic nongyrotropic) and $\infty\infty$ (isotropic gyrotropic). The existence of the third-rank tensor $\Gamma^{(1)}_{ijn}$ is forbidden by symmetry in $\infty\infty m$ media, but can be nonzero in $\infty\infty$ media. Such media consist of randomly oriented chiral molecules (e.g. a sugar solution in water). In isotropic gyrotropic media the nonlocality tensor is totally antisymmetric, $\Gamma^{(1)}_{ijn} = -\Gamma^{(1)}_{jin} = \Gamma^{\mathrm{a}} e_{ijn}$.

To derive the specific form of the equations (3.30) for isotropic gyrotropic media, we shall calculate vector $\mathbf{\Omega}^{\mathrm{L}}$, but before that, the material parameters u_{ij}. By substituting $\epsilon_{ij} = n^2\delta_{\mathrm{ij}}$ and $\Gamma^{(1)}_{ijn} = \Gamma^{\mathrm{a}} e_{ijn}$ into equation (3.19) we get

$$u_{ij} = i\frac{\omega}{c}\Gamma^{\mathrm{a}} e_{ijz} \tag{3.43}$$

Now we calculate the components of vectors $\mathbf{\Omega}^{\mathrm{L}}$ and Ω^{L}_0 as given by formula (3.21):

$$\Omega^{\mathrm{L}}_{\alpha} = \sigma^{(\alpha)}_{ji} u_{ij} = i\frac{\omega\Gamma^{\mathrm{a}}}{c}\sigma^{(\alpha)}_{ji} e_{ijz} \tag{3.44}$$

From here, resorting equation (3.4) we readily obtain

$$\begin{aligned} \Omega^{\mathrm{L}}_1 &= \Omega^{\mathrm{L}}_2 = 0 \\ \Omega^{\mathrm{L}}_3 &= -\frac{2\omega\Gamma^{\mathrm{a}}}{c} \\ \Omega^{\mathrm{L}}_0 &= 0 \end{aligned} \tag{3.45}$$

The polarization evolution analysis can now be based on equations (3.30) with $\mathbf{\Omega}^{\mathrm{L}}$ and Ω^{L}_0 given by formulae (3.35).

The evolution of the polarization state in isotropic gyrotropic media may be easily visualized in Stokes space if the medium is absorption-less. In this case $\mathrm{Im}\{n\} = 0$ and $\mathrm{Im}\{\Gamma^{\mathrm{a}}\} = 0$; therefore Ω^{L} is real and the evolution equations (3.30) take the form (3.36). In the case of optical activity equation (3.36) describes a precession of the Stokes vector $\mathbf{S}$ around vector $\mathbf{\Omega}^{\mathrm{L}}$. In an absorption-less medium vector $\mathbf{\Omega}^{\mathrm{L}}$ in Stokes space is directed along the S_3 axis (see Fig. 3.5). Thus, from any initial position the end of the Stokes vector of a propagating wave will trace out a horizontal circle on the Poincaré sphere. Since the longitude of a point on the Poincaré sphere gives the azimuth of light polarization, the azimuth Θ will steadily change along the direction of propagation. This is optical activity. It is only seen when $\Gamma^{\mathrm{a}} \neq 0$; otherwise the initial Stokes vector remains unchanged on propagation. Note that in the dissipation-

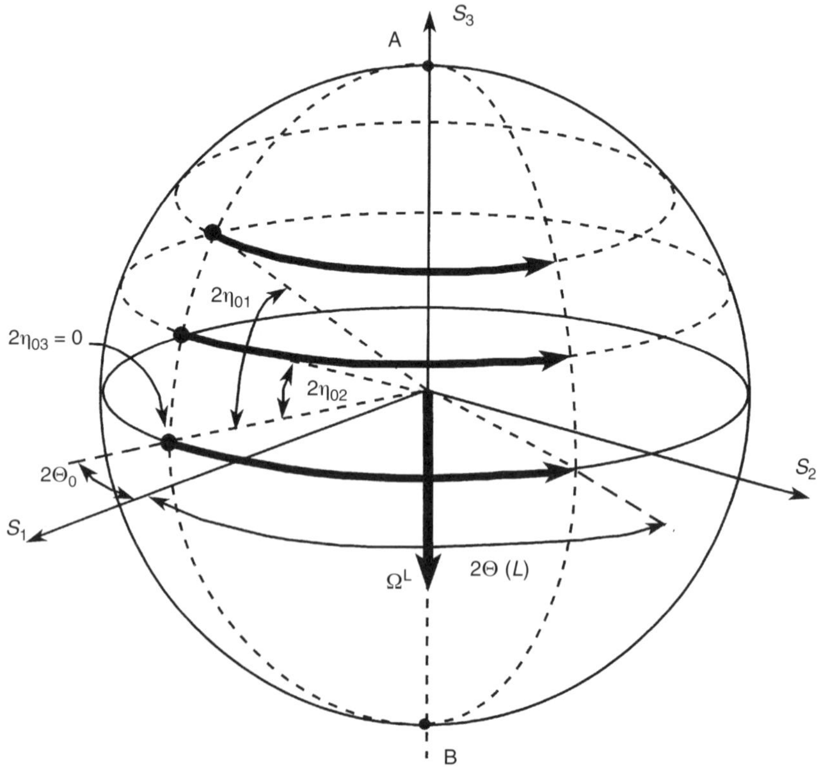

Fig. 3.5 The Poincaré sphere presentation of optical activity in a dissipation-less, isotropic, gyrotropic medium. Different initial polarization states with ellipticities η_{01} and η_{02} and $\eta_{03} = 0$ all corresponding to the same initial polarization azimuth Θ_0 are considered. The vector $\mathbf{\Omega}^{\mathrm{L}}$ is directed along S_3. When a light wave propagates through the medium, the point on the Poincaré sphere which represents the polarization state traces out a circular orbit parallel to the S_1S_2 plane. For initially linearly polarized light the point moves along the equator, i.e. light remains linearly polarized but the polarization azimuth changes steadily. If the initial polarization state is represented by either point A or point B on the poles (RCP or LCP), the polarization state does not change: right-handed and left-handed circular polarizations are the medium's eigenpolarizations

less media, the trajectory of the end of the Stokes vector always remains on the same altitude; therefore the initial ellipticity angle η does not change. In particular, the plane of polarization of linearly polarized light rotates, but the light remains linearly polarized.

Let us now include dissipation. In coordinate form the set of equations (3.30) will be

$$\begin{aligned}\frac{dS_1}{dz} &= -2\,\mathrm{Im}\,\{k\}S_1 - \frac{\omega^2\,\mathrm{Re}\,\{\Gamma^a\}}{c^2}S_2\\ \frac{dS_2}{dz} &= -2\,\mathrm{Im}\,\{k\}S_2 + \frac{\omega^2\,\mathrm{Re}\,\{\Gamma^a\}}{c^2}S_1\\ \frac{dS_3}{dz} &= -2\,\mathrm{Im}\,\{k\}S_3 + \frac{\omega^2\,\mathrm{Im}\,\{\Gamma^a\}}{c^2}S_0\\ \frac{dS_0}{dz} &= -2\,\mathrm{Im}\,\{k\}S_0 + \frac{\omega^2\,\mathrm{Im}\,\{\Gamma^a\}}{c^2}S_3\end{aligned} \tag{3.46}$$

According to equations (3.46) S_1 and S_2 change independently of S_3 and S_0. Thus, the evolution of the polarization azimuth $\Theta = \frac{1}{2}\tan^{-1}(S_2/S_1)$ is independent of the evolution of the ellipticity $\eta = \frac{1}{2}\sin^{-1}(S_3/S_0)$. If the initial Stokes parameters of the wave at $z = 0$ were $S_{\alpha 0} = \{S_{00}, S_{10}, S_{20}, S_{30}\}$, solving the set of equations (3.46) would yield the following result:

$$\begin{aligned}S_1(L) &= \left[S_{10}\cos\left(\frac{\omega^2}{c^2}\,\mathrm{Re}\,\{\Gamma^a\}L\right) - S_{20}\sin\left(\frac{\omega^2}{c^2}\,\mathrm{Re}\,\{\Gamma^a\}L\right)\right]\mathrm{e}^{-2\mathrm{Im}\{k\}L}\\ S_2(L) &= \left[S_{20}\cos\left(\frac{\omega^2}{c^2}\,\mathrm{Re}\,\{\Gamma^a\}L\right) + S_{10}\sin\left(\frac{\omega^2}{c^2}\,\mathrm{Re}\,\{\Gamma^a\}L\right)\right]\mathrm{e}^{-2\mathrm{Im}\{k\}L}\\ S_3(L) &= \left[S_{30}\cosh\left(\frac{\omega^2}{c^2}\,\mathrm{Im}\,\{\Gamma^a\}L\right) + S_{00}\sinh\left(\frac{\omega^2}{c^2}\,\mathrm{Im}\,\{\Gamma^a\}L\right)\right]\mathrm{e}^{-2\mathrm{Im}\{k\}L}\\ S_0(L) &= \left[S_{00}\cosh\left(\frac{\omega^2}{c^2}\,\mathrm{Im}\,\{\Gamma^a\}L\right) + S_{30}\sinh\left(\frac{\omega^2}{c^2}\,\mathrm{Im}\,\{\Gamma^a\}L\right)\right]\mathrm{e}^{-2\mathrm{Im}\{k\}L}\end{aligned} \tag{3.47}$$

From here we obtain the formulae for the polarization azimuth and ellipticity of the propagating wave at $z = L$:

$$\Delta\Theta = \Theta(L) - \Theta(0) = \frac{\omega^2}{2c^2}\,\mathrm{Re}\,\{\Gamma^a\}L \tag{3.48}$$

$$\eta(L) = \tfrac{1}{2}\sin^{-1}\left(\frac{\sin[2\eta(0)] + \tanh[(\omega^2/c^2)\,\mathrm{Im}\,\{\Gamma^a\}L]}{1 + \sin[2\eta(0)]\tanh[(\omega^2/c^2)\,\mathrm{Im}\,\{\Gamma^a\}L]}\right) \tag{3.49}$$

The real part of Γ^a gives the polarization azimuth rotation, while its imaginary part controls the evolution of the ellipticity. The azimuth rotation is proportional to the length of the medium L. Here $G = \Delta\Theta/L = \omega^2/(2c^2)\,\mathrm{Re}\,\{\Gamma^a\}$ is known as the specific polarization rotatory power. The azimuth rotates due to the difference in refractive indices for left and right circularly polarized waves. Such a difference is often referred to as circular birefringence. Depending on the sign of $\mathrm{Im}\,\{\Gamma^a\}$, irrespective of the initial polarization, an elliptically polarized wave will tend towards the right or left circular polarization. The polarization states of the left- and right-handed circularly polarized waves remain unchanged; i.e. these polarization states are the eigenpolarizations of isotropic gyrotropic media. The evolution of the ellipticity is due to the difference in absorption coefficients for the left- and right-handed circularly polarized eigenwaves. This is often referred to as circular dichroism. Sometimes both, the effects of polarization azimuth rotation and ellipticity change associated with Γ^a are referred to as optical activity.

Formulae (3.48) and (3.49) have been derived for isotropic media; however, they can also be used for describing optical activity in anisotropic crystals along directions of high symmetry, such as the optical axes where no birefringence, due to anisotropy of the dielectric tensor ϵ_{ij}, is seen. For example, crystalline quartz (SiO_2) is optically active along its optical axis. At a wavelength $\lambda = 0.63\ \mu$m its specific polarization rotatory power is $G = 18$ deg/mm, or in the esu system of units, $G = 3.15$ rad/cm. Therefore $\mathrm{Re}\,\{\Gamma^a\} = 6.3 \times 10^{-10}$ cm. Some materials show huge specific rotatory power which normally increases close to the absorption resonances. For instance, the popular crystal of nonlinear optics, lithium iodate ($LiIO_3$) gives a rotation of 199 deg/mm, or 34.7 rad/cm at $\lambda = 0.448\ \mu$m ($\mathrm{Re}\,\{\Gamma^a\} = 3.6 \times 10^{-9}$ cm). The optically active organic liquid ethyl tartrate (tartaric acid) shows, at wavelength $\lambda = 0.63\ \mu$m, specific rotatory power of approximately 0.09 deg/mm or 1.6×10^{-2} rad/cm ($\mathrm{Re}\,\{\Gamma^a\} = 3 \times 10^{-12}$ cm).

From these examples we can clearly see that in the constitutive equation (3.42) the nonlocal contribution to the optical response from the post term does not exceed 10^{-3} of the local contribution from the dielectric tensor. Therefore, the first-order spatial dispersion approximation works very well here.

Let us now examine light propagation in a birefringent crystal with optical activity. We will see below that under certain circumstances the combined action of optical activity and birefringence leads to a very interesting polarization sensitive effect useful in narrow-band spectral filters.

To describe light propagation in an optically active birefringent crystal we assume the laboratory coordinate frame installation presented in Fig. 3.3 with $\phi = 0$. The dielectric tensor of the crystal will be given by equation (3.42) and the optical activity of the crystal will be considered to be isotropic media, $\Gamma^a = \Gamma^{(1)}_{xyz} = -\Gamma^{(1)}_{yxz}$. The local and nonlocal terms are independent additives in formula (3.19) for u_{ij}. Consequently, the vector $\mathbf{\Omega}^L$ for the birefringent, optically active medium is a sum of the corresponding vectors for an isotropic optically active medium (3.45) and for a birefringent crystal (3.35):

$$
\begin{aligned}
\Omega_1^{\mathrm{L}} &= \frac{\Delta\epsilon}{n} \\
\Omega_2^{\mathrm{L}} &= 0 \\
\Omega_3^{\mathrm{L}} &= -\frac{2\omega}{c}\Gamma^{\mathrm{a}} \\
\Omega_0^{\mathrm{L}} &= \frac{\epsilon_{11} + \epsilon_{33} - 2n^2}{n}
\end{aligned}
\tag{3.50}
$$

Here we have defined $\Delta\epsilon = \epsilon_{11} - \epsilon_{33}$. If absorption is negligible ($\mathrm{Im}\{\Omega^{\mathrm{L}}\} = \mathrm{Im}\{\Omega_0^{\mathrm{L}}\} = 0$), the evolution equations (3.30) for the Stokes parameters take the following form:

$$
\begin{aligned}
\frac{\mathrm{d}S_1}{\mathrm{d}z} &= -2GS_2 \\
\frac{\mathrm{d}S_2}{\mathrm{d}z} &= 2GS_1 + 2\delta' S_3 \\
\frac{\mathrm{d}S_3}{\mathrm{d}z} &= -2\delta' S_2 \\
\frac{\mathrm{d}S_0}{\mathrm{d}z} &= 0
\end{aligned}
\tag{3.51}
$$

where $G = [\omega^2/(2c^2)]\Gamma^{\mathrm{a}}$ is the specific optical rotatory power of the medium (see formula (3.48)) and $\delta' = \Delta\epsilon\omega/(4cn)$ is a measure of the birefringence in an absorption-less medium (see formula (3.38)).

Let us examine the case when the incident light is linearly polarized along the X direction by a polarizer, and after passing through the crystal it goes through an analyser oriented along the Y direction (see Fig. 3.6a). Therefore $S_0(0) = S_1(0) = S_{00}$, $S_2(0) = S_3(0) = 0$ and for $z = L$ equations (3.51) give

$$
\begin{aligned}
S_0(L) &= S_{00} \\
S_1(L) &= S_{00}\left[1 - \frac{G^2}{\mu^2}(1 - \cos 2\mu L)\right] \\
S_2(L) &= S_{00}\frac{G}{\mu}\sin 2\mu L \\
S_3(L) &= -S_{00}\frac{G\delta'}{\mu^2}(1 - \cos 2\mu L)
\end{aligned}
\tag{3.52}
$$

where $\mu^2 = G^2 + (\delta')^2$. From the definition of the Stokes parameters (3.7) it follows that the intensity of light transmitted through an ideal polarizer oriented

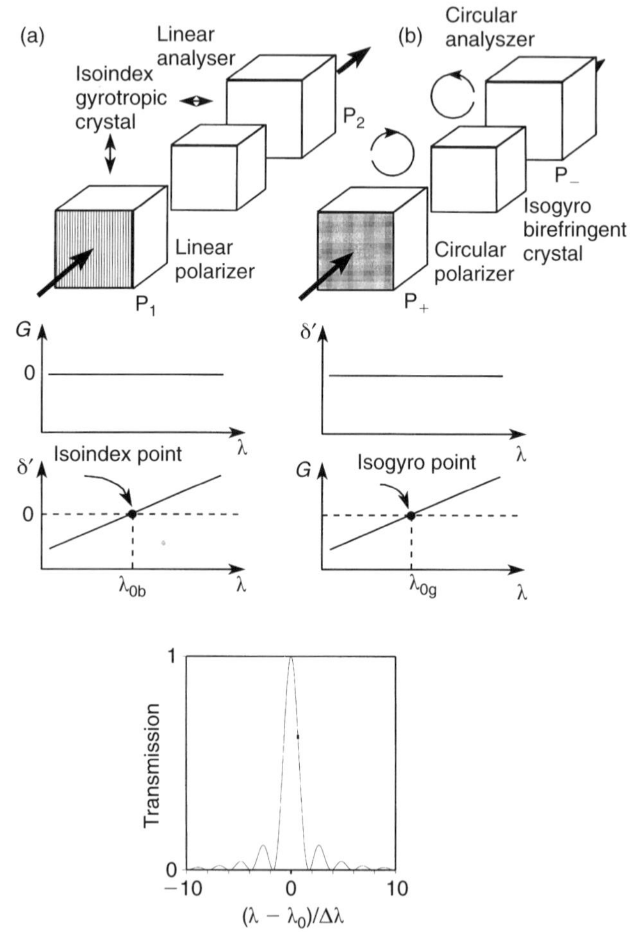

Fig. 3.6 Isoindex gyrotropic spectral filter (a) and isogyration birefringent filter (b). In the isoindex gyrotropic spectral filter an optically active birefringent crystal is sandwiched between two orthogonal linear polarizers P_1 and P_2. The crystal birefringence changes its sign at λ_{0b}, while the rotatory power G merely depends on λ in the vicinity of λ_{0b}. Near the wavelength of birefringence zero-crossing the filter has a narrow spectral transmission peak. In the isogyration birefringent spectral filter an optically active birefringent crystal is sandwiched between right and left circular polarizers P_+ and P_-. The crystal specific rotatory power changes its sign at λ_{0g}, while birefringence merely depends on λ in the vicinity of λ_{0g}. Near the isogyro point the filter has a narrow spectral transmission peak. The inset at the bottom of the figure shows the transmission coefficient T as a function of the wavelength near the birefringence or gyrotropy zero-crossing point. For the iso-birefringent filter $\Delta\lambda = (\lambda_{0b}/2L)(\mathrm{d}n/\mathrm{d}\lambda)^{-1}$ while for the isogyrotropic filter $\Delta\lambda = (\pi/\lambda)(\mathrm{d}G/\mathrm{d}\lambda)^{-1}$

along the Y direction is $c(S_0 - S_1)/(4\pi)$. Therefore, neglecting reflection losses we obtain the overall transmission coefficient T for the gyrotropic birefringent crystal sandwiched between two crossed linear polarizers:

$$T = \frac{S_0(L) - S_1(L)}{2S_0(0)} = (\mathrm{GL})^2 \left(\frac{\sin \mu L}{\mu L}\right)^2 \tag{3.53}$$

One can see from equation (3.53) that if the crystal length L is chosen in such a way that $GL = \pi/2$, the transmission of 100% ($T = 1$) will be achieved for zero birefringence, i.e. for $\delta' = 0$. With increasing birefringence the transmission will drop according to formula (3.53), becoming negligible for $\delta' \gg G$.

In some birefringent crystals, such as CdS, the birefringence becomes zero and changes its sign at a particular wavelength λ_{0b} (isoindex point). This is known as birefringence zero-crossing. If the crystal is also gyrotropic, such as $CuAlSe_2$ [23], the dependence of transmission on birefringence near the isoindex point is used in narrow-spectral-band filters. In the region near λ_{0b} we can expand

$$\delta' \simeq \pi \left(\frac{\mathrm{d}\Delta n}{\mathrm{d}\lambda}\right) \frac{\lambda - \lambda_{0b}}{\lambda_{0b}} \tag{3.54}$$

Therefore a zero-crossing, birefringent, gyrotropic crystal sandwiched between two orthogonal linear polarizers shows a sharp peak of transmission in the proximity of λ_{0b}:

$$T = \frac{\sin^2 \dfrac{\pi}{2} \sqrt{1 + \left(2L \dfrac{\mathrm{d}\Delta n}{\mathrm{d}\lambda} \dfrac{\lambda - \lambda_{0b}}{\lambda_{0b}}\right)^2}}{1 + \left(2L \dfrac{\mathrm{d}\Delta n}{\mathrm{d}\lambda} \dfrac{\lambda - \lambda_{0b}}{\lambda_{0b}}\right)^2} \tag{3.55}$$

This sort of filter was suggested in Ref. [24] and is known as an isoindex gyrotropic spectral filter.

Similarly, a narrow-band spectral filter may be made with a gyrotropic birefringent crystal near the wavelength λ_{0g}, where the specific optical rotatory power G of the crystal crosses zero (isogyration point):

$$G \simeq \left(\frac{\mathrm{d}G}{\mathrm{d}\lambda}\right)(\lambda - \lambda_{0g}) \tag{3.56}$$

For example, crystals of nickel sulfate hexahydrate, α-$NiSO_4{\cdot}6H_2O$, and nickel selenate hexahydrate α-$NiSeO_4{\cdot}6H_2O$, have an isogyration point at 1150 nm, and crystals of $CsCuCl_3$ [18] have a corresponding point at 950 nm. To make a spectral filter the crystal must be sandwiched between left and right circular polarizers (see Fig. 3.6b). If the thickness of the crystal is chosen in such a way that $\delta' L = \pi/2$, i.e. if at zero rotatory power the crystal has birefringence corresponding to a $\lambda/2$-wave plate, the transmission coefficient of the filter has the following wavelength dependence:

$$T = \frac{\sin^2 \dfrac{\pi}{2}\sqrt{1 + \left(\dfrac{2L}{\pi}\dfrac{\mathrm{d}G}{\mathrm{d}\lambda}\right)^2 (\lambda - \lambda_{0g})^2}}{1 + \left(\dfrac{2L}{\pi}\dfrac{\mathrm{d}G}{\mathrm{d}\lambda}\right)^2 (\lambda - \lambda_{0g})^2} \tag{3.57}$$

Such a filter is known as an isogyration birefringent filter and was first described in Ref. [25]. The dependence of birefringence and specific optical rotatory power on the light intensity (see the next chapter) gives rise to dependence of the transmission characteristics of the zero-crossing filters on light intensity [23].

3.4 GENERAL SOLUTIONS TO THE STOKES PARAMETERS EVOLUTION EQUATIONS, EIGENPOLARIZATIONS AND PROPAGATION OF PARTIALLY POLARIZED LIGHT

In early sections we looked at some elementary manifestations of optical polarization phenomena such as circular dichroism and birefringence (optical activity) and linear birefringence and dichroism (optical anisotropy). In this section we will consider the general case of wave propagation in linear media taking account of anisotropy, nonlocality and dissipation. All terms in the constitutive equation (2.16) will be accounted for and the analytical formulae for the Stokes parameters of the propagating wave will be obtained for an arbitrary incident polarization state.

However, we will first consider a situation where the self-action vector is complex, $\mathbf{\Omega}^{\mathrm{L}} = \mathbf{\Omega}^{\mathrm{L}'} + i\mathbf{\Omega}^{\mathrm{L}''}$, but its imaginary and real parts are parallel to one another in Stokes space, that is $\mathbf{\Omega}^{\mathrm{L}} = (\Omega^{\mathrm{L}'} + i\Omega^{\mathrm{L}''})\mathbf{m}$, where $\mathbf{m}$ is a unit vector in Stokes space. This is a particularly important case, because the self-action vector for the simple manifestations of circular and linear dichroism and birefringence considered above fall into this category (see formulae (3.34) and (3.44). With $\Omega^{\mathrm{L}} = (\Omega^{\mathrm{L}'} + i\Omega^{\mathrm{L}''})\mathbf{m}$ the Stokes vector evolution equations (3.30) reduce to

$$\begin{aligned} \frac{\mathrm{d}\mathbf{S}}{\mathrm{d}z} &= -\Delta\mathbf{S} - 2\delta''\mathbf{m}S_0 - 2\delta'[\mathbf{m} \times \mathbf{S}] \\ \frac{\mathrm{d}S_0}{\mathrm{d}z} &= -\Delta S_0 - 2\delta''(\mathbf{m}\cdot\mathbf{S}) \end{aligned} \tag{3.58}$$

where

$$\Delta = \frac{\omega}{2c}\,\mathrm{Im}\,\{4n + \Omega_0^{\mathrm{L}}\}$$
$$\delta' + i\delta'' = \frac{\omega}{4c}(\boldsymbol{\Omega}^{\mathrm{L}}\cdot\mathbf{m}) \tag{3.59}$$

By multiplying the second equation of (3.58) by $\mathbf{m}$ and taking into account the vector identity $(\mathbf{m}\cdot[\mathbf{m}\times\mathbf{S}]) = 0$, we obtain the following set of equations for $(\mathbf{m}\cdot\mathbf{S})$ and S_0:

$$\frac{\mathrm{d}}{\mathrm{d}z}(\mathbf{m}\cdot\mathbf{S}) = -\Delta(\mathbf{m}\cdot\mathbf{S}) - 2\delta'' S_0$$
$$\frac{\mathrm{d}S_0}{\mathrm{d}z} = -\Delta S_0 - 2\delta''(\mathbf{m}\cdot\mathbf{S}) \tag{3.60}$$

This set of two linear differential equations can easily be solved to give S_0 and $(\mathbf{m}\cdot\mathbf{S})$:

$$(\mathbf{m}\cdot\mathbf{S}) = [(\mathbf{m}\cdot\mathbf{S}_0)\cosh(2\delta'' z) - S_{00}\sinh(2\delta'' z)]\,\mathrm{e}^{-\Delta z}$$
$$S_0 = [S_{00}\cosh(2\delta'' z) - (\mathbf{m}\cdot\mathbf{S}_0)\sinh(2\delta'' z)]\,\mathrm{e}^{-\Delta z} \tag{3.61}$$

Now, the second equation of (3.58) and the vector product of its both sides with $\mathbf{m}$ give us a set of equations for $[\mathbf{m}\times\mathbf{S}]$ and $\mathbf{S}$:

$$\frac{\mathrm{d}\mathbf{S}}{\mathrm{d}z} + \Delta\mathbf{S} + 2\delta'[\mathbf{m}\times\mathbf{S}] = -2\delta'' S_0\mathbf{m}$$
$$\frac{\mathrm{d}}{\mathrm{d}z}[\mathbf{m}\times\mathbf{S}] + \Delta[\mathbf{m}\times\mathbf{S}] + 2\delta'\mathbf{S} = -2\delta'\mathbf{m}(\mathbf{m}\cdot\mathbf{S}) \tag{3.62}$$

Here we have used the identity $[\mathbf{m}\times[\mathbf{m}\times\mathbf{S}]] = \mathbf{m}(\mathbf{m}\cdot\mathbf{S}) - \mathbf{S}$. The right-hand sides of equations (3.62) may be found from equations (3.61). Therefore, equations (3.62) are a set of two linear differential equations that can be solved to yeld an expression for $\mathbf{S}$:

$$\mathbf{S}(L) = \{\mathbf{S}(0)\cosh(2\delta'' L) - \mathbf{m}S_{00}\sinh(2\delta'' L) + [\mathbf{m}\times\mathbf{S}(0)]\sin(2\delta' L)$$
$$+ [\mathbf{m}\times[\mathbf{m}\times\mathbf{S}(0)]][\cosh(2\delta'' L) - \cos(2\delta' L)]\}\,\mathrm{e}^{-\Delta L} \tag{3.63}$$

If the incident polarization is such that the corresponding Stokes vector $\mathbf{S}(0)$ is parallel (or antiparallel) to the vector $\mathbf{m}$, then $[\mathbf{m}\times\mathbf{S}(0)] = 0$, and the last two terms in equation (3.63) vanish. The remaining two terms in the equation are vectors parallel to the initial Stokes vector $\mathbf{S}(0)$. Consequently, if $\mathbf{m}\|\mathbf{S}(0)$, the light polarization does not change with propagation. Therefore, the condition

$$[\boldsymbol{\Omega}^{\mathrm{L}}\times\mathbf{S}(0)] = 0 \tag{3.64}$$

is, in fact, the condition for eigenpolarizations of the medium where the real and imaginary parts of vector $\mathbf{\Omega}^{\mathrm{L}}$ are parallel: there are two orthogonal eigenpolarizations in the medium with Stokes vectors parallel and antiparallel to the vector $\mathbf{\Omega}^{\mathrm{L}}$ (see Fig. 3.7).

The parallelism of the real and imaginary parts of the vector $\mathbf{\Omega}^{\mathrm{L}}$ is a common situation in media of high symmetry but not necessary the general case. For example, in birefringent absorptive media with optical activity the vector $\mathbf{\Omega}^{\mathrm{L}}$ will typically have a structure like $\mathbf{\Omega}^{\mathrm{L}} = n^{-1}(\epsilon_{11} - \epsilon_{33};\ 0;\ -2k\Gamma^{\mathrm{a}})$, where the real and imaginary parts of the vector are not necessarily parallel and the eigenwave condition (3.64) cannot be used. In such a case the medium does not support any eigenpolarization. Nevertheless, in dissipation-less media the vector $\mathbf{\Omega}^{\mathrm{L}}$ is purely real and formula (3.64) may be used for finding the eigenpolarizations of a medium of arbitrary symmetry.

Even if the real and imaginary parts of the vector $\mathbf{\Omega}^{\mathrm{L}}$ are not parallel, the

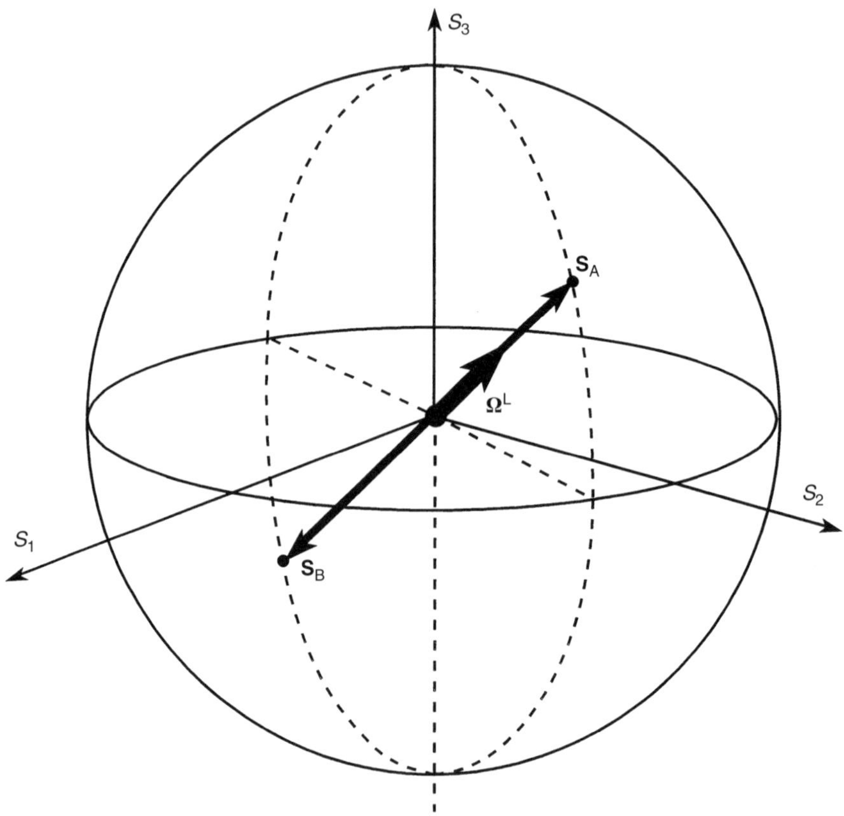

Fig. 3.7 Dissipation-less medium: the Stokes vectors $\mathbf{S}_{\mathrm{A}}$ and $\mathbf{S}_{\mathrm{B}}$ represent orthogonal eigenpolarizations of the medium on the Poincaré sphere. They are correspondingly parallel and anti-parallel to vector $\mathbf{\Omega}^{\mathrm{L}}$

solutions to the evolution equations for the Stokes vector parameters is obtainable in a general form:

$$
\begin{aligned}
\mathbf{S} = {} & \{[(\boldsymbol{\delta}''(\boldsymbol{\delta}'\cdot\boldsymbol{\delta}'') - \boldsymbol{\delta}''\mu_1^2)A_1 + \mu_1[\boldsymbol{\delta}' \times \boldsymbol{\delta}'']B_1] \sin 2\mu_1 L \\
& + [(\boldsymbol{\delta}''(\boldsymbol{\delta}'\cdot\boldsymbol{\delta}'') - \boldsymbol{\delta}''\mu_1^2)B_1 + \mu_1[\boldsymbol{\delta}' \times \boldsymbol{\delta}'']A_1] \cos 2\mu_1 L \\
& \times [(\boldsymbol{\delta}''(\boldsymbol{\delta}'\cdot\boldsymbol{\delta}'') + \boldsymbol{\delta}''\mu_2^2)A_2 - \mu_2[\boldsymbol{\delta}' \times \boldsymbol{\delta}'']B_2] \sinh 2\mu_2 L \\
& + [(\boldsymbol{\delta}''(\boldsymbol{\delta}'\cdot\boldsymbol{\delta}'') + \boldsymbol{\delta}''\mu_2^2)B_2 - \mu_2[\boldsymbol{\delta}' \times \boldsymbol{\delta}'']A_2] \cosh 2\mu_2 L\}\, \mathrm{e}^{-\boldsymbol{\delta}L} \\
S_0 = {} & -\{\mu_1[(\boldsymbol{\delta}'\cdot\boldsymbol{\delta}') - \mu_1^2](A_1 \cos 2\mu_1 L - B_1 \sin 2\mu_2 L) \\
& + \mu_2[(\boldsymbol{\delta}'\cdot\boldsymbol{\delta}') + \mu_2^2](A_2 \cosh 2\mu_2 L + B_2 \sin 2\mu_2 L)\}\, \mathrm{e}^{-\boldsymbol{\delta}L}
\end{aligned}
\tag{3.65}
$$

where $\boldsymbol{\delta}' = [\omega/(4c)]\,\mathrm{Re}\,\{\boldsymbol{\Omega}^{\mathrm{L}}\}$, $\boldsymbol{\delta}'' = [\omega/(4c)]\,\mathrm{Im}\,\{\boldsymbol{\Omega}^{\mathrm{L}}\}$, and we have introduced

$$
\mu_{1,2}^2 = \tfrac{1}{2}\{\sqrt{[(\boldsymbol{\delta}'\cdot\boldsymbol{\delta}') - (\boldsymbol{\delta}''\cdot\boldsymbol{\delta}'')]^2 + 4(\boldsymbol{\delta}'\cdot\boldsymbol{\delta}'')^2} \pm [(\boldsymbol{\delta}'\cdot\boldsymbol{\delta}') - (\boldsymbol{\delta}''\cdot\boldsymbol{\delta}'')]\} \tag{3.66}
$$

$$
\begin{aligned}
A_{1,2} &= \mp \frac{([\boldsymbol{\delta}' \times \boldsymbol{\delta}'']\cdot\mathbf{S}_0) - [(\boldsymbol{\delta}'\cdot\boldsymbol{\delta}') \mp \mu_{1,2}^2]S_{00}}{\mu_{1,2}(\mu_1^2 + \mu_2^2)[(\boldsymbol{\delta}'\cdot\boldsymbol{\delta}') \mp \mu_{1,2}^2]} \\
B_{1,2} &= \frac{\mu_{2,1}^2(\boldsymbol{\delta}'\cdot\mathbf{S}_0) \mp (\boldsymbol{\delta}'\cdot\boldsymbol{\delta}'')(\boldsymbol{\delta}''\cdot\mathbf{S}_0)}{(\boldsymbol{\delta}'\cdot\boldsymbol{\delta}'')(\mu_1^2 + \mu_2^2)[(\boldsymbol{\delta}'\cdot\boldsymbol{\delta}') \mp \mu_{1,2}^2]}
\end{aligned}
\tag{3.67}
$$

Therefore equations (3.65) give general solutions for all four Stokes parameters of waves propagating in anisotropic dissipative media with nonlocal optical response (on this issue also see Refs. [8] and [9]).

In the analysis above, and particularly in the derivation of the evolution equations (3.30) and their general solutions (3.65), we did not presume that the light was totally polarized. In fact equations (3.30) are suitable for describing the propagation of partially polarized light in nonlocal, dissipative, anisotropic media. For partially polarized light we are primarily interested in the evolution of the degree of polarization r (see formula (3.9)):

$$
r = \sqrt{\frac{(\mathbf{S}\cdot\mathbf{S})}{S_0^2}} \tag{3.68}
$$

Here the scalar product $(\mathbf{S}\cdot\mathbf{S})$ may be found from the differential equation which we obtain by multiplying the first equation of (3.30) by vector $\mathbf{S}$. Similarly, the equation for S_0^2 may be obtained by multiplying the second equation of (3.30) by S_0:

$$
\begin{aligned}
\frac{\mathrm{d}(\mathbf{S}\cdot\mathbf{S})}{\mathrm{d}z} &= -\frac{\omega}{c}\,\mathrm{Im}\,\{(4n+\Omega_0^{\mathrm{L}})(\mathbf{S}\cdot\mathbf{S})+(\mathbf{\Omega}^{\mathrm{L}}\cdot\mathbf{S})S_0\} \\
\frac{\mathrm{d}S_0^2}{\mathrm{d}z} &= -\frac{\omega}{c}\,\mathrm{Im}\,\{(4n+\Omega_0^{\mathrm{L}})S_0^2+(\mathbf{\Omega}^{\mathrm{L}}\cdot\mathbf{S})S_0\}
\end{aligned}
\tag{3.69}
$$

For totally polarized light, when $(\mathbf{S}\cdot\mathbf{S}) = S_0^2$, the two equations of (3.69) are identical; however, this is not the case when light is partially polarized, i.e. when $r \neq 1$. Differentiating equation (3.68) we obtain

$$
\frac{\mathrm{d}r^2}{\mathrm{d}z} = \frac{1}{S_0^2}\frac{\mathrm{d}(\mathbf{S}\cdot\mathbf{S})}{\mathrm{d}z} - \frac{(\mathbf{S}\cdot\mathbf{S})}{S_0^4}\frac{\mathrm{d}S_0^2}{\mathrm{d}z}
\tag{3.70}
$$

Then, substituting the derivatives from equations (3.69) into (3.70) we arrive at the following evolution equation for the degree of polarization r:

$$
\frac{\mathrm{d}(r^2)}{\mathrm{d}z} = -\frac{\omega}{c}\,\mathrm{Im}\left\{\frac{(\mathbf{\Omega}^{\mathrm{L}}\cdot\mathbf{S})}{S_0}\right\}(1-r^2)
\tag{3.71}
$$

Equation (3.71) leads to two important conclusions concerning the propagation of partially polarized light. First, totally polarized light ($r = 1$) remains totally polarized under any circumstances. This is because the right-hand side of equation (3.71) is zero and so the initial degree of polarization does not change. Second, if there is a change in degree of polarization, it is due to the imaginary part of Ω^{L}, which is related to the dissipation in optical anisotropy or nonlocality. Such a change in the degree of polarization results from differential absorption of different polarization states. Therefore, in a dissipation-free medium the initial degree of polarization does not change.

REFERENCES

1. E. Collett, *Polarization of Light: Fundamentals and Applications*, Marcel Dekker, New York, 1993.
2. S. Huard, *Polarization of light*, John Wiley & Sons, Chichester, 1997; M. Born and E. Wolf, *Principles of Optics*, Pergamon Press, Oxford, 1970.
3. R.M.A. Azzam and N.M. Bashara, *Ellipsometry and Polarized Light*, North-Holland, Amsterdam, 1977.
4. D.S. Kliger, J.W. Lewis and G.E. Randall, *Polarized Light in Optics and Spectroscopy*, Academic Press, San Diego, 1990.
5. G.V. Rozenberg, *Soviet Physics Uspekhi*, **121**, 97 (1977).
6. G.G. Stokes, *Trans. Camb. Phil. Soc.*, **9**, 399 (1852). Reprinted in *Mathematical and Physical Papers*, Vol. 3, Cambridge University Press, London, 1901, p. 233.
7. L.D. Landau and E.M. Lifshitz, *Quantum Mechanics. Non-relativistic Theory*, Pergamon Press, Oxford, 1977.
8. H. Poincaré, *Théorie Mathematique de la Lumierè*, Vol. 2, Cauthiers-Villars, Paris, 1892, Ch. 12.
9. H. Kubo and R. Nagata, *Opt. Commun.*, **34**, 306 (1980).

10. H. Kubo and R. Nagata, *J. Opt. Soc. Am. A*, **2**, 30 (1985).
11. K.L. Sala, *Phys. Rev. A*, **29**, 1944 (1984).
12. M.V. Tratnik and J. E. Sipe, *Phys. Rev. A*, **35**, 2965 (1987).
13. D. David, D.D. Holm and M.V. Tratnik, *Physics Reports*, **187**, 281 (1990).
14. N.I. Zheludev and Yu.P. Svirko, Nonlinear polarization spectroscopy of semiconductors, in S.A. Akhmanov and N.I. Zheludev (eds), Modern Problems of Laser Physics, Vol. 2, VINITI, Moscow, 1990 (in Russian).
15. Y.P. Svirko and N.I. Zheludev, *Phys. Rev. A*, **50**, 709 (1994).
16. A.R. Hall, *All Was Light*, Oxford University Press, 1993, p. 131.
17. V.G. Dmitriev, G.G. Gurzadyan and D.N. Nikogosyan, *Handbook of Nonlinear Optical Crystals*, Springer-Verlag, Berlin, 1991.
18. V.A. Kizel and V.I. Burkov, *Gyrotropy of Crystals*, Nauka, Moscow, 1980 (in Russian).
19. V.M. Agranovich and V.L. Ginsburg, *Spatial Dispersion in Crystal Optics and Theory of Excitons*, Interscience Publishers, New York, 1966.
20. L.D. Barron, *Molecular Light Scattering and Optical Activity*, Cambridge University Press, London, 1982.
21. F. Ciardelli and P. Salvadori (eds.), *Fundamental Aspects and Recent Developments in Optical Rotatory Dispersion and Circular Dichroism*, Heyden & Son Ltd, London, 1973.
22. D.J. Caldwell and H. Eyring, *The Theory of Optical Activity*, Wiley-Interscience, New York, 1971.
23. S.V. Popov, A.S. Semenikhin, V.A. Tarasenko, N.I. Zheludev, Yu.P. Svirko and L.A. Makovetskaya, *Opt. Lett.*, **18**, 993 (1990).
24. P. Yeh, *Appl. Opt.*, **21**, 4054 (1982).
25. N.I. Zheludev, *Soviet J. Quantum. Electron.*, **19**, 993 (1989).

4 The Evolution Equations for the Stokes Parameters in Nonlinear Media and Self-Action Polarization Phenomena

The propagation of intense light in nonlinear optical media results in a change in the medium s refraction, absorption, anisotropy and nonlocality. In return this alters the conditions of propagation for the wave itself and affects the light s intensity, phase and polarization state. This is known as self-action. The evolution equations for the Stokes parameters in nonlinear media will be derived in this chapter. Basic polarization self-action effects will be analysed and relevant pioneering experiments reviewed.

4.1 THE EVOLUTION EQUATIONS FOR THE STOKES PARAMETERS IN NONLINEAR MEDIA

In this section the Stokes parameter evolution equations (3.30) will be generalized to include the cubic optical nonlinearity of a medium as the first nonvanishing nonlinear term in the constitutive equation contributing to the self-action of light. It will be shown that the evolution equations for the Stokes parameters in nonlinear media are similar to those used in linear media, but with the vector $\mathbf{\Omega}$ being itself a function of the wave's Stokes parameters. The derived nonlinear evolution equations for the Stokes parameters will be suitable for the description of self-action nonlinear optical effects such as polarization ellipse self-rotation, nonlinear optical activity, nonlinear anisotropic effect and nonlinear nonlocal birefringence.

We will consider the propagation of a narrow-band plane wave with frequency ω and wave vector $\mathbf{k}$ in the z direction in a medium with cubic nonlinearity, using the slow envelope approach. Quadratic nonlinearities will be ignored because they do not affect nonlinear response directly on the frequency of excitation and, therefore, do not contribute to self-action. It may be shown, however, that quadratic nonlinearities could give rise to so-called 'cascade' cubic nonlinearities [29]. These nonlinearities result from a cascade of nonlinear interaction of propagating waves in the nonlinear medium. For example concessive second harmonic generation $\omega + \omega \rightarrow 2\omega$ and differential frequency generation $2\omega - \omega \rightarrow \omega$ give, similarly to the cubic nonlinearity, a nonlinear response at frequency ω. If no special measures are taken to optimize the phase matching of the interacting waves the efficiency of the cascade processes is usually small. Here and below we will assume that this 'cascade' contribution is already accounted for in the cubic nonlinearity. Therefore, we will use the following nonlinear constitutive equation which accounts for the nonlinearity and nonlocality of the medium's response (see equations (2.16) and (2.56)):

$$P_i(\omega, \mathbf{k}) = P_{\mathrm{i}}^{(1)}(\omega, \mathbf{k}) + P_{\mathrm{i}}^{(3)}(\omega, \mathbf{k})$$
$$= \frac{1}{4\pi}(\epsilon_{ij} - \delta_{ij} + ik\Gamma_{ijz}^{(1)})A_j + (\chi_{ijkl} + ik\Gamma_{ijklz})A_j A_k A_l^* \qquad (4.1)$$

where i, j, k and l may be either x or y. Substituting $\mathbf{D} = \mathbf{E} + 4\pi\mathbf{P}$ into the wave equation (3.15) and using condition equation (3.16) we obtain the following equation for the amplitude of the electric field in the nonlinear medium:

$$2ik\frac{\mathrm{d}A_i}{\mathrm{d}z} + \frac{\omega^2}{c^2}(\epsilon_{ij} + ik\Gamma_{ijz}^{(1)} - n^2\delta_{ij})A_j$$
$$+ \frac{4\pi\omega^2}{c^2}(\chi_{ijkl} + ik\Gamma_{ijklz})A_j A_k A_l^* \; e^{-2\mathrm{Im}\{k\}z} = 0 \quad (4.2)$$

or, after simple rearrangement:

$$\frac{\mathrm{d}A_i}{\mathrm{d}z} = \frac{i\omega}{2c}(u_{ij}A_j + \kappa_{ijkl}A_j A_k A_l^* \; e^{-2\mathrm{Im}\{k\}z}) \qquad (4.3)$$

The above equation is similar to equation (3.18) but includes a nonlinear term in the right-hand side. Here coefficients u_{ij} are as defined in equation (3.19) and we have introduced new material parameters responsible for the medium nonlinearity:

$$\kappa_{ijkl} = \frac{4\pi}{n}(\chi_{ijkl} + ik\Gamma_{ijklz}) \qquad (4.4)$$

In both u_{ij} and κ_{ijkl} indices i, j, k and l may be either x or y. The parameters u_{ij} and κ_{ijkl} may be expanded over the Pauli matrices as follows:

$$\begin{aligned} u_{ij} &= \tfrac{1}{2}v_\alpha \sigma_{ij}^{(\alpha)} \\ \kappa_{ijkl} &= \tfrac{1}{4}w_{\alpha\beta}\sigma_{ij}^{(\alpha)}\sigma_{lk}^{\beta} \end{aligned} \qquad (4.5)$$

Summations are presumed over α and $\beta = 0, \ldots, 3$, and the expansion coefficients are

$$\begin{aligned} v_\alpha &= \sigma_{ji}^{(\alpha)}u_{ij} \\ w_{\alpha\beta} &= \sigma_{ji}^{(\alpha)}\kappa_{ijkl}\sigma_{kl}^{(\beta)} \end{aligned} \qquad (4.6)$$

From now on we will consider totally polarized light. The case of partially polarized light will be studied in the next chapter. When expansions (4.5) are substituted into equation (4.3), the latter may be rearranged as follows:

$$\frac{\mathrm{d}A_i}{\mathrm{d}z} = \frac{i\omega}{4c}\Omega_\alpha \sigma_{ik}^{(\alpha)}A_k \qquad (4.7)$$

or for complex-conjugated amplitudes

$$\frac{dA_j^*}{dz} = -\frac{i\omega}{4c}\Omega_\alpha^* \sigma_{kj}^{(\alpha)} A_k^* \tag{4.8}$$

Equations (4.7) and (4.8) are similar to equations (3.22) and (3.23). However, in nonlinear media $\Omega_\alpha^{\mathrm{L}}$ is replaced with $\Omega_\alpha = \{\Omega_0; \Omega_1; \Omega_2; \Omega_3\}$, which is now itself dependent on the Stokes parameters of the wave:

$$\Omega_\alpha = v_\alpha + \tfrac{1}{2}w_{\alpha\beta}S_\beta = \Omega_\alpha^{\mathrm{L}} + \tfrac{1}{2}w_{\alpha\beta}S_\beta \tag{4.9}$$

where α and $\beta = 0, \ldots, 3$ and a summation is presumed over the repeated index β. The vector $\mathbf{\Omega}$ with projections Ω_1, Ω_2 and Ω_3 is a vector in Stokes space. From now on this will be called the self-action vector. If nonlinearity is neglected the self-action vector coincides with the vector $\mathbf{\Omega}^{\mathrm{L}}$ defined in equation (3.21).

Now, using the same routine as was used to derive the evolution equations in linear optics (see equations (3.22) to (3.29)), we readily obtain the evolution equations for the Stokes parameters in nonlinear medium:

$$\begin{aligned}\frac{d\mathbf{S}}{dz} &= -\frac{\omega}{2c}\,\mathrm{Im}\,\{(4n + \Omega_0)\mathbf{S} + \mathbf{\Omega}S_0 + i[\mathbf{\Omega}\times\mathbf{S}]\}\\ \frac{dS_0}{dz} &= -\frac{\omega}{2c}\,\mathrm{Im}\,\{(4n + \Omega_0)S_0 + (\mathbf{\Omega}\cdot\mathbf{S})\}\end{aligned} \tag{4.10}$$

These equations are structurally identical to the evolution equations for the Stokes parameters (3.30) in linear optics. However, in nonlinear medium $\Omega_\alpha = \{\Omega_1; \Omega_2; \Omega_3; \Omega_4\}$ does not only depend on the material characteristics of the medium, but is also a function of the current values of the Stokes parameters of the wave. Recall that in linear media $\Omega_\alpha^{\mathrm{L}}$ depends only on the material parameters of the media. This reflects the fact that in nonlinear media the propagation conditions are changed by the wave itself. Therefore, in nonlinear media the evolution equations (4.10) are nonlinear differential equations. Such evolution equations may be applied to nonlinear media of arbitrary symmetry. However, a low anisotropy along the direction of propagation shall be assumed. From the second equation in (4.10) we can conclude that a dissipation-less medium will correspond to $\mathrm{Im}\,\{n\} = 0$, $\mathrm{Im}\,\{\mathbf{\Omega}\} = 0$ and $\mathrm{Im}\,\{\Omega_0\} = 0$.

4.2 POLARIZATION SELF-ACTION IN ISOTROPIC DISSIPATION-LESS MEDIA

In this section we will illustrate how the evolution equations (4.10) may be used for the analysis of elementary nonlinear polarization effects in an isotropic medium. In Chapter 3 we saw that the only reason for a change in the light polarization in an isotropic linear medium was optical activity. The effect of optical activity was described by the antisymmetric tensor of the nonlocal susceptibility $\Gamma_{ijn}^{(1)} = -\Gamma_{jin}^{(1)}$. To study the nonlinear polarization effects we will need to account for the local and nonlocal nonlinearities, χ_{ijkl} and Γ_{ijkln} and identify the relevant nonzero components of the nonlinearity tensors. In an isotropic medium the tensor of cubic nonlinearity has the following nonzero components relevant to our consideration:

$$\begin{aligned} \chi_{yxyx} &= \chi_{yyxx} = \chi_{xxyy} = \chi_{yxyx} \\ \chi_{yyyy} &= \chi_{xxxx} = 2\chi_{yxyx} + \chi_{xyyx} \end{aligned} \tag{4.11}$$

(these may be derived from formula (2.45) taking into account the frequency permutation symmetry of the tensor components in dissipation-less media). If the isotropic medium is also gyrotropic ($\infty\infty$ class of symmetry) it will show linear and nonlinear nonlocal optical responses:

$$\begin{aligned} &\Gamma^{(1)}_{xyz} = -\Gamma^{(1)}_{yxz} \\ &\Gamma_{xyxxz} = \Gamma_{xxyxz} = -\Gamma_{yyxyz} = -\Gamma_{yxyyz} \\ &\Gamma_{yyyxz} = -\Gamma_{xxxyz} = \Gamma_{yxxxz} + \Gamma_{xyxxz} + \Gamma_{xxyxz} \\ &\Gamma_{xyyyz} = -\Gamma_{yxxxz} = \Gamma_{xyxxz} + \Gamma_{xxyxz} + \Gamma_{xxxyz} \end{aligned} \tag{4.12}$$

To specify the evolution equation (4.10) for isotropic nonlinear media we first need to calculate Ω_α as it is given by equation (4.9). Its linear part, $\Omega^{\mathrm{L}}_\alpha = v_\alpha$ for isotropic gyrotropic medium have already been calculated (see equation (3.45)):

$$\begin{aligned} v_1 &= v_2 = 0 \\ v_3 &= -\frac{2\omega\Gamma^{\mathrm{a}}}{c} \\ v_0 &= 0 \end{aligned} \tag{4.13}$$

where $\Gamma^{(1)}_{ijn} = \Gamma^{\mathrm{a}} e_{ijn}$. The parameters w_{ij} responsible for the intensity-dependent part of Ω_α need to be found using formula (4.6) and tensor components (4.11) and (12):

$$\begin{aligned} &w_{11} = w_{22} = \frac{8\pi}{n}(\chi_{xxyy} + \chi_{xyyx}) \\ &w_{33} = \frac{8\pi}{n}(\chi_{xxyy} - \chi_{xyyx}), \qquad w_{00} = \frac{8\pi}{n}(\chi_{xxxx} + \chi_{xxyy}) \\ &w_{03} = \frac{8\pi\omega}{c}(\Gamma_{xxxyz} - \Gamma_{xxyxz}), \qquad w_{30} = \frac{8\pi\omega}{c}(\Gamma_{yxxxz} - \Gamma_{xxyxz}) \\ &w_{12} = \frac{8\pi i\omega}{c}(\Gamma_{xxxyz} + \Gamma_{xxyxz}), \qquad w_{21} = \frac{8\pi i\omega}{c}(\Gamma_{yxxxz} + \Gamma_{xxyxz}) \\ &w_{01} = w_{10} = w_{02} = w_{20} = w_{13} = w_{31} = w_{23} = w_{32} = 0 \end{aligned} \tag{4.14}$$

Thus, in isotropic gyrotropic nonlinear media Ω_α has the following components:

$$
\begin{aligned}
\Omega_1 &= \frac{4\pi}{n}[(\chi_{xxyy} + \chi_{xyyx})S_1 + ik(\Gamma_{xxxyz} + \Gamma_{xxyxz})S_2] \\
\Omega_2 &= \frac{4\pi}{n}[(\chi_{xxyy} + \chi_{xyyx})S_2 + ik(\Gamma_{yxxxz} + \Gamma_{xxyxz})S_1] \\
\Omega_3 &= \frac{1}{n}[-2k\Gamma^\alpha + 4\pi(\chi_{xxyy} - \chi_{xyyx})S_3 + 4\pi k(\Gamma_{yxxxz} - \Gamma_{xxyxz})S_0] \\
\Omega_0 &= \frac{4\pi}{n}[2(\chi_{xxxx} + \chi_{xxyy})S_0 + k(\Gamma_{xxxyz} - \Gamma_{xxyxz})S_3]
\end{aligned}
\tag{4.15}
$$

Formulae (4.15) may be used in absorptive media. However, for the sake of simplicity, we will continue the analysis of polarization self-action effects in isotropic gyrotropic media with no dissipation. In this case all components of Ω_α are real. Indeed, for a self-action process when all the frequency arguments are equal, the frequency permutation relations (2.37) and (2.38) yield

$$
\begin{aligned}
\mathrm{Im}\,\{\chi_{xxyy}\} &= \mathrm{Im}\,\{\chi_{xyyx}\} = 0 \\
\Gamma_{xxxyz} &= \Gamma_{yxxxz} = -\{\Gamma_{xxyxz}\}^*
\end{aligned}
\tag{4.16}
$$

Similarly, condition (2.36) gives $\mathrm{Im}\,\{\Gamma^{\mathrm{a}}\} = 0$ and $\mathrm{Im}\,\{\epsilon\} = 0$ and, therefore, $\mathrm{Im}\,\{n\} = 0$ in the absence of dissipation. Direct inspection shows that with such material tensors $\mathrm{Im}\,\{\Omega_0\} = 0$ and $\mathrm{Im}\,\{\mathbf{\Omega}\} = 0$ which simplifies formulae (4.15):

$$
\begin{aligned}
\Omega_1 &= \frac{4\pi}{n}(\chi_{xxyy} + \chi_{xyyx})S_1 \\
\Omega_2 &= \frac{4\pi}{n}(\chi_{xxyy} + \chi_{xyyx})S_2 \\
\Omega_3 &= \frac{1}{n}[4\pi(\chi_{xxyy} - \chi_{xyyx})S_3 - 2k\ \mathrm{Re}\,\{\Gamma^\alpha - 4\pi\Gamma_{yxxxz}S_0\}] \\
\Omega_0 &= \frac{8\pi}{n}[(\chi_{xxxx} + \chi_{xxyy})S_0 + k\,\mathrm{Re}\,\{\Gamma_{yxxxz}\}S_3]
\end{aligned}
\tag{4.17}
$$

The nonlinear evolution equations for the Stokes parameters in isotropic dissipation-less media can now be written as follows:

$$\begin{aligned}
\frac{\mathrm{d}S_1}{\mathrm{d}z} &= -\frac{\omega^2}{c^2}\,\mathrm{Re}\left\{\Gamma^\alpha - 4\pi\Gamma_{yxxxz}S_0 + 4\pi\frac{\chi_{xyyx}}{k}S_3\right\}S_2 \\
\frac{\mathrm{d}S_2}{\mathrm{d}z} &= \frac{\omega^2}{c^2}\,\mathrm{Re}\left\{\Gamma^\alpha - 4\pi\Gamma_{yxxxz}S_0 + 4\pi\frac{\chi_{xyyx}}{k}S_3\right\}S_1 \\
\frac{\mathrm{d}S_3}{\mathrm{d}z} &= 0 \\
\frac{\mathrm{d}S_0}{\mathrm{d}z} &= 0
\end{aligned} \tag{4.18}$$

(compare with equations (3.46) for linear isotropic dissipative medium). From the last two equations we immediately observe that in a dissipation-less isotropic medium the wave ellipticity $\eta = \sin^{-1}(S_3/S_0)$ does not change. With the initial conditions $S_{\alpha 0} = \{S_{00};\, S_{10};\, S_{20};\, S_{30}\}$ the evolution equations (4.18) have the following solutions:

$$\begin{aligned}
S_1(L) &= S_{10}\cos\left(\frac{\omega^2}{c^2}\,\mathrm{Re}\,\{\Gamma^{\mathrm{eff}}\}L\right) - S_{20}\sin\left(\frac{\omega^2}{c^2}\,\mathrm{Re}\,\{\Gamma^{\mathrm{eff}}\}L\right) \\
S_2(L) &= S_{20}\cos\left(\frac{\omega^2}{c^2}\,\mathrm{Re}\,\{\Gamma^{\mathrm{eff}}\}L\right) + S_{10}\sin\left(\frac{\omega^2}{c^2}\,\mathrm{Re}\,\{\Gamma^{\mathrm{eff}}\}L\right) \\
S_3(L) &= S_{30} \\
S_0(L) &= S_{00}
\end{aligned} \tag{4.19}$$

where

$$\Gamma^{\mathrm{eff}} = \Gamma^\alpha - 4\pi\Gamma_{yxxxz}S_{00} + \frac{4\pi}{k}\chi_{xyyx}S_{00}\sin(2\eta_0) \tag{4.20}$$

By resorting equations (4.19) the polarization azimuth $\Theta(L)$ and ellipticity $\eta(L)$ of a propagating wave can readily be obtained:

$$\Theta(L) = \Theta_0 + \frac{\omega^2}{2c^2}\,\mathrm{Re}\,\{\Gamma^{\mathrm{eff}}\}L \tag{4.21}$$

$$\eta(L) = \eta_0 \tag{4.22}$$

Here L is the length of the medium and Θ_0 and η_0 are the polarization azimuth and ellipticity of the incident wave. Equation (4.21) is similar to equation (3.48) for polarization azimuth rotation in linear media. The difference between equations (3.48) and (4.21) is obviously in the fact that in nonlinear media Γ^{eff} depends on the light wave intensity and initial ellipticity η_0.

Using formula (4.21) we can introduce the two most important nonlinear self-action polarization phenomena. The first one is called nonlinear optical activity

(NLOA) which is a nonlinear analog of conventional optical activity (OA) (see Fig. 4.1). One can clearly identify this effect if the incident light is linearly polarized, that is $\eta_0 = 0$, in which case the last term in equation (4.20) vanishes and we are left with $\Theta(L) = \Theta_0 + \Theta^{\mathrm{OA}}(L) + \Theta^{\mathrm{NLOA}}(L)$, where

$$\begin{aligned} \Theta^{\mathrm{OA}}(L) &= \frac{2\pi^2 L}{\lambda^2}\,\mathrm{Re}\,\{\Gamma^{\alpha}\} \\ \Theta^{\mathrm{NLOA}}(L) &= -\frac{64\pi^4 L}{c\lambda^2(1+n)^2}\,\mathrm{Re}\,\{\Gamma_{yxxxz}\}I \end{aligned} \tag{4.23}$$

Here the angles are in radians, I is the intensity of light with wavelength λ incident, from a vacuum, on a nonlinear medium of refractive index n. In equations (4.23) $\Theta^{\mathrm{OA}}(L)$ describes the polarization plane rotation due to natural optical activity and $\Theta^{\mathrm{NLOA}}(L)$ gives the polarization azimuth rotation due to intensity-dependent nonlinear optical activity. In isotropic dissipation-less media nonlinear optical activity is simply an addendum, proportional to $\mathrm{Re}\,\{\Gamma_{yxxxz}\}$, to the natural optical rotation (see Figs 4.1 and 4.2).

The term 'nonlinear optical activity' (NLOA) is also used to refer to polarization effects due to the real and imaginary parts of the cubic nonlocal susceptibility Γ_{ijkln} in media of lower symmetries including crystals. Although we do not

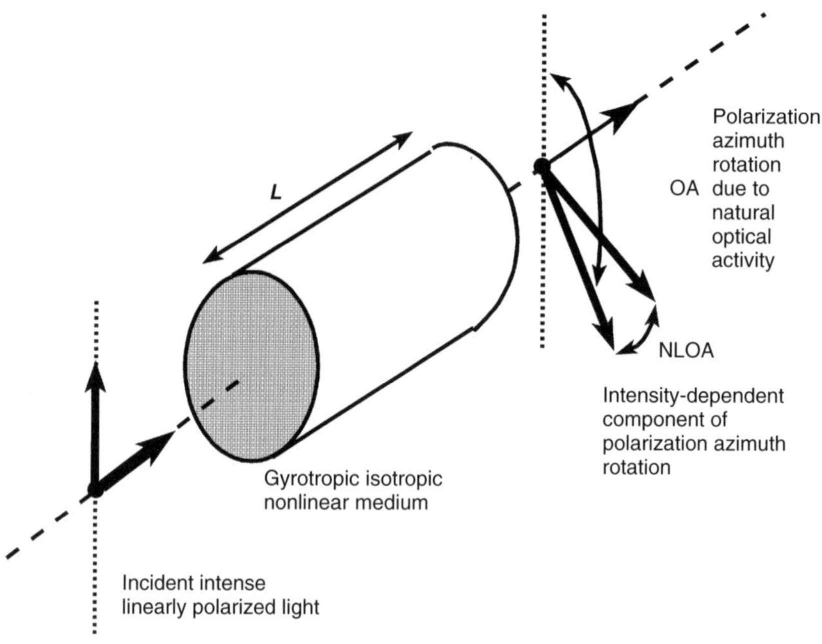

Fig. 4.1 Linear (OA) and nonlinear (NLOA) optical activity in isotropic gyrotropic medium

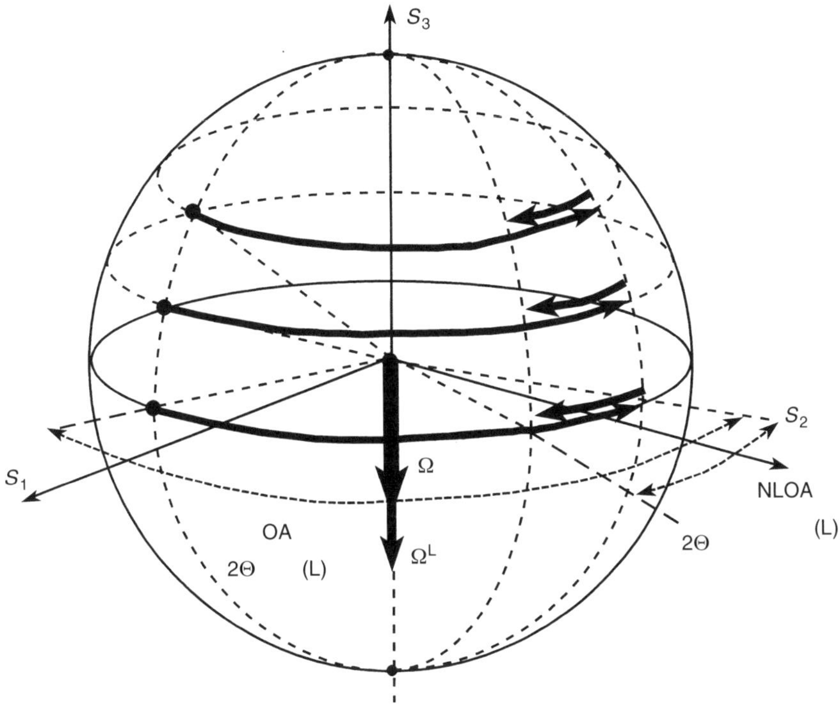

Fig. 4.2 Nonlinear optical activity (NLOA): the Poincaré sphere presentation of the evolution of the polarization state of light in a dissipation-less nonlinear isotropic gyrotropic medium. The self-action vector $\mathbf{\Omega}$ is directed along S_3. When a light wave propagates through the medium, the point on the Poincaré sphere representing the polarization state traces of an arc parallel to the equator: the light does not change its ellipticity but the polarization azimuth changes steadily. The length of the arc is proportional to the length of the medium. The arc consists of two segments. The first one is due to natural optical activity (OA) and the second is due to nonlinear optical activity (NLOA). The length of the second segment increases with intensity.

produce the relevant formulae here, it can be shown that if Re $\{\Gamma_{ijkln}\}$ gives polarization azimuth rotation, Im $\{\Gamma_{ijkln}\}$ will contribute to an intensity-dependent change in the light wave ellipticity. For example, in isotropic media with Im $\{\Gamma_{xxyxz}\} \neq 0$, light initially linearly polarized will become elliptically polarized, with ellipticity increasing with intensity. This can be compared to circular dichroism in linear optics where linearly polarized light becomes elliptically polarized in optically active media with Im $\{\Gamma^{\alpha}\} \neq 0$ (see formula (3.49)).

Nonlinear optical activity or 'nonlinear gyrotropy' was first predicted and discribed by Akhmanov and Zharikov in Ref. [1]. Papers [2–7] provided theoretical analysis of the wave picture and of the different mechanisms of the effect. An account of work on the microscopic theory of nonlocal nonlinearities responsible for NLOA will be given in Chapters 8 and 9. NLOA has also been intensively studied experimentally with the most effort concentrated on the search for the fast 'electronic' mechanism of NLOA. However, initially an intensity-dependent optical activity due to the thermal mechanism of nonlinearity was seen. The experiment was performed using a radiation coloured quartz crystal, SiO_2 [8], which showed considerable natural rotatory power along its optical axis (see Chapter 3, page 54). Rotatory power depends on the crystal temperature. The NLOA observed in Ref. [8] resulted from heating of the crystal by an intense laser pulse. Forty degrees of laser-induced polarization azimuth rotation corresponding to 475 K increase of the crystal temperature were seen. This experiment clearly indicated the main problem when trying to observe the electronic mechanisms of nonlinear optical activity: it must be distinguished from the thermally induced rotation and from the huge background of natural optical activity. An elegant experiment where nonlinear optical activity was separated from the strong natural rotation was undertaken with a lithium iodate crystal ($LiIO_3$) [9]. Using a dedicated polarimeter presented in Fig. 4.3 and a

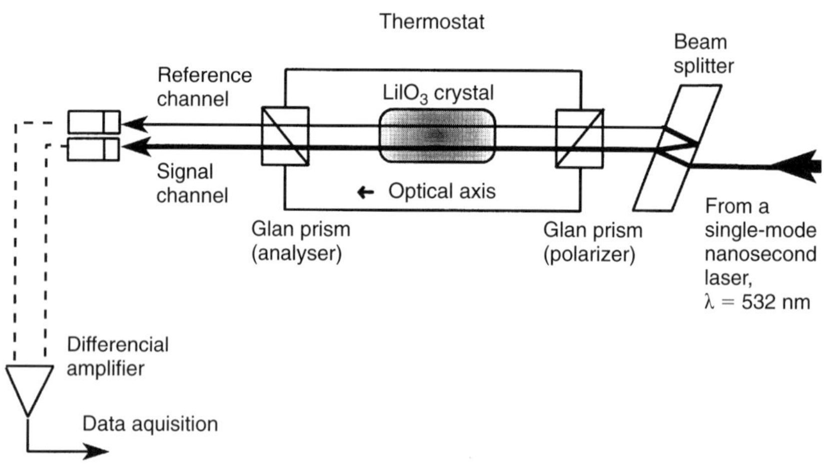

Fig. 4.3 Experimental arrangement for detecting nonlinear optical activity in crystals with a strong natural rotatory power ($LiIO_3$). The laser beam is split by a beamsplitter into reference and signal beams with an intensity ratio of 1:10. The reference and signal beams are identically linearly polarized by a Glan prism. They then pass through the crystal and an analyser, installed so as to compensate for the natural optical rotation. The analyser is set nearly, but not exactly, at the crossed position to the polarizer. Fluctuations of the laser frequency and variations in crystal temperature, which lead to changes in the natural optical activity, affect the signal and reference channels equally. Thus the polarimeter balance is insensitive to such variations. However, with increasing laser power, the ratio of the currents in the signal and reference channels will change due to the nonlinear polarization azimuth rotation which occures predominantly in the signal beam. (Reproduced by permission from American Institute of Physics from Ref. [9])

nanosecond, pulsed, single-frequency, stabilized laser ($\lambda = 532$ nm), electronic nonlinear optical activity was identified for the first time in the background of the thermal effect. At the laser wavelength the lithium iodate crystal gave a natural rotation of 1250 grad/cm along its optical axis. The nonlincar rotation of 1.8×10^{-2} deg at $I = 300\ \mathrm{MW^t/cm^2}$ was seen in a 6.2 cm long crystal. According to formula (4.23) this gives $|\mathrm{Re}\,\Gamma_{yyxxz}| = 3.10^{-21}$ esu.

Other experiments on nonlinear optical activity included the study of $B_{12}GeO_{20}$ [10], $B_{12}SiO_{20}$ [11,12], ZnP_2 [13,14] and CdP_2 [13] crystals where mainly the thermal effect was observed. Thermal nonlinear optical activity may reach considerable levels. For instance, in ZnP_2 the specific constant of thermal nonlinear optical activity reaches 2000 deg cm/W [14] (see Fig. 4.4). For the observation of NLOA in organic liquids see Ref. [15].

The second fundamental nonlinear polarization effect observable in isotropic media is that of polarization ellipse self-rotation (see Fig. 4.5). The effect can be seen in isotropic nongyrotropic media ($\Gamma^a = 0$ and $\Gamma_{xxyxz} = 0$), but only if the light wave is initially elliptically polarized. In such a medium it follows from equations (4.20) and (4.21) that the polarization azimuth rotates as the wave propagates along the propagation direction:

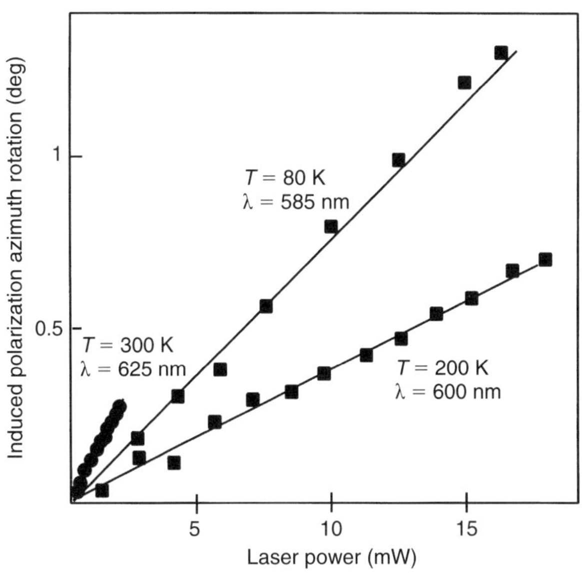

Fig. 4.4 The intensity-dependent component of polarization azimuth rotation along the optical axis of a ZnP_2 crystal of thickness 0.55 mm as a function of beam power for different wavelengths of a CW dye laser and different crystal temperatures. The thermal mechanism of nonlinear optical activity dominates (after Ref. [14])

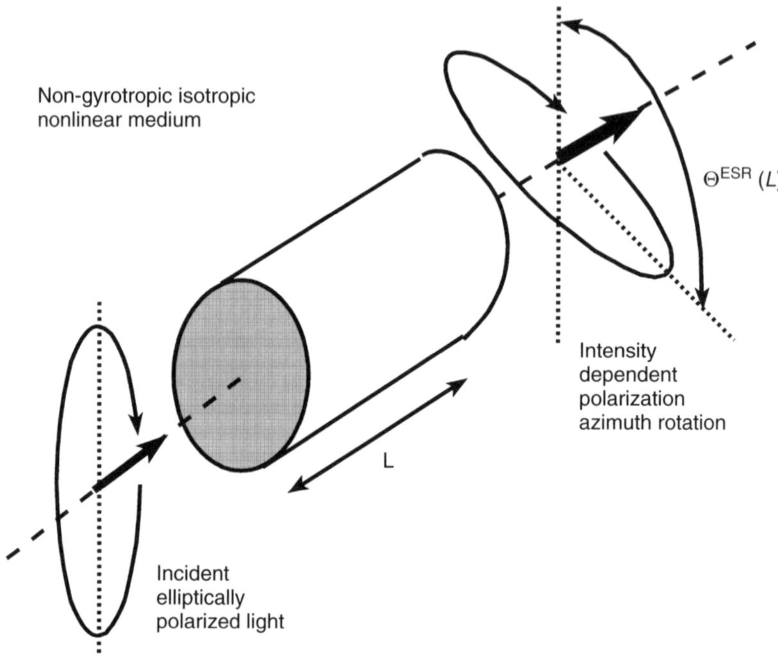

Fig. 4.5 Polarization ellipse self-rotation effect in an isotropic medium

$$\begin{aligned} \Theta^{\mathrm{ESR}}(L) - \Theta_0 &= \frac{32\pi^3 L}{\lambda c n |1+n|^2} \operatorname{Re}\{\chi_{xyyx}\} I \sin(2\eta_0) \\ \eta(L) &= \eta_0 \end{aligned} \tag{4.24}$$

Here I is the intensity of a light wave approaching a nonlinear medium of refractive index n from a vacuum. Since the rotation requires an initial ellipticity η_0, which does not change on propagation, the effect is often referred to as polarization ellipse self-rotation. On the Poincaré sphere this effect is represented by a point tracing out a horizontal arc above or below the equator (see Fig 4.6). More general analysis based on the evolution equations (4.10), with the self-action vector (4.15), shows that in dissipative isotropic media where the nonlinearity χ_{ijkl} is not purely real, the azimuth rotation is accompanied by an intensity-dependent change in ellipticity.

Polarization ellipse self-rotation is one of the most easily observed nonlinear optical effects. It was first seen by Maker, Terhune and Savage who produced

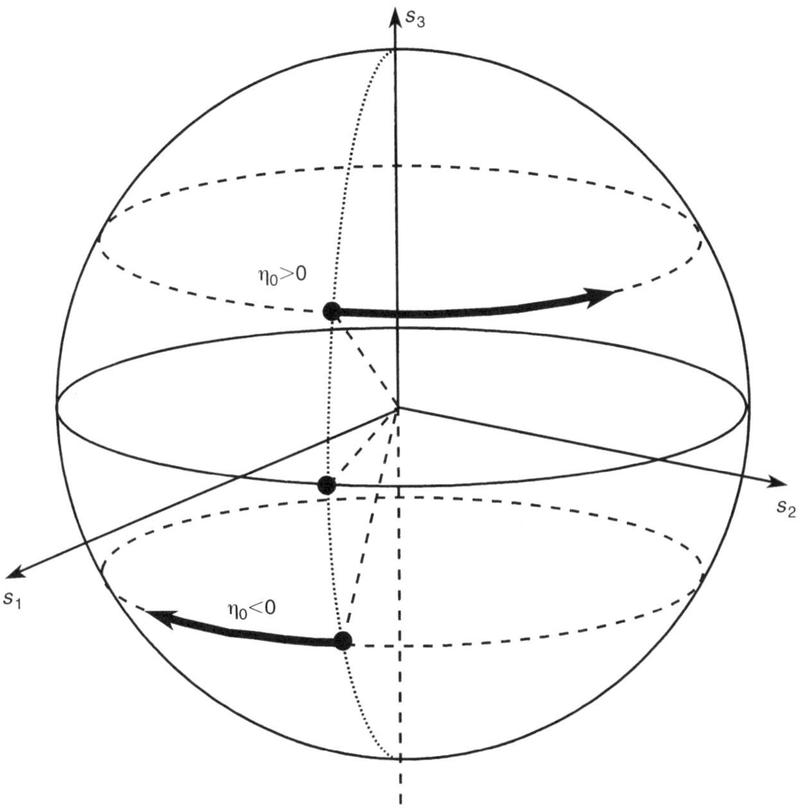

Fig. 4.6 Polarization ellipse self-rotation: the Poincaré sphere presentation of the evolution of the polarization state of elliptically polarized light in a dissipation-less, nonlinear, isotropic, non-gyrotropic medium. When a light wave propagates through the medium, the point on the Poincaré sphere which represents the polarization state traces out an arc parallel to the equator. The length of the arc is proportional to the length of the medium and increases with intensity and initial ellipticity. Depending on the sign of the initial ellipticity, the azimuth rotation will be positive or negative

probably the most acclaimed experimental paper on nonlinear polarization effects [16]. In this experiment a pulsed ruby laser ($\lambda = 0.69\,\mu$m) was used to observe polarization ellipse self-rotation in various non-gyrotropic liquids including water, methanol, bromoform and carbon disulfide. The experimental set-up is presented in Fig. 4.7. In the 1 m long cell polarization azimuth rotation of several degrees was seen at intensities of approximately $100\,\mathrm{MW/cm^2}$. In

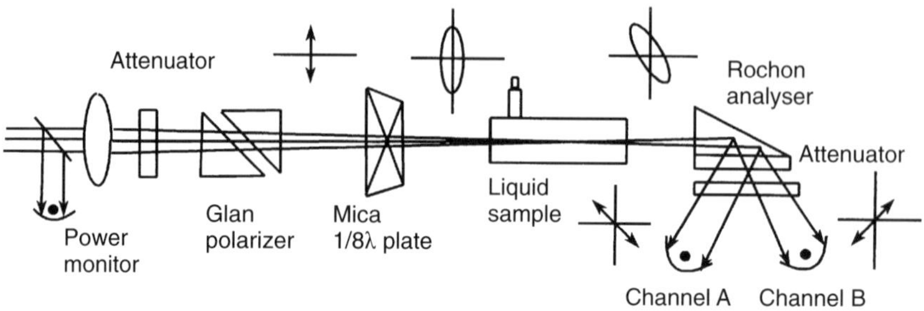

Fig. 4.7 Experimental arrangement used in the first experiment for detecting intensity-dependent rotation of the polarization ellipse. The polarization of the beam, as seen when facing the laser, is indicated at several positions. The initial elliptical polarization state is achieved using $\lambda/8$-wave plates (Reproduced by permission of the American Physical Society from Ref. [16])

bromoform, for instance, this corresponds to the cubic nonlinearity $\mathrm{Re}\,\{\chi_{xyyx}\} \simeq 10^{-14}$ esu [16] (see Fig. 4.8).

Polarization ellipse self-rotation is used for the routine measurement of cubic optical nonlinearities and in spectroscopic applications [17].

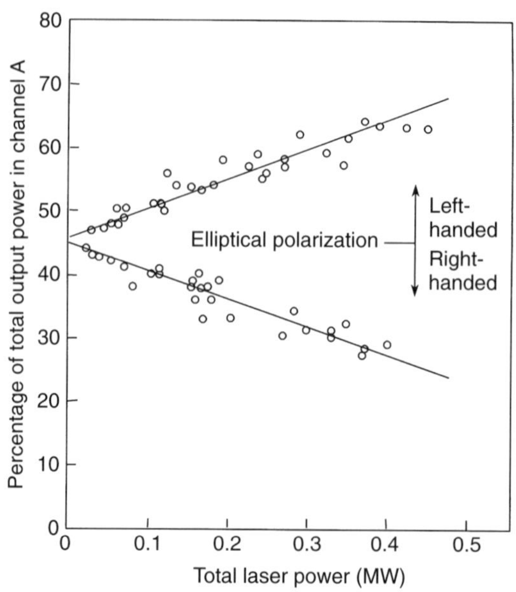

Fig. 4.8 Measured percentage of total output power in channel A (see Fig. 4.7) versus laser power for a bromoform sample. The upper and lower data correspond to the beam entering the liquid sample with left and right elliptical polarizations, indicating polarization ellipse rotations in opposite directions compare with Fig. 4.6. (Reproduced by permission of the American Physical Society from Ref. [16])

4.3 POLARIZATION SELF-ACTION EFFECTS IN CUBIC CRYSTALS

In this section we will consider the effects of polarization self-action in crystals of cubic symmetry. We will introduce two major nonlinear polarization effects: the nonlinear anisotropic effect and nonlinear nonlocal birefringence.

There are five different cubic crystal symmetry point groups; 23, $m3$, 432, $\bar{4}3m$ and $m3m$. All cubic crystals have isotropic dielectric tensors $\epsilon_{ij} = \epsilon\delta_{ij}$. Therefore without taking into account nonlocality, their linear optical properties are isotropic. Even within the first-order spatial dispersion approximation, the linear optical properties of crystals of $m3$ and $m3m$ point groups are indistinguishable from those of isotropic nongyrotropic media and these crystal show no nonlocal effects since $\Gamma^{(1)}_{ijn} = 0$ by symmetry. Some of the cubic crystals, namely the 23 and 432 classes, show natural and nonlinear optical activity similar to that of isotropic gyrotropic media (departure from isotropic optical activity is discussed in Appendix 4). Here in more detail we will examine crystals of the $\bar{4}3m$ point group. These crystals show no conventional optical activity as it is forbidden by symmetry, $\Gamma^{\alpha}_{ijn} = -\Gamma^{\alpha}_{jin} = 0$. However, because this point group supports the symmetric nonlocality tensor $\Gamma^{s}_{ijn} = \Gamma^{s}_{jin}$, in certain circumstances they may exhibit nonlocal birefringence (see Appendix 4). This choice of crystal symmetry point group will allow us to illustrate, as specific cases, both main nonlinear polarization effects in cubic crystals.

We will consider light propagating along the cubic axis [001] in the crystal's natural Cartesian frame. In cubic crystals, including those of the $\bar{4}3m$ point group symmetry, the natural frame coincides with the edges of the cubic crystal cell. In cubic crystals of the $\bar{4}3m$ group the tensors of local and nonlocal cubic nonlinearity have the following non-zero components which are important for our analysis:

$$\begin{aligned} &\chi_{yxyx} = \chi_{yyxx} = \chi_{xxyy} = \chi_{yxyx} \\ &\chi_{yyyy} = \chi_{xxxx} \\ &\chi_{yxxy} = \chi_{xyyx} \end{aligned} \tag{4.25}$$

$$\begin{aligned} &\Gamma_{xyxxz} = \Gamma_{xxyxz} = \Gamma_{yyxyz} = \Gamma_{yxyyz} \\ &\Gamma_{yyyxz} = \Gamma_{xxxyz} \\ &\Gamma_{xyyyz} = \Gamma_{yxxxz} \end{aligned} \tag{4.26}$$

The parameters $v_\alpha = \{v_0, v_1, v_2, v_3\}$ in this crystal point group may be found in Appendix 4:

$$v_1 = 0, \qquad v_2 = \frac{2i\omega\Gamma^{s}}{c} \qquad v_3 = 0 \qquad v_0 = 0 \tag{4.27}$$

where $\Gamma^{(1)}_{xyz} = \Gamma^{(1)}_{yxz} = \Gamma^{s}$. Resorting equations (4.25) and (4.26) we can calculate the parameters w_{ij} as given by formula (4.6):

$$\begin{aligned}
w_{00} &= \frac{8\pi}{n}(\chi_{xxxx} + \chi_{xxyy}), \qquad & w_{11} &= \frac{8\pi}{n}(\chi_{xxxx} - \chi_{xxyy}) \\
w_{22} &= \frac{8\pi}{n}(\chi_{xxyy} + \chi_{xyyx}), \qquad & w_{33} &= \frac{8\pi}{n}(\chi_{xxyy} - \chi_{xyyx}) \\
w_{13} &= \frac{8\pi\omega}{c}(\Gamma_{xxxyz} - \Gamma_{xxyxz}), \qquad & w_{31} &= \frac{8\pi\omega}{c}(\Gamma_{yxxxz} - \Gamma_{xxyxz}) \\
w_{02} &= \frac{8\pi i\omega}{c}(\Gamma_{xxxyz} + \Gamma_{xxyxz}), \qquad & w_{20} &= \frac{8\pi i\omega}{c}(\Gamma_{yxxxz} + \Gamma_{xxyxz}) \\
w_{01} &= w_{10} = w_{12} = w_{21} = w_{03} = w_{30} = w_{23} = w_{32} = 0
\end{aligned} \tag{4.28}$$

Therefore the self-action vector for light propagating along the [001] direction in the crystals of the $\bar{4}3m$ point group symmetry will be

$$\begin{aligned}
\Omega_1 &= \frac{4\pi}{n}[(\chi_{xxxx} - \chi_{xxyy})S_1 + k(\Gamma_{xxxyz} - \Gamma_{xxyxz})S_3] \\
\Omega_2 &= \frac{4\pi}{n}\left[ik\frac{\Gamma^{s}}{2\pi} + (\chi_{xxyy} + \chi_{xyyx})S_2 + ik(\Gamma_{yxxxz} + \Gamma_{xxyxz})S_0\right] \\
\Omega_3 &= \frac{4\pi}{n}[(\chi_{xxyy} - \chi_{xyyx})S_3 + k(\Gamma_{yxxxz} - \Gamma_{xxyxz})S_1] \\
\Omega_0 &= \frac{4\pi}{n}[(\chi_{xxxx} + \chi_{xxyy})S_0 + ik(\Gamma_{xxxyz} + \Gamma_{xxyxz})S_2]
\end{aligned} \tag{4.29}$$

The nonlinear evolution equations for the Stokes parameters (4.10) will become

$$\begin{aligned}
\frac{\mathrm{d}S_1}{\mathrm{d}z} = &-\frac{4\pi\omega}{c}\left(\mathrm{Im}\left\{\frac{\chi_{xxxx}}{n}\right\}S_0S_1 + \mathrm{Re}\left\{\frac{\chi_{xyyx}}{n}\right\}S_2S_3\right) \\
&+ \frac{2\pi\omega^2}{c^2}\left(\mathrm{Im}\left\{\frac{\Gamma^{s}}{2\pi}\right\}S_3 - \mathrm{Im}\{\Gamma_{xxxyz} - \Gamma_{yxxxz} - 2\Gamma_{xxyxz}\}S_0S_3\right. \\
&\left. - \mathrm{Re}\{\Gamma_{xxxyz} - \Gamma_{yxxxz} + 2\Gamma_{xxyxz}\}S_1S_2\right)
\end{aligned}$$

$$\frac{dS_2}{dz} = -\frac{2\pi\omega}{c}\left(\mathrm{Im}\left\{\frac{\chi_{xxxx}+2\chi_{xxyy}+\chi_{xyyx}}{n}\right\}S_0S_2+\mathrm{Re}\left\{\frac{2\chi_{xxyy}-\chi_{xxxx}-\chi_{xyyx}}{n}\right\}S_1S_3\right)$$
$$-\frac{2\pi\omega^2}{c^2}\left(\mathrm{Re}\left\{\frac{\Gamma^{\mathrm{s}}}{2\pi}\right\}S_0+\mathrm{Re}\{\Gamma_{xxxyz}(S_2^2-S_3^2)+\Gamma_{yxxxz}(S_0^2+S_1^2)+2\Gamma_{xxyxz}(S_2^2+S_3^2)\}\right)$$

$$\frac{dS_3}{dz} = -\frac{2\pi\omega}{c}\left(\mathrm{Im}\left\{\frac{\chi_{xxxx}+2\chi_{xxyy}-\chi_{xyyx}}{n}\right\}S_0S_3+\mathrm{Re}\left\{\frac{\chi_{xxxx}-2\chi_{xxyy}-\chi_{xyyx}}{n}\right\}S_1S_2\right)$$
$$-\frac{4\pi\omega^2}{c^2}\left(\mathrm{Im}\left\{\frac{\Gamma^{\mathrm{s}}}{4\pi}\right\}S_1+\mathrm{Im}\{\Gamma_{yxxxz}\}S_0S_1+\mathrm{Re}\{\Gamma_{xxxyz}\}S_2S_3\right)$$

$$\frac{dS_0}{dz} = -\frac{2\pi\omega}{c}\mathrm{Im}\left\{2\frac{\chi_{xxxx}}{n}S_1^2+\frac{\chi_{xxxx}+2\chi_{xxyy}+\chi_{xyyx}}{n}S_2^2+\frac{\chi_{xxxx}+2\chi_{xxyy}-\chi_{xyyx}}{n}S_3^2\right\}$$
$$-\frac{2\pi\omega^2}{c^2}\left(\mathrm{Re}\left\{\frac{\Gamma^{\mathrm{s}}}{\pi}\right\}S_2+\mathrm{Re}\{\Gamma_{yxxxz}+\Gamma_{xxxyz}+2\Gamma_{xxyxz}\}S_0S_2\right.$$
$$\left.+\,\mathrm{Im}\{\Gamma_{yxxxz}+\Gamma_{xxxyz}-2\Gamma_{xxyxz}\}S_1S_3\right) \tag{4.30}$$

Using equations (4.30) the two main nonlinear polarization effects of light self-action in cubic crystals can be introduced. The first is the nonlinear anisotropic effect which is due to the local cubic nonlinearity χ_{ijkl}. To isolate this effect we need to drop the linear and nonlinear nonlocal susceptibilities from equations (4.30). Note that linear and nonlinear nonlocal susceptibilities vanish in crystals with an inversion centre. Thus, this simplification is effectively an increase in the crystal symmetry up to the centrosymmetric $m3m$ point group.

As mentioned above, the linear optical properties of crystals in this group are identical to those of isotropic nongyrotropic liquids. It would, however, be incorrect to assume that nonlinear polarization effects occurring in such crystals would manifest themselves exactly as in the isotropic liquids. Indeed, the nonlinearity tensor χ_{ijkl} in cubic crystals has three independent components while in isotropic media it has only two (compare formulae (4.11) and (4.25)). In particular, in isotropic nongyrotropic liquid the only possible nonlinear polarization effect was the polarization ellipse self-rotation which required non-zero initial ellipticity of the light beam. In cubic crystals, even when nonlocality is neglected, the polarization azimuth of initially linearly polarized light will rotate as shown below.

When we drop the terms containing Γ^{s} and Γ_{ijkln} the nonlinear evolution equations simplify to:

$$
\begin{aligned}
\frac{\mathrm{d}S_1}{\mathrm{d}z} &= -\frac{4\pi\omega}{c}\left(\mathrm{Im}\left\{\frac{\chi_{xxxx}}{n}\right\}S_0S_1 + \mathrm{Re}\left\{\frac{\chi_{xyyx}}{n}\right\}S_2S_3\right) \\
\frac{\mathrm{d}S_2}{\mathrm{d}z} &= -\frac{2\pi\omega}{c}\left(\mathrm{Im}\left\{\frac{\chi_{xxxx}+2\chi_{xxyy}+\chi_{xyyx}}{n}\right\}S_0S_2\right. \\
&\qquad\left. + \mathrm{Re}\left\{\frac{2\chi_{xxyy}-\chi_{xxxx}-\chi_{xyyx}}{n}\right\}S_1S_3\right) \\
\frac{\mathrm{d}S_3}{\mathrm{d}z} &= -\frac{2\pi\omega}{c}\left(\mathrm{Im}\left\{\frac{\chi_{xxxx}+2\chi_{xxyy}-\chi_{xyyx}}{n}\right\}S_0S_3\right. \\
&\qquad\left. + \mathrm{Re}\left\{\frac{\chi_{xxxx}-2\chi_{xxyy}-\chi_{xyyx}}{n}\right\}S_1S_2\right) \\
\frac{\mathrm{d}S_0}{\mathrm{d}z} &= \\
&-\frac{2\pi\omega}{c}\mathrm{Im}\left\{2\frac{\chi_{xxxx}}{n}S_1^2 + \frac{\chi_{xxxx}+2\chi_{xxyy}+\chi_{xyyx}}{n}S_2^2 + \frac{\chi_{xxxx}+2\chi_{xxyy}-\chi_{xyyx}}{n}S_3^2\right\}
\end{aligned} \tag{4.31}
$$

Let the incident light to be linearly polarized with polarization azimuth Θ_0 be

$$S_\alpha = S_{00}(1,\ \cos 2\Theta_0,\ \sin 2\Theta_0,\ 0) \tag{4.32}$$

We will reasonably assume only small polarization changes with propagation, $\Theta(L) - \Theta_0 \ll \pi$ and $\eta(L) \ll \pi$. Eqautions (4.31) thus yield the following formulae for the polarization azimuth and ellipticity of a wave propagating through a crystal along the [001] axis of the natural frame:

$$\begin{pmatrix}\Theta(L)-\Theta_0 \\ \eta(L)\end{pmatrix} = \frac{8\pi^3 L}{\lambda c|1+n|^2}\begin{pmatrix}\mathrm{Im} \\ -\mathrm{Re}\end{pmatrix}\left\{\frac{\Delta\chi}{n}\right\} I\sin 4\Theta_0 \tag{4.33}$$

Here L is length of the medium, I is the intensity of the light wave with wavelength λ approaching the nonlinear medium of core refractive index n from vacuum and

$$\Delta\chi = \chi_{xxxx} - 2\chi_{xxyy} - \chi_{xyyx} \tag{4.34}$$

Thus, the polarization azimuth of an intense light beam rotates even if the initial ellipticity of the light is zero (see Fig. 4.9). The sign of the nonlinear anisotropic polarization effect depends on the initial polarization azimuth Θ_0. If the light is polarized along the [100] direction (cubic X axis), the [010] direction (cubic Y axis), the bisector [110] between X and Y, or perpendicular to this bisector, i.e. if

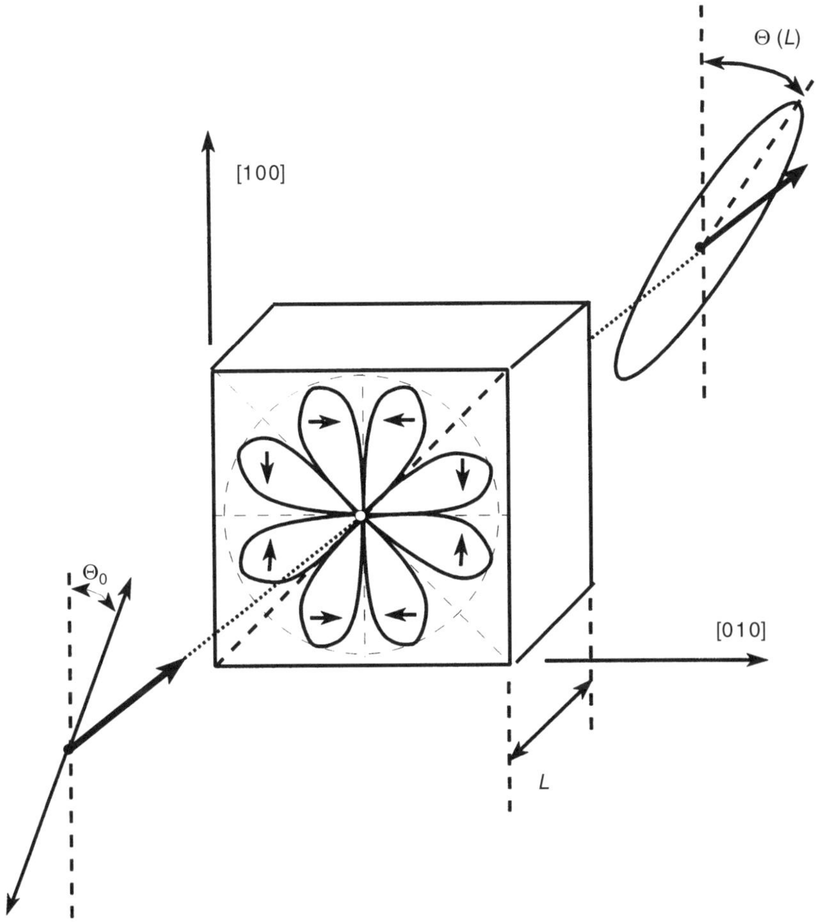

Fig. 4.9 The nonlinear anisotropic polarization effect in a crystal of cubic symmetry. Intense, initially linearly polarized light with polarization azimuth Θ_0 propagates along the [001] crystal direction. On emerging from the crystal the light is elliptically polarized with a different polarization azimuth $\Theta(L)$. The $|\sin 4\Theta_0|$ rosette shows the magnitude of the nonlinear anizotropic polarization effect as a function of the incident polarization azimuth Ω_0. Arrows indicate the alternating direction of the intensity dependent polarization azimuth rotation

$\sin 4\Theta_0 = 0$, then no polarization change occurs. For any other initial polarization the polarization state of the wave will change.

In dissipation-less media, where the refractive index n and cubic optical nonlinearity χ_{ijkl} are real, the light will become eliptically polarized, but the main axis of the ellipse will retain its initial azimuth. The self-induced elipticity will be proportional to the intensity and the length of the crystal. In dissipative media where n or χ_{ijkl} is complex, the self-induced ellipticity will be accompanied by rotation of the polarization azimuth. Depending on the sign of $\mathrm{Im}\,\{\Delta\chi/n\}$, the azimuth will tend towards the cubic axes or their bisectors. Since for small induced polarization azimuth rotation the nonlinear anisotropic effect manifests itself in a very similar way to nonlinear optical activity, in early works [5, 18] it was called NOA-II or NLOA-II (i.e. the second type of nonlinear optical activity). Therefore, the nonlinear anisotropic polarization effect reveals hidden anisotropy of third-order local optical nonlinearity of cubic crystals. This anisotropic effect is represented by $\Delta\chi = \chi_{xxxx} - 2\chi_{xxyy} - \chi_{xyyx}$. This combination of components of the cubic optical nonlinearity tensor is zero by symmetry in isotropic media (see Appendix 1, point groups $\infty\infty$ and $\infty\infty$m), and thus shows the difference between the nonlinearity tensor in isotropic and cubic media.

Intensity-dependent polarization plane rotation due to the nonlinear anisotropic effect was first seen in a GaAs monocrystal of the $\bar{4}3m$ point group [18]. Self-induced ellipticity due to the nonlinear anisotropic effect was first observed in a range of highly transparent alkali–halid crystals of the $m3m$ point group, namely NaCl, KBr, KCl, LiF [19]. In these crystals $\mathrm{Im}\{\Delta\chi\} = 0$ in the visible optical range, and no polarization azimuth rotation can be seen. However, $\mathrm{Re}\{\Delta\chi\}$ is not necessary zero. As described by formula (4.33), initially linearly polarized light becomes eliptically polarized if $\mathrm{Re}\{\Delta\chi\} \neq 0$. If a crystal with non-zero $\mathrm{Re}\{\Delta\chi\}$ is sandwiched between crossed polarizers, the transmitted light intensity depends on the self-induced ellipticity and is proportional to $|\eta(L)|^2$ (see Fig. 4.10). The anisotropic nonlinearity seen in the alkali–halid crystals was rather small. In NaCl, for instance, the observed value of $\eta(L)/(IL) \simeq 1.7 \times 10^{-19}$ cm/erg corresponds to $\mathrm{Re}\{\Delta\chi\} \simeq 10^{-14}$ esu. Spectral dependencies of the polarization rotation due to the anisotropic effect were studied in Ref. [20].

A polarization effect similar to the nonlinear anisotropic effect was seen in cubic crystals of KCl containing F_A(Li) colour centres [21]. The optical properties of these crystals depend upon the orientation of the F_A(Li) centres with respect to the light beam polarization. Formula (4.33) is not directly applicable to this situation (the nonlinearity is very slow), nevertheless, the effect is due to the cubic anisotropic nonlinearity and formula (4.33) correctly predicts the orientational dependence of the effect on Θ_0.

The nonlinear anisotropic effect has very interesting spectroscopic applications because it is sensitive to departures of the crystal electronic structure from isotropy. In fact, the nonlinear anisotropic effect depends on the same anisotropic nonlinearity as the polarization-dependent component of two-photon absorbtion in cubic crystals [22]. The quantum mechanical [23–25] and classical oscillator [26] theories of this anisotropy have been developed. The anisotropic polarization effect is most pronounced along cubic directions, e.g. [001]. Polarization self-action along other directions in cubic crystals has been studied in Refs. [27] and [28].

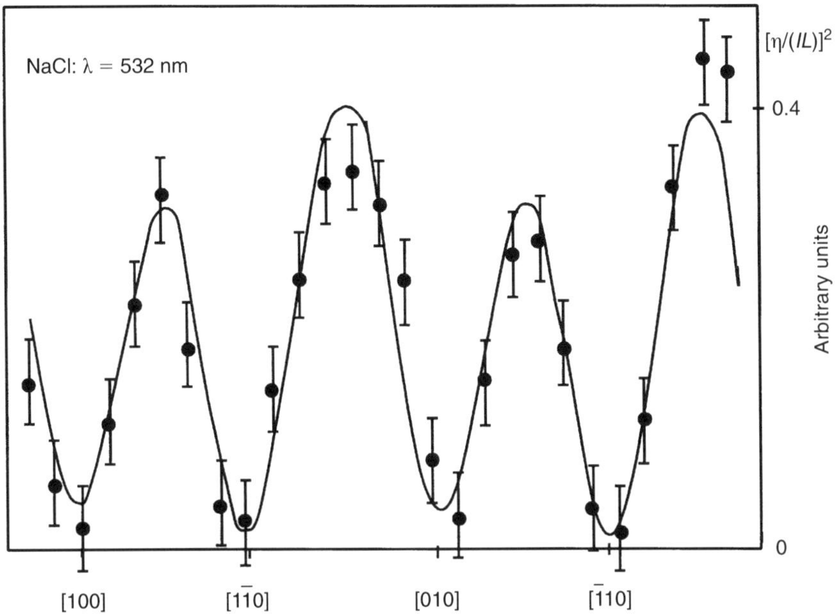

Fig. 4.10 Self-induced ellipticity due to the nonlinear anisotropic polarization effect as a function of the crystal orientation Θ_0. Initially linearly polarized picosecond optical pulses, at $\lambda = 532$ nm, propagate along the [001] direction in a 5 mm long NaCl crystal. The photodetector signal is proportional to η^2 and therefore oscillates with Θ_0 twice as fast as η. The inequality of the peaks is related to small residual birefringence in the sample (Reprinted with kind permission of Elsevier Science from Ref. [19])

The second major polarization effect in the nonlinear optics of cubic crystals is due to the nonlocality of nonlinear response. In crystals of the $\bar{4}3m$ symmetry group the nonlocal susceptibilities Γ^{s} and Γ_{ijkln} cannot be ignored. Small polarization changes of initially linearly polarized light will depend both on local and nonlocal optical nonlinearities:

$$\begin{pmatrix} \Theta(L) - \Theta_0 \\ \eta(L) \end{pmatrix} = \frac{8\pi^3 L}{c\lambda|1+n|^2} \begin{pmatrix} \mathrm{Im} \\ -\mathrm{Re} \end{pmatrix} \left\{ \frac{\Delta\chi}{n} \right\} I \sin 4\Theta_0 - \frac{2\pi^2 L}{\lambda^2} \begin{pmatrix} \mathrm{Re} \\ \mathrm{Im} \end{pmatrix} \left\{ \Gamma^{\mathrm{s}} + \Gamma_{yxxxz} \frac{32\pi^2}{c|1+n|^2} I \right\} \cos 2\Theta_0 \quad (4.35)$$

One can see that the first term on the right-hand side of equation (4.35) produces the nonlinear anisotropic effect already examined above. The second term has two components. The first, which is proportional to Γ^{s}, does not depend on intensity. This component describes natural nonlocal birefringence (gyrotropic linear dichroism) (see equation (A4.4) in Appendix 4). The second component is proportional to the intensity. It describes nonlinear nonlocal birefringence. Polarization effects due to the nonlocality can be clearly distinguished from those due to nonlinear anisotropy. The effect due to $\Delta\chi$ depends on the initial polarization azimuth as $\sin 4\Theta_0$, while the polarization effect associated with nonlocality depends on the initial polarization as $\cos 2\Theta_0$ (compare Figs. 4.9 and 4.11).

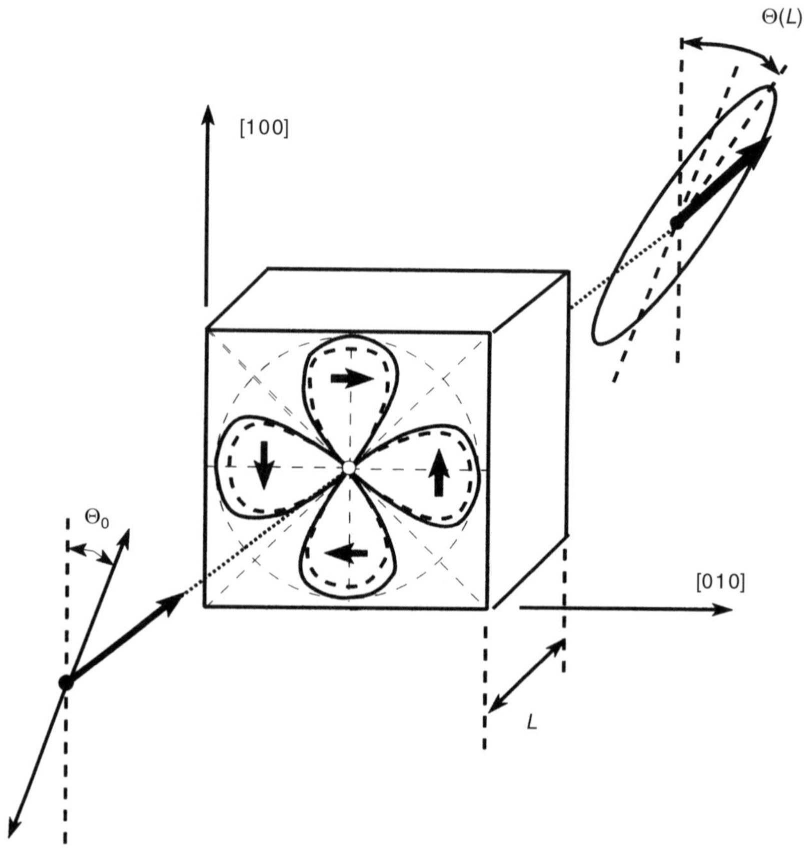

Fig. 4.11 Natural and intensity-dependent nonlocal birefringence in a cubic crystal of the $\bar{4}3m$ point group. The nonlinear anisotropic effect is not shown. Intense, initially linearly polarized light with polarization azimuth Θ_0 propagates along the [001] direction. On emerging from the crystal the light is elliptically polarized with a different polarization azimuth $\Theta(L)$. The dashed line of the $|\cos 2\Theta_0|$ rosette shows the magnitude of the polarization effect due to natural gyrotropic linear dichroism. This does not depend on intensity. The slightly bigger solid line rosette shows the polarization effect including the intensity-dependent gyrotropic linear dichroism. Arrows indicate the alternating direction of induced polarization rotation.

In crystals of the $\bar{4}3m$ symmetry point group the nonlocal and anisotropic polarization self-action effects were first documented in Ref. [18]. They are generally very small effects. Rotation of the polarization azimuth of initially linearly polarized light was studied, as a function of the crystal orientation and light intensity, in a GaAs crystal along the [001] direction. A nanosecond pulsed laser operating at $\lambda = 1.064\ \mu$m was used as the optical source and sensitivity the rotation of 0.02° was achieved. Only the intensity-dependent component of azimuth relation was detected. Contributions from the nonlinear anisotropic and nonlocal effects were separated by Fourier analysis of the orientational dependence of the azimuth rotation. The anisotropic effect due to $\Delta\chi$, which depends on the crystal orientation as $\sin 4\Theta_0$, gave the fourth harmonic in the Fourier

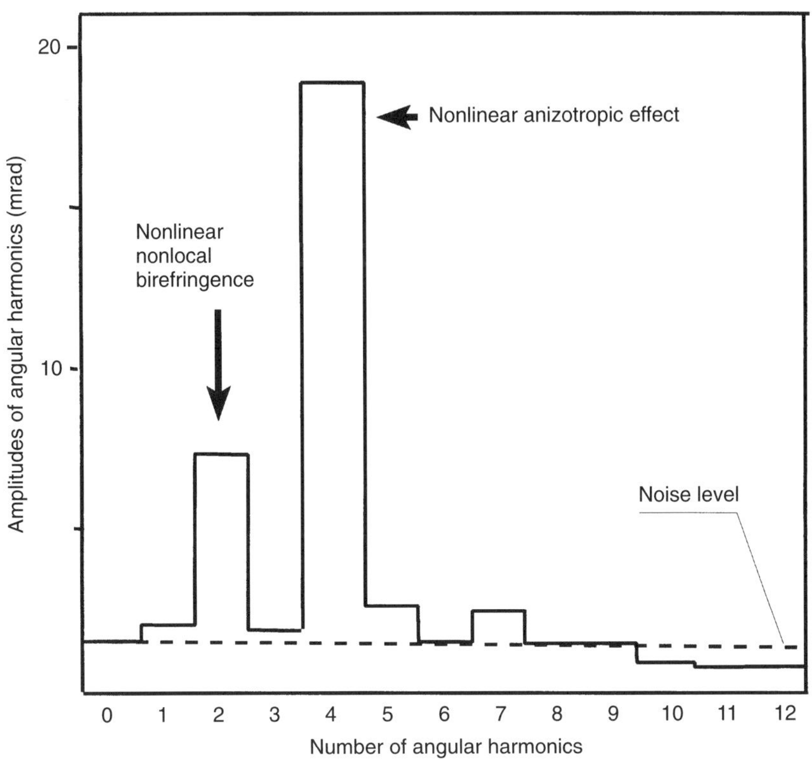

Fig. 4.12 Observation of nonlinear nonlocal birefringence in GaAs, a crystal of the $\bar{4}3m$ point group symmetry. The angular Fourier spectrum of the orientational dependence of the polarization azimuth rotation is presented (see formula (4.35)). Two distinct peaks correspond to the contribution of the nonlinear anisotropic effect ($N = 4$) and nonlinear nonlocal birefringence ($N = 2$) The intensity was approximately 3 MW/cm^2 (Reprinted with kind permission of Elsevier Science from Ref. [18])

spectrum. The nonlinear nonlocal effect due to Γ_{yxxxz}, which depends on the crystal orientation as $\cos 2\Theta_0$, gave the second harmonic (see Fig. 4.12). The following values of the nonlinear susceptibilities were derived from the experimental data: $\mathrm{Im}\{\Delta\chi\} = 6 \times 10^{-11}$ esu and $|\mathrm{Re}\{\Gamma_{yxxxz}\}| = 10^{-15}$ esu. The spectral dependence of the nonlinear anisotropic effect and nonlinear nonlocal birefringence were studied in Ref. [20] using tunable picosecond lasers.

REFERENCES

1. S.A. Akhmanov and V.I. Zharikov, *JETP Lett*, **6**, 137 (1967).
2. P.A. Atkins and L.D. Barron, *Proc. R. Soc.*, **304**, 303 (1968).
3. S. Kielich, *Phys. Lett. A*, **25**, 517 (1967).
4. S. Kielich, *Opt. Commun.*, **1**, 129 (1969).
5. A.D. Petrenko and N.I. Zheludev, *Optica Acta*, **31**, 1177 (1984).
6. A.A. Golubkov and V.A. Makarov, *J. Modern Opt.*, **37**, 1531 (1990).
7. G. Rivoire, Nonlinear propagation of laser light of different polarizations, in M. Evans and S. Kielich (eds.), *Modern Nonlinear Optics*, Part 1, John Wiley & Sons, New York, 1993.
8. D.V. Vlasov and V.P. Zaitseva, *JETP Lett.*, **13**, 112 (1971).
9. S.A. Akhmanov, B.V. Zhdanov, N.I. Zheludev, N.I. Kovrigin and V.I. Kuznetsov, *JETP Lett.*, **29**, 264 (1979).
10. B.Kh. Bairamov, B.P. Zakharchenya and Z.M. Khashhozhev, *Soviet Phys. Solid State*, **14**, 2357 (1973).
11. B.Kh. Bairamov, B.P. Zakharchenya, V.V. Toporov and Z.M. Khashkhozhev, *Soviet Phys. Solid State*, **15**, 1245 (1973).
12. N.I. Zheludev, I.S. Ruddock and R. Illingworth, *Appl. Phys. B*, **49**, 65 (1989).
13. V.V. Borshch, M.P. Lisitsa, P.E. Mozol and I.V. Fekeshgazi, *Soviet J. Quantum Electron*, **8**, 393 (1978).
14. N.I. Zheludev, New nonlinear polarization effects and polarization coding of optical information, in S.A. Akhmanov and M.A. Vorontsov (eds.), *New Physical Principles for Optical Data Processing*, Nauka, Moscow, 1990 (in Russian).
15. R. Cameron and G.C. Tabisz, *Molec. Phys.*, **90**, 159 (1997).
16. P.D. Maker, R.W. Terhune and C.M. Savage, *Phys. Rev. Lett.*, **12**, 507 (1964).
17. M. Kuwata and N. Nagasawa, *J. Phys. Soc. Japan*, **51**, 2591 (1982).
18. A.I. Kovrighin, D.V. Yakovlev, B.V. Zhdanov and N.I. Zheludev, *Opt. Commun.*, **35**, 92 (1980).
19. R.S. Zadoyan, N.I. Zheludev and L.B. Meysner, *Solid State Commun.*, **55**, 713 (1985).
20. M.G. Dubenskaya, R.S. Zadoya and N.I. Zheludev, *J. Opt. Soc. Am. B*, **2**, 1174 (1985).
21. M.Ya. Valakh, M.I. Dyknam, M.P. Lisitsa, G.Yu. Rudko and G.G. Tarasov, *Solid State Commun.*, **30**, 133 (1979).
22. S.J. Bepko, *Phys. Rev.*, **12B**, 669 (1975).
23. M.I. Dykman and Yu.G. Rubo, *Phys. Rev.*, **45B**, 5926 (1992).
24. M.G. Dubenskaya, T.M. Ilinova, A.V. Trukhov and A.A. Fortygin, *Izv. Akad. Nauk SSSR, Ser. Fiz.*, **56**, 95 (1992).
25. D.C. Hutchings and B.S. Wherrett, *Phys. Rev.*, **B52**, 8150 (1995).
26. N.I. Zheludev and Yu.P. Svirko, *Kristallographia*, **35**, 279 (1990).
27. N.R. Belashenkov, S.A. Kozlov, *SPIE-The Int. Soc. for Opt. Engng Proc.*, **1841**, 169 (1992).

28. D.C. Hutchings, *Opt. Lett.*, **20**, 1607 (1995).
29. N.I. Zheludev, A.D. Petrenko, Yu.P. Svirko and G.S. Filippova, *Izv. Akad. Nauk SSSR, Ser. Fiz.*, **48**, 603 (1984).

FURTHER READING

M. Evance and S. Kielich (eds.), *Modern Nonlinear Optics*, Vols. 1–3, John Wiley & Sons, New York, 1994.

5 Polarization Self-Action, Instability and Propagation of Partially Polarized Light

In this section we will find the general solution of the Stokes parameter evolution equations for small polarization changes. We will investigate polarization stability in low birefringence media and cubic crystals and find the evolution equations for partially polarized light in nonlinear media.

5.1 POLARIZATION SELF-ACTION IN NONLINEAR MEDIA: SMALL POLARIZATION CHANGES

It was not paticularly difficult to solve the Stokes parameter evolution equations (4.10) in order to illustrate the elementary nonlinear polarization effects considered in Chapter 4. In more complex cases, especially when many more components of the nonlinearity tensors are non-zero, solving equations (4.10) analytically can become impossible or exceptionally laborious. Fortunately the evolution equations for the Stokes parameters have the advantage of being easy to solve numerically. However, for weakly nonlinear media the solutions may be obtained analytically. The acquisition of these solutions is not dependent on the number of nonzero components of the linear and nonlinear response tensor, but low anisotropy along the direction of light propagation is assumed.

Let the light wave at $z = 0$ to be totally polarized with Stokes parameters $S_0(z = 0) = S_{00}$ and $\mathbf{S}(z = 0) = \mathbf{S}_0$ corresponding to the initial polarization azimuth Θ_0 and ellipticity η_0. If the changes in the Stokes parameters of the wave after travelling a distance L inside the medium are small:

$$\mathbf{S}(L) = \mathbf{S}_0 + L\left.\frac{\mathrm{d}\mathbf{S}}{\mathrm{d}z}\right|_{z=0}$$

$$S_0(L) = S_{00} + L\left.\frac{\mathrm{d}S_0}{\mathrm{d}z}\right|_{z=0} \tag{5.1}$$

Here the derivatives are the right-hand sides of the evolution equations (4.10) which need to be calculated at $z = 0$. Substituting equations (5.1) into the definitions of the polarization azimuth (3.12) and the ellipticity angle (3.13) we readily obtain the following formulae:

$$\Theta(L) - \Theta_0 = \frac{L}{2} \times \frac{\mathrm{d}S_2/\mathrm{d}z|_{z=0} S_{10} - \mathrm{d}S_1/\mathrm{d}z|_{z=0} S_{20}}{S_{00}^2 \cos^2 2\eta_0}$$

$$\eta(L) - \eta_0 = \frac{L}{2} \times \frac{\mathrm{d}S_3/\mathrm{d}z|_{z=0} S_{00} - \mathrm{d}S_0/\mathrm{d}z|_{z=0} S_{30}}{S_{00}^2 \cos 2\eta_0} \tag{5.2}$$

or, combining formulae (4.10) with (5.2), we arrive at

$$\begin{pmatrix} [\Theta(L) - \Theta_0] \cos^2 2\eta_0 \\ [\eta(L) - \eta_0] \cos 2\eta_0 \end{pmatrix}$$

$$= \frac{\omega L}{4cS_{00}^2} \begin{pmatrix} \mathrm{Re} \\ \mathrm{Im} \end{pmatrix} \{ S_{30}(\mathbf{\Omega}_0 \cdot \mathbf{S}_0) - \Omega_{30} S_{00}^2 + i(S_{10}\Omega_{20} - S_{20}\Omega_{10}) S_{00} \} \tag{5.3}$$

where we have introduced $\Omega_{\alpha 0}$ at $z = 0$:

$$\Omega_{\alpha 0} = v_\alpha + \tfrac{1}{2} w_{\alpha\beta} S_{\beta 0} \tag{5.4}$$

or, in terms of u_{ij} and $\boldsymbol{\kappa}_{ijkl}$,

$$\begin{aligned} \Omega_{10} = u_{xx} - u_{yy} &+ \tfrac{1}{2}(\kappa_{xxxx} - \kappa_{yyxx} + \kappa_{xxyy} - \kappa_{yyyy}) S_{00} \\ &+ \tfrac{1}{2}(\kappa_{xxxx} - \kappa_{yyxx} - \kappa_{xxyy} + \kappa_{yyyy}) S_{10} \\ &+ \tfrac{1}{2}(\kappa_{xxxy} - \kappa_{yyxy} + \kappa_{xxyx} - \kappa_{yyyx}) S_{20} \\ &+ \frac{i}{2}(-\kappa_{xxxy} + \kappa_{yyxy} + \kappa_{xxyx} - \kappa_{yyyx}) S_{30} \end{aligned} \tag{5.5}$$

$$\begin{aligned} \Omega_{20} = u_{xy} + u_{yx} &+ \tfrac{1}{2}(\kappa_{xyxx} + \kappa_{yxxx} + \kappa_{xyyy} + \kappa_{yxyy}) S_{00} \\ &+ \tfrac{1}{2}(\kappa_{xyxx} + \kappa_{yxxx} - \kappa_{yxxx} - \kappa_{yxyy}) S_{10} \\ &+ \tfrac{1}{2}(\kappa_{yxxy} + \kappa_{xyxy} + \kappa_{yxyx} + \kappa_{xyyx}) S_{20} \\ &+ \frac{i}{2}(-\kappa_{yxxy} - \kappa_{xyxy} + \kappa_{xyyx} - \kappa_{yxyx}) S_{30} \end{aligned} \tag{5.6}$$

$$\Omega_{30} = i(u_{xy} - u_{yx}) + \frac{i}{2}(-\kappa_{yxxx} + \kappa_{xyxx} + \kappa_{yxyy} - \kappa_{xyyy})S_{00}$$
$$+ \frac{i}{2}(-\kappa_{yxxx} + \kappa_{xyxx} + \kappa_{yxyy} - \kappa_{xyyy})S_{10}$$
$$+ \frac{i}{2}(-\kappa_{yxxy} + \kappa_{xyxy} - \kappa_{yxyx} + \kappa_{xyyx})S_{20}$$
$$+ \tfrac{1}{2}(-\kappa_{yxxy} + \kappa_{xyxy} + \kappa_{yxyx} - \kappa_{xyyx})S_{30} \quad (5.7)$$

$$\Omega_{00} = u_{xx} + u_{yy} + \tfrac{1}{2}(\kappa_{xxxx} + \kappa_{xxyy} + \kappa_{yyxx} + \kappa_{yyyy})S_{00}$$
$$+ \tfrac{1}{2}(\kappa_{xxxx} - \kappa_{xxyy} + \kappa_{yyxx} - \kappa_{yyyy})S_{10}$$
$$+ \tfrac{1}{2}(\kappa_{xxxy} + \kappa_{xxyx} + \kappa_{yyxy} + \kappa_{yyyx})S_{20}$$
$$+ \frac{i}{2}(-\kappa_{xxxy} + \kappa_{xxyx} - \kappa_{yyxy} + \kappa_{yyyx})S_{30} \quad (5.8)$$

Equation (5.3) is the desired formula for small changes in the polarization state of the light wave propagating in a weakly nonlinear medium of arbitrary symmetry. It may be rewritten in terms of the polarization azimuth Θ_0 and ellipticity η_0 of the incident wave:

$$\begin{pmatrix} [\Theta(L) - \Theta_0]\cos 2\eta_0 \\ \eta(L) - \eta_0 \end{pmatrix}$$
$$= \frac{\omega L}{4c}\begin{pmatrix} \mathrm{Re} \\ \mathrm{Im} \end{pmatrix}\{(\Omega_{10}\cos 2\Theta_0 + \Omega_{20}\sin 2\Theta_0)\sin 2\eta_0 - \Omega_{30}\cos 2\eta_0$$
$$+ i[\Omega_{20}\cos 2\Theta_0 - \Omega_{10}\sin 2\Theta_0]\} \quad (5.9)$$

Thus, small changes of polarization state in nonlinear media are determined by the components of the self-action vectors Ω_{10}, Ω_{20} and Ω_{30} at the point of incidence. An elegant and useful formula for small changes to initially linearly polarized light ($\eta_0 = 0$) in a dissipation-free medium ($\mathrm{Im}\{\Omega_{\alpha 0}\} = 0$) may easily be derived from (5.9):

$$\begin{pmatrix} \Theta(L) - \Theta_0 \\ \eta(L) - \eta_0 \end{pmatrix} = -\frac{\omega L}{4c}\begin{pmatrix} \Omega_{30} \\ \Omega_{10}\sin 2\Theta_0 - \Omega_{20}\cos 2\Theta_0 \end{pmatrix} \quad (5.10)$$

It should be noted that simplified formulae (5.3), (5.9) and (5.10) give polarization effects due to natural anizotropy ($\Delta\epsilon$), natural nonlocality (Γ_{ijk}) and local and nonlocal nonlinearity (χ_{ijkl} and Γ_{ijklm}). They do not account for 'crossing' polarization phenomena such as resulting from the interference of nonlinearity and natural anizotropy where polarization changes will be proportional to

$\Delta\epsilon \times \chi$, or from the interference of nonlinearity and natural nonlocality where polarization changes will depend on $\Gamma_{ijk} \times \chi$.

5.2 POLARIZATION INSTABILITY

In linear optics, the polarization of an electromagnetic wave in a medium is independent of the intensity of light and is uniquely related to the polarization incident on the medium. Importantly, the 'output' wave polarization is a smooth function of the 'input' polarization. Waves of a particular polarization type (eigenwaves) retain their polarization with propagation. For example, as we saw in Chapter 3, in an isotropic optically active medium, the eigenwaves are circularly polarized. In crystals, eigenwaves are, in general, elliptically polarized, although in the simple case of birefringent nongyrotropic crystals they are orthogonally linearly polarized. In dissipative media exhibiting linear or circular dichroism, arbitrarily polarized waves tend to one of the eigenwaves of the medium as they propagate through it.

In nonlinear optics the situation is radically different. The refraction, absorption, anisotropy and nonlocality of a medium are all functions of wave intensity. The following specifically nonlinear polarization effects have emerged (see review [1]):

(a) When the 'input' intensity or polarization of light undergoes a slow change, the 'output' polarization is found to be a multivalued function containing both stable and unstable branches. The output polarization state depends on the 'prehistory' of the excitation (polarization hysteresis). Given a particular combination of the parameters of the radiation and of the nonlinear medium, with steady state 'input' radiation, the 'output' polarization may oscillate in time or even change randomly with a continuous frequency spectrum creating 'polarization chaos'.

(b) In linear optics the polarization of the wave either oscillates along the direction of propagation or steadily tends towards one of the eigenpolarizations (eigenwaves remain unchanged). In nonlinear propagation the polarization parameters along the propagation coordinate may change ubruptly can depend strongly and 'unpredictably' on the initial conditions.

Often, polarization instability and chaos appears in nonlinear interactions between two waves; however, simple examples of polarization instability can be seen in self-action effects. In this section we will show how instability of polarization develops for intense light in birefringent and cubic crystals. For instance, we will see that when the intensity of initially polarized light propagating in a birefringent crystal reaches a certain threshold value, the polarization state become unstable and switches to a right- or left-handed elliptical polarization. This is known as spontaneous polarization symmetry breaking, a term justified by the fact that the linearly polarized wave has a plane of symmetry, while the elliptically polarized wave does not. Similar examples of spontaneous symmetry breaking are well known in mechanics. For example, consider a gravitational pendulum whose point of suspension is forced to oscillate horizontally with amplitude A and frequency ω as $A\cos(\omega t)$, where $\omega^2 \gg g/l$ (g is the gravitational acceleration and l is the length of the pendulum). When the amplitude of the suspension point oscillations is small, the steady state position of the pendulum is vertical. However, when the oscillation amplitude A reaches the value $A_0 = (2gl/\omega^2)^{1/2}$, the vertical equilibrium position becomes unstable. Instead the pendulum's stable equilibrium position is at an angle $+\psi$ or $-\psi$ to the vertical, where $\psi = \cos^{-1}[2gl/(A^2\omega^2)]$ (see Ref. [2] and Fig. 5.1).

Now let us see how spontaneous polarization symmetry breaking happens in a nonlinear transparent birefringent medium. We will consider light propagating along the Z axis of the laboratory Cartesian frame in a birefringent nonlinear medium with local optical response. We assume that the X and Y axes of the Cartesian frame are directed along the eigenpolarizations of the crystal. This situation is similar to that presented in Fig. 3.3 for $\phi = 0$. The Stokes vector

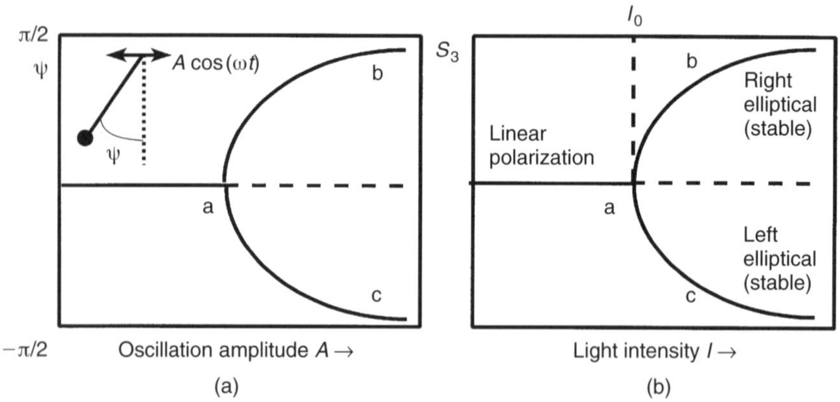

Fig. 5.1 Spontaneous symmetry breaking in mechanical and optical systems (following Ref. [1]). (a) A gravitational pendulum with a rapidly oscillating suspension point. When the amplitude of oscillation of the suspension point is small, the pendulum has only one, vertical, equilibrium position (a). When this amplitude reaches a threshold value, two new equilibrium positions (b and c) appear. These new equilibrium positions are stable and are deviated from the vertical by equal angles to the left and right. The vertical position becomes one of unstable equilibrium: a small fluctuation will result in a 'jump' of the pendulum from the unstable vertical to one of the stable deviated positions. (b) Spontaneous polarization symmetry breaking in a birefringent crystal. A linearly polarized eigenwave (a) with a smaller refractive index is stable unless the intensity reaches a threshold value I_0. For intensities above the threshold value the linear polarization becomes unstable. A small fluctuation of the initial polarization state will lead to a 'jump' of the output polarization to either right (b) or left (c) elliptical polarization (Reproduced by permission of American Institute of Physics from Ref. [1])

evolution equations (4.10) are very well suited for a qualitative analysis of such instability. A more detailed description of the effect may be found in Refs [3] and [4]. In this case the components of the self-action vector and Ω_0 are:

$$
\begin{aligned}
\Omega_1 &= \frac{\epsilon_{11} - \epsilon_{33}}{n} + \frac{8\pi\chi_{xxyy}}{n} S_1 \\
\Omega_2 &= \frac{8\pi\chi_{xxyy}}{n} S_2 \\
\Omega_3 &= 0 \\
\Omega_0 &= \frac{16\pi\chi_{xxyy}}{n} S_0
\end{aligned}
\tag{5.11}
$$

For the sake of simplicity we have accounted for anisotropy only in the dielectric tensor and presumed that the cubic nonlinearity is isotropic, as given by equations (4.11). With these, and given that in transparent media $\operatorname{Im}\{\chi_{ijkl}\} = 0$ and $\operatorname{Im}\{\epsilon_{ij}\} = 0$, the nonlinear evolution equations (4.10) become

$$
\begin{aligned}
\frac{dS_1}{dz} &= -\frac{4\pi\omega}{cn}\chi_{xxyy} S_3 S_2 \\
\frac{dS_2}{dz} &= \frac{\omega}{2cn}(\epsilon_{11} - \epsilon_{33} + 8\pi\chi_{xxyy} S_1) S_3 \\
\frac{dS_3}{dz} &= -\frac{\omega}{2cn}(\epsilon_{11} - \epsilon_{33}) S_2 \\
\frac{dS_0}{dz} &= 0
\end{aligned}
\tag{5.12}
$$

The evolution equations (5.12) can easily be rearranged into equations for the polarization azimuth Θ and ellipticity angle η of the wave:

$$
\begin{aligned}
\frac{d\Theta}{dz} &= \delta' \tan 2\eta \cos 2\Theta + \delta^{NL} \sin 2\eta \\
\frac{d\eta}{dz} &= -\delta' \sin 2\Theta
\end{aligned}
\tag{5.13}
$$

where we have introduced the following measures of anisotropy and nonlinearity:

$$
\begin{aligned}
\delta' &= \frac{\omega(\epsilon_{11} - \epsilon_{33})}{4cn} \\
\delta^{NL} &= \frac{2\pi\omega\chi_{xxyy} S_{00}}{cn}
\end{aligned}
\tag{5.14}
$$

Spontaneous polarization symmetry breaking may be clearly seen if the incident light wave is launched into the birefringent medium as an eigenpolarization. As we saw in Chapter 3, a birefringent linear medium supports two orthogonal linear eigenpolarizations with Stokes vectors parallel and antiparallel to the vector $\mathbf{\Omega}$ as given by equations (5.11) with $\chi_{ijkl} = 0$. In the laboratory Cartesian frame such waves are polarized along the X and Y directions. When the intensity of light is small, these polarization states are stable, i.e. the propagating wave will remain polarized as it was when it was launched into the crystal.

Let us now consider the stability of the eigenpolarizations with increasing wave intensity. The stability analysis suggests that we need to watch the development of small fluctuations, $\Delta\Theta$ and $\Delta\eta$, in the polarization azimuth and ellipticity. If these small fluctuations will lead to a drastic departure of the 'output' polarization from the incident one, the polarization state is said to be 'unstable'. Let the incident wave be linearly polarized close to the X direction, but not strictly along it, so that we introduce an initial 'fluctuation' of the light polarization azimuth $\Delta\Theta = \Theta - \Theta_0$ at $z = 0$. A wave strictly polarized along the X direction would have $\Theta = \Theta_0 = 0$ and $\eta_0 = 0$. For our wave polarized close to the X direction, equations (5.12) may be simplified:

$$\begin{aligned} \frac{\mathrm{d}\Theta}{\mathrm{d}z} &= 2(\delta' + \delta^{\mathrm{NL}})\eta \\ \frac{\mathrm{d}\eta}{\mathrm{d}z} &= -2\delta'\Theta \end{aligned} \tag{5.15}$$

By differentiating the first of these equations with respect to z, and substituting $\mathrm{d}\eta/\mathrm{d}z$ from the second, we arrive at the following second-order differential equation for the polarization azimuth:

$$\frac{\mathrm{d}^2\Theta}{\mathrm{d}z^2} + 4\delta'(\delta' + \delta^{\mathrm{NL}})\Theta = 0 \tag{5.16}$$

The solution to this equation will be critically dependant on the sign of $\delta'(\delta' + \delta^{\mathrm{NL}})$. In dissipation-less media χ_{xxyy} is real and, as a rule, positive, that is δ^{NL} is real and positive. Therefore $\delta'(\delta' + \delta^{\mathrm{NL}}) > 0$ for all values of δ' and δ^{NL} except when $-\delta^{\mathrm{NL}} < \delta' < 0$. For $\delta'(\delta' + \delta^{\mathrm{NL}}) > 0$ equation (5.16) has the following solution:

$$\Theta = \Delta\Theta \cos\left[2\sqrt{\delta'(\delta' + \delta^{\mathrm{NL}})}z\right] \tag{5.17}$$

In this case, the polarization azimuth oscillates within strict limits imposed by the initial fluctuation $\Delta\Theta$, and so linear polarization along the X direction is stable. When $\delta'(\delta' + \delta^{\mathrm{NL}}) < 0$ the solution to equation (5.16) is

$$\Theta = \Delta\Theta \cosh\left[2\sqrt{-\delta'(\delta' + \delta^{\mathrm{NL}})}z\right] \tag{5.18}$$

For $z^2 \geqslant 1/|4\delta'(\delta' + \delta^{\mathrm{NL}})|$

$$\Theta \approx \frac{\Delta\Theta}{2} \exp\left[\sqrt{-\delta'(\delta' + \delta^{\mathrm{NL}})}z\right] \tag{5.19}$$

Here, a small initial fluctuation $\Delta\Theta$ of the polarization azimuth will exponentially increase with propagation. Therefore, for $\delta'(\delta' + \delta^{\mathrm{NL}}) < 0$, linear polarization along the X direction is unstable.

What does the condition $\delta'(\delta' + \delta^{\mathrm{NL}}) < 0$ for polarization instability mean and how can it be realistically achieved? In terms of the material parameters this condition is achieved, if

$$\chi_{xxyy} S_{00} > \frac{\epsilon_{33} - \epsilon_{11}}{8\pi} > 0 \tag{5.20}$$

Primarily, polarization instability requires ϵ_{11} to be smaller than ϵ_{33}; i.e. the refractive index for a wave polarized along the X direction, $n_x = \epsilon_{11}^{1/2}$, should be smaller than the refractive index $n_y = \epsilon_{33}^{1/2}$ for a wave polarized along the Y direction: only the eigenpolarization with the smaller refractive index can be unstable. The left-hand side of inequality (5.20) also imposes a condition on the light intensity. Polarization instability occur if $8\pi\chi_{xxyy}\mathrm{S}_{00} > \epsilon_{33} - \epsilon_{11}$ or, in terms of the threshold light intensity I_0,

$$I > I_0 = \frac{cn(n_y - n_x)(1 + n)^2}{32\pi^2 \chi_{xxyy}} \tag{5.21}$$

This is the polarization instability criterion for the intensity of a wave approaching a crystal with core refractive coefficient n from a vacuum. It takes into account the reflection losses at the boundary. If the intensity exceeds the threshold value (5.21) a small fluctuation of the incident polarization state will lead to a very considerable change in the output polarization. More detailed analysis shows that, depending on the sign of the initial fluctuation, the wave will become right or left elliptically polarized.

The question of polarization instability in birefringent media was purely academic when Sala was the first to draw attention to the 'trigger' behaviour of the polarization of light in a birefringent crystal [3]. The intensity threshold for polarization instability is easily achieved in optical fibres with small birefringence. Quartz glass optical fibres typically have nonlinearities of $\chi_{xxyy} \simeq 10^{-15}$ esu. For a fibre with birefringence of $n_y - n_x \simeq 10^{-7}$ and $n \simeq 1.5$ the polarization instability threshold intensity is about $I_0 = 10^{17}$ erg/cm^2, which is equivalent to $I_0 = 10$ GW/cm^2. With a fibre core of 5 μm diameter the threshold power required for the observation of polarization instability is of the order of several kilowatts. A great number of theoretical and experimental works devoted to this problem have been published (see review [1] and papers [4–7] in particular). A similar effect to polarization instability occurs with circularly polarized light in birefringent media. When the light intensity exceeds a threshold value, circular polarization becomes unstable. In fact, polarization instability was first seen in a birefringent fibre with circularly polarized light [8] (see Fig. 5.2). It will also be interesting to observe spontaneous polarization symmetry breaking are crystals with an 'isotropic point', i.e. crystals for which the frequency dependence of the two main refractive indices cross one another (see Section 3.3). Many of the popular birefringent materials, such as ZnO, ZnS, CdSe, SnO_2, MgF_2, Al_2O_3 and CdS, have just such a point. In such crystals the threshold for the observation of spontaneous symmetry breaking could be very small.

Polarization instability can also be seen along the [001] direction in cubic crystals. This type of instability, however, does not require the intensity to exceed a threshold value but also develops as an exponential instability. Indeed, if the initial polarization state is linear, at an angle Θ_0 to the X direction in the natural coordinate frame, i.e.

$$\mathbf{S} = S_{00}(\cos 2\Theta_0, \sin 2\Theta_0, 0) \tag{5.22}$$

in a dissipation-less crystal of $m3m$ symmetry the formula (4.31) may be reduced to

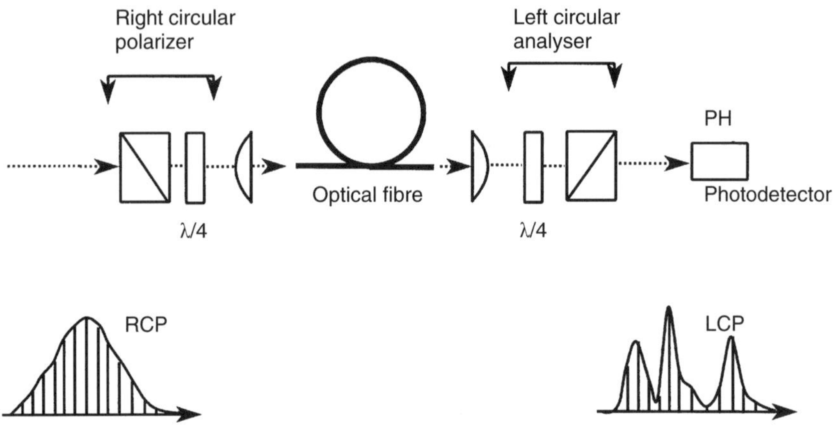

Fig. 5.2 Experimental observation of polarization instability in a weakly birefringent optical fibre with retardation approximately equal to the wavelength. A bell-shaped train of picosecond light pulses with right circular polarization is launched into the fibre. For low power levels the fibre works as a λ-waveplate: the output wave retains its initial polarization. The photodetector, equipped with a left circular analyser, does not detect any signal. When the intensity reaches a threshold value, the circular polarization ceases to be stable from pulse to pulse inside the train. This is detected by the photodetector as a fragmentation of the pulse train envelope (Reproduced by permission of American Institute of Physics from Ref. [8])

$$\begin{aligned}
\frac{dS_1}{dz} &= -\frac{4\pi\omega}{cn}\chi_{xxyy}S_2S_3 \\
\frac{dS_2}{dz} &= \frac{2\pi\omega}{cn}(2\chi_{xxyy} + \Delta\chi)S_1S_3 \\
\frac{dS_3}{dz} &= -\frac{2\pi\omega}{cn}\Delta\chi S_1S_2 \\
\frac{dS_0}{dz} &= 0
\end{aligned} \tag{5.23}$$

where $\Delta\chi = \chi_{xxxx} - 3\chi_{xxyy}$. Since these evolution equations are rather simple in form, they may be rearranged directly into equations for the polarization azimuth Θ and ellipticity angle η:

$$\begin{aligned}
\frac{d\Theta}{dz} &= \tfrac{1}{2}(2\delta^{NL} + \delta_a^{NL}\cos^2 2\Theta)\sin 2\eta \\
\frac{d\eta}{dz} &= -\tfrac{1}{4}\delta_a^{NL}\sin 4\Theta\cos 2\eta
\end{aligned} \tag{5.24}$$

Here we have introduced the following measures of optical nonlinearity and anisotropy of the nonlinearity:

$$\begin{aligned}
\delta^{NL} &= \frac{2\pi\omega}{cn}\chi_{xxyy} \\
\delta_a^{NL} &= \frac{2\pi\omega}{cn}\Delta\chi
\end{aligned} \tag{5.25}$$

From equations (5.24) it follows that polarizations which exactly satisfy the conditions $\sin 4\Theta = 0$ and $\eta = 0$ will not change since the right-hand sides of the evolution equations are zero. To study the stability of these linear polarizations we will consider a small fluctuation $\Delta\Theta$ in the polarization azimuth of the incident wave allowing $\Theta(0) = \Delta\Theta + \Theta_0$ with $\sin 4\Theta_0 = 0$ and $\eta(0) = 0$. In this case equations (5.24) may be simplified as follows for $\Theta - \Theta_0 \ll \pi$ and $\eta \ll \pi$:

$$\begin{aligned}
\frac{d\Theta}{dz} &= (2\delta^{NL} + \delta_a^{NL}\cos^2 2\Theta_0)\eta \\
\frac{d\eta}{dz} &= -\delta_a^{NL}\cos 4\Theta_0
\end{aligned} \tag{5.26}$$

Solving equations (5.26) we obtain

$$\Theta = \theta_0 + \Delta\Theta \cos\sqrt{\mu}z$$
$$\eta = \frac{\sqrt{\mu}\Delta\Theta}{2\delta^{\rm NL} + \delta_{\rm a}^{\rm NL}\cos^2 2\Theta_0}\sin\sqrt{\mu}z \tag{5.27}$$

where

$$\mu = (2\delta^{\rm NL} + \delta_{\rm a}^{\rm NL}\cos^2 2\Theta_0)\delta_{\rm a}^{\rm NL}\cos 4\Theta_0 \tag{5.28}$$

The character of this solution depends on the sign of μ. For positive μ the polarization azimuth and ellipticity will oscillate within strict limits imposed by the initial fluctuation $\Delta\Theta$, and the incident linear polarization will be stable. For negative μ the solution will become exponential in nature: $\cos(\mu^{1/2}z) = {\rm ch}(|\mu|^{1/2}z) \simeq \frac{1}{2}\exp(|\mu|^{1/2}z)$. Therefore for $\mu < 0$ and $|\mu|^{1/2}z \gg 1$, the output polarization will depart exponentially from the input one. As usual, anisotropy is small, $|\Delta\chi| \ll |\chi_{xxyy}|$, and therefore $|\delta^{\rm NL}| \ll |\delta_{\rm a}^{\rm NL}|$. Since the nonlinearity is often positive, that is $\chi_{xxyy} > 0$ and $\delta^{\rm NL} > 0$, the term in brackets in the first equation of (5.26) is positive. Therefore the sign of μ and the stability of the polarization state will depend on the sign of $\delta_{\rm a}^{\rm NL}\cos 4\Theta_0$, i.e. on the sign of $\Delta\chi\cos 4\Theta_0$. For

$$\Delta\chi\cos 4\Theta_0 < 0 \tag{5.29}$$

the initial linear polarization corresponding to $\sin 4\Theta_0 = 0$ is unstable. If, however,

$$\Delta\chi\cos 4\Theta_0 > 0 \tag{5.30}$$

the linear polarization corresponding to $\sin 4\Theta_0 = 0$ is stable. For example, for positive $\Delta\chi$, eigenpolarizations along the cubic crystal axes [100] and [010] are stable. In contrast, polarization along bisectors, [110] and [$\bar{1}$10], of these axes are unstable: small departures of the initially linearly polarized light from pure linear polarization will grow exponentially with propagation. This instability does not have a threshold intensity and is known as soft polarization symmetry breaking. Importantly, it was found that the interaction of two counter-propagating beams in a cubic crystal can create a chaotic polarization distribution along the propagation coordinate [9–16]. The first experimental research into counter-propagation polarization instability was performed using mirrors deposited on the cubic crystal [17].

5.3 PROPAGATION OF PARTIALLY POLARIZED LIGHT IN NONLINEAR MEDIA

In nonlinear optics the evolution of the degree of polarization is evidently more complicated than in the linear case (see Section 3.4). In this section we will derive evolution equations suitable for describing partially polarized light propagating in a medium with cubic nonlinearity. Partially polarized light may be created, for example, by scattering radiation from an ideal single-mode laser. Narrow-band partially polarized light may also be prepared using a thermal light source (the Sun or a light bulb, for instance), by polarization-sensitive reflection or scattering combined with passing through a narrow-band spectral filter. A realistic single-mode laser whose output is spoilt by spontaneous emission in the active medium is one more example of an intense partially polarized light source.

The nonlinear evolution equations (4.10) were derived for totally polarized light. These equations are also suitable for describing narrow-band partially polarized light if, for S_0, S_1, S_2, S_3, we take their instantaneous, but not the time averaged values (3.7). To make the notation for the instantaneous quadratic forms distinctively different from the proper Stokes vectors, we will use small letters $\{s_0, s_1, s_2, s_3\}$. They will be referred to as the s-parameters in what follows:

$$\begin{aligned} s_0 &= (A_x A_x^* + A_y A_y^*)\,\mathrm{e}^{-2\,\mathrm{Im}\,\{k\}z} \\ s_1 &= (A_x A_x^* - A_y A_y^*)\,\mathrm{e}^{-2\,\mathrm{Im}\,\{k\}z} \\ s_2 &= (A_x A_y^* + A_y A_x^*)\,\mathrm{e}^{-2\,\mathrm{Im}\,\{k\}z} \\ s_3 &= i(A_x A_y^* - A_y A_x^*)\,\mathrm{e}^{-2\,\mathrm{Im}\,\{k\}z} \end{aligned} \tag{5.31}$$

Evidently, the s-parameters are similar to the Stokes parameters except that the Stokes parameters are time-averaged over the integration time of the photodetector. (compare with definitions (3.7) of the proper Stokes parameters). To obtain the evolution equations for the s parameters, Ω_0 and $\boldsymbol{\Omega}$, in the evolution equations (4.10), will be replaced with their instantaneous values Ξ_0 and $\boldsymbol{\Xi}$:

$$\Xi_\alpha = v_\alpha + \tfrac{1}{2} w_{\alpha\beta} s_\beta \tag{5.32}$$

(compare with equations (4.9)). Thus, we arrive at the following evolution equations for the s-parameters:

$$\begin{aligned} \frac{\mathrm{d}s_\alpha}{\mathrm{d}z} &= -\frac{\omega}{2c}\,\mathrm{Im}\,\{(4n + \Xi_0)s_\alpha + \Xi_\alpha s_0 + i e_{\alpha\beta\gamma}\Xi_\beta s_\gamma\} \\ \frac{\mathrm{d}s_0}{\mathrm{d}z} &= -\frac{\omega}{2c}\,\mathrm{Im}\,\{(4n + \Xi_0)s_0 + \Xi_\alpha s_\alpha)\} \end{aligned} \tag{5.33}$$

It is desirable to adapt equations (5.33) so as to make them suitable for describing the evolution of the Stokes parameters of partially polarized light. We therefore need to average them in time. Time-averaging of the left-hand sides of (5.33) does not present any problems: $\langle \mathbf{s} \rangle = \mathbf{S}$ and $\langle s_0 \rangle = S_0$ (where $\langle \cdot \rangle$ denotes a time average); therefore $\langle d\mathbf{s}/dz \rangle = d\mathbf{S}/dz$ and $\langle ds_0/dz \rangle = dS_0/dz$. However, in nonlinear media, Ξ and Ξ_0 depend on $\mathbf{s}$ and s_0, and so the right-hand side terms of these equations may hide biquadratic structures like $\langle s_\alpha s_\beta \rangle$, which must be treated with care. In order to time-average such biquadratic terms, we decompose the amplitude of the electric field of partially polarized light into a totally polarized coherent component with amplitude $\mathbf{A}_0$ and a totally unpolarized, randomly fluctuating component with amplitude $\delta\mathbf{A}$:

$$\mathbf{A} = \mathbf{A}_0 + \delta\mathbf{A} \tag{5.34}$$

In the totally unpolarized component the X and Y projections of the electric field vector fluctuate independently with mean values of zero; therefore:

$$\begin{gathered} \langle \delta A_x \rangle = \langle \delta A_y \rangle = 0 \\ \langle \delta A_i^* \delta A_j \rangle = \tfrac{1}{2} \langle (\delta\mathbf{A} \cdot \delta\mathbf{A}^*) \rangle \delta_{ij} \end{gathered} \tag{5.35}$$

Now the components of the Stokes vector (3.7) may be written as follows:

$$S_\alpha = \langle (A_{0i}^* + \delta A_i^*) \sigma_{ij}^{(\alpha)} (A_{0j} + \delta A_j) > \mathrm{e}^{-2\,\mathrm{Im}\,\{k\} z} = A_{0i}^* \sigma_{ij}^{(\alpha)} A_{0j}\, \mathrm{e}^{-2\,\mathrm{Im}\,\{k\} z} \tag{5.36}$$

for $\alpha = 1$, 2 and 3, and

$$S_0 = \langle (A_{0i}^* + \delta A_i^*) \sigma_{ij}^{(0)} (A_{0j} + \delta A_j) \rangle\, \mathrm{e}^{-2\,\mathrm{Im}\,\{k\} z} = [(\mathbf{A}_0 \cdot \mathbf{A}_0^*) + \langle (\delta\mathbf{A} \cdot \mathbf{A}^*) \rangle]\, \mathrm{e}^{-2\,\mathrm{Im}\,\{k\} z} \tag{5.37}$$

Equation (5.36) shows that the Stokes parameters S_1, S_2 and S_3 depend only on the totally polarized component of the light wave. The Stokes parameter S_0, however, being a measure of the light intensity, depends on the intensities of both the polarized and unpolarized components.

To average the biquadratic structure $s_\alpha s_\beta$ we use the definition of the s-parameters (5.31):

$$\langle s_\alpha s_\beta \rangle = \langle (A_{0i}^* + \delta A_i^*) \sigma_{ij}^{(\alpha)} (A_{0j} + \delta A_j)(A_{0k}^* + \delta A_k^*) \sigma_{kl}^{(\beta)} (A_{0l} + \delta A_{0l}) \rangle\, \mathrm{e}^{-4\,\mathrm{Im}\,\{k\} z} \tag{5.38}$$

Now combining equations (5.36) and (5.38) we obtain

$$\begin{aligned} \langle s_\alpha s_\beta \rangle = {} & S_\alpha S_\beta + \sigma_{ij}^{(\alpha)} \sigma_{kl}^{(\beta)} \langle (A_{0i}^* A_{0l} \delta A_k^* \delta A_j + A_{0k}^* A_{0j} \delta A_i^* \delta A_l \\ & + \delta A_i^* \delta A_j \delta A_k^* \delta A_l \rangle\, \mathrm{e}^{-4\,\mathrm{Im}\,\{k\} z} \end{aligned} \tag{5.39}$$

Since the degree of polarization

$$r = \sqrt{\frac{S_1^2 + S_2^2 + S_3^2}{S_0^2}} \tag{5.40}$$

represents the ratio of the intensity of the polarized part of a light wave to its total intensity

$$(\delta\mathbf{A}\cdot\delta\mathbf{A}^*) = S_0(1-r)\,\mathrm{e}^{+2\,\mathrm{Im}\,\{k\}z} \tag{5.41}$$

and formula (5.39) may be rearranged:

$$\langle s_\alpha s_\beta\rangle = S_\alpha S_\beta + (1-r)rS_0^2\delta_{\alpha\beta} + \sigma_{ij}^{(\alpha)}\sigma_{kl}^{(\beta)}\langle\delta A_i^*\delta A_j\delta A_k^*\delta A_l\rangle\ \mathrm{e}^{-4\,\mathrm{Im}\,\{k\}z} \tag{5.42}$$

In order to find the time-averaged value $\langle\delta A_i^*\delta A_j\delta A_k^*\delta A_l\rangle$ we need to know the statistical properties of the unpolarized component of the light wave. We will presume that it obeys Gaussian statistics (a common assumption in statistical optics [18, 19]). For a signal obeying Gaussian statistics the following relationship is, by definition, valid:

$$\langle\delta A_i\delta A_j^*\delta A_k\delta A_l^*\rangle = \langle\delta A_i\delta A_j^*\rangle\langle\delta A_k\delta A_l^*\rangle + \langle\delta A_i\delta A_l^*\rangle\langle\delta A_k\delta A_j^*\rangle \tag{5.43}$$

Propagation in linear media does not affect the statistical properties of light. However, the statistical properties of the unpolarized component may change in a nonlinear medium. Indeed, the intensity-dependent change in the amplitude δA of the unpolarized component's electric field may be evaluated as $\delta A(L) \approx \delta A(0)[1+(\kappa L/\lambda)A^2]$. Here κ is of the order of magnitude of κ_{ijkl} in equation (4.3). This nonlinear parameter incorporates the local and nonlocal contributions to the third-order nonlinearity. Therefore, changes in the statistical properties of light may be taken into account by including a nonlinearity-proportional correction in equation (5.43):

$$\begin{aligned}\langle\delta A_i\delta A_j^*\delta A_k\delta A_l^*\rangle &= \langle\delta A_i\delta A_j^*\rangle\langle\delta A_k\delta A_l^*\rangle + \langle\delta A_i\delta A_l^*\rangle\langle\delta A_k\delta A_j^*\rangle \\ &\quad + O\left(|\kappa|\,|\mathbf{A}|^2\frac{L}{\lambda}\right)O(\delta\mathbf{A}^4)\end{aligned} \tag{5.44}$$

Here $O(x)$ stands for terms of the order of x or higher. The last term in equation (5.44) reflects the difference between the Gaussian statistics of the incident light's unpolarized component and the statistical properties of the light as it propagates in the nonlinear medium. Thus, the statistical properties of the unpolarized components inside the nonlinear medium will be close to Gaussian statistics as long as

$$|\kappa|\,|\mathbf{A}|^2\frac{L}{\lambda} \ll 1 \tag{5.45}$$

Condition (5.45) requires the intensity-dependent changes of the phase and amplitude of the light wave propagating in the nonlinear medium to be small. Therefore relation (5.43) may be used as long as the intensity dependent changes in the polarization state of the wave Θ^{NL}, $\eta^{\mathrm{NL}} \ll \pi$. From now on we will assume this to be the case and say that Gaussian statistics hold everywhere in the medium. With this assumption, equation (5.42) may be rearranged:

$$\begin{aligned}\langle s_\alpha s_\beta\rangle &= S_\alpha S_\beta + \tfrac{1}{2}S_0^2(1-r^2)\delta_{\alpha\beta}, \qquad \alpha, \beta = 1, 2, 3\\ \langle s_0^2\rangle &= S_0^2 + \tfrac{1}{2}S_0^2(1-r^2)\\ \langle s_0 s_\alpha\rangle &= S_0 S_\alpha(2-r)\end{aligned} \tag{5.46}$$

and we may now continue with our averaging of the evolution equations (5.33) for the s-parameters. First, using (5.46) we calculate

$$\begin{aligned}\langle \Xi_0 s_0 + \Xi_\alpha s_\alpha\rangle &= \Omega_0 S_0 + \Omega_\alpha S_\alpha + \tfrac{1}{4}(w_{0\alpha} + w_{\alpha 0})S_0^2(1-r^2)\\ &\quad + \tfrac{1}{2}(w_{\alpha 0} + w_{0\alpha})S_\alpha S_0(1-r)\\ \langle \Xi_\alpha s_0 + \Xi_0 s_\alpha\rangle &= \Omega_\alpha S_0 + \Omega_0 S_\alpha + \tfrac{1}{4}(w_{0\alpha} + w_{\alpha 0})S_0^2(1-r^2)\\ &\quad + \tfrac{1}{2}(w_{00}S_\alpha + w_{\alpha\beta}S_\beta)S_0(1-r)\\ \langle e_{\alpha\beta\gamma}\Xi_\beta s_\gamma\rangle &= e_{\alpha\beta\gamma}\Omega_\beta S_\gamma + \tfrac{1}{4}e_{\alpha\beta\gamma}w_{\beta\gamma}S_0^2(1-r^2) + \tfrac{1}{2}e_{\alpha\beta\gamma}w_{\beta 0}S_\gamma S_0(1-r)\end{aligned} \tag{5.47}$$

Using equations (5.47), equations (5.33) may be readily converted into the evolution equations for the Stokes parameters of partially polarized light:

$$\begin{aligned}\frac{dS_\alpha}{dz} &= -\frac{\omega}{2c}\,\mathrm{Im}\,\{(4n + \Omega_0)S_\alpha + \Omega_\alpha S_0 + ie_{\alpha\beta\gamma}\Omega_\beta S_\gamma\\ &\quad + \tfrac{1}{4}(w_{0\alpha} + w_{\alpha 0})S_0^2(1-r^2) + \tfrac{1}{2}(w_{\alpha\beta}S_\beta + w_{00}S_\alpha)S_0(1-r)\\ &\quad + \frac{i}{2}e_{\alpha\beta\gamma}w_{\beta 0}S_\gamma S_0(1-r) + \frac{i}{4}e_{\alpha\beta\gamma}w_{\beta\gamma}S_0^2(1-r^2)\}\\ \frac{dS_0}{dz} &= -\frac{\omega}{2c}\,\mathrm{Im}\,\{(4n + \Omega_0)S_0 + \Omega_\alpha S_\alpha + \tfrac{1}{4}(w_{00} + w_{\alpha\alpha})S_0^2(1-r^2)\\ &\quad + \tfrac{1}{2}(w_{\alpha 0} + w_{0\alpha})S_\alpha S_0(1-r)\}\end{aligned} \tag{5.48}$$

Clearly, for totally polarized light ($r = 1$) equations (5.48) become equations (4.10). In a manner similar to that used to derive equation (3.71) for the evolution of the degree of polarization in linear media, the equation for the degree of polarization r in nonlinear media may be deduced from equations (5.48):

$$\begin{aligned}\frac{d(r^2)}{dz} &= -\frac{\omega}{c}\,\mathrm{Im}\left\{\frac{(\mathbf{\Omega}\cdot\mathbf{S})}{S_0}(1-r^2) + \tfrac{1}{4}[(w_{0\alpha} + w_{\alpha 0})S_\alpha + w_{00}S_0](1-r)^2(1+2r)\right.\\ &\quad \left. + \tfrac{1}{2}w_{\alpha\beta}\frac{S_\alpha S_\beta}{S_0}(1-r) + \tfrac{1}{4}w_{\alpha\alpha}S_0 r^2(1-r^2) + \frac{i}{4}e_{\alpha\beta\gamma}w_{\beta\gamma}S_\alpha(1-r^2)\right\}\end{aligned} \tag{5.49}$$

Equation (3.71) is returned from equations (5.49) when nonlinearity is ignored, that is $w_{\alpha\beta} = 0$. From equations (5.49) it can be seen that totally polarized light ($r = 1$) remains totally polarized, since its right-hand side becomes zero. Concerning theoretical study of the propagation of partially polarized light and light depolarization in nonlinear media also see Refs. [20] to [22].

REFERENCES

1. N.I. Zheludev, *Soviet Phys. Usp.*, **32**, 357 (1989).
2. L.D. Landau and E.M. Lifshitz, *Mechanics*, Pergamon Press, Oxford, 1960.
3. K.L. Sala, *Phys. Rev. A*, **29**, 1944 (1984).
4. H.G. Winful, *Appl. Phys. Lett.*, **47**, 213 (1985).
5. H.G. Winful, *Opt. Lett.*, **11**, 33 (1986).
6. B. Daino, G. Gregori and S. Wabnitz, *Opt. Lett.*, **11**, 42 (1986).
7. M. Roman, S. Kielich and W. Gadomski. *Phys. Rev. A*, **34**, 351 (1986).
8. S. Trillo, S. Wabnitz, R.H. Stolen, G. Assanto, C.T. Seaton and G.I. Stegeman, *Appl. Phys. Lett.*, **49**, 1224 (1986).
9. J. Yumoto and K. Otsuka, *Phys. Rev. Lett.*, **54**, 1806 (1985).
10. G. Gregori and S. Wabnitz, *Phys. Rev. Lett.*, **56**, 600 (1986).
11. M.V. Tratnik and J.E. Sipe, *Phys. Rev. A*, **35**, 2965 (1987).
12. M.V. Tratnik and J.E. Sipe, *Phys. Rev. A*, **35**, 2976 (1987).
13. M.V. Tratnik and J.E. Sipe. *Phys. Rev. A*, **36**, 4817 (1987).
14. J. Yumoto and K. Otsuka, *Phys. Rev. A*, **34**, 4445 (1986).
15. M.I. Dykman and G.G. Tarasov, *Soviet Phys. JETP*, **59**, 1180 (1984).
16. M.I. Dykman and G.G. Tarasov. *Opt. Commun.*, **49**, 79 (1984).
17. V.A. Makarov, S.M. Pershin, A.A. Podshivalov, R.S. Zadoyan and N.I. Zheludev. *Opt. Lett.*, **8**, 557 (1983).
18. S.A. Akhmanov, Yu.E. D'yakov and A.S. Chirkin, *Introduction to Statistical Radiophysics and Optics* (in Russian), Nauka, Moscow, 1981.
19. L. Mandel and E. Wolf, *Optical Coherence and Quantum Optics*, Cambridge University Press, Cambridge, 1995.
20. B. Crosignani, B. Daino and P. Di Porto, *JOSA-B*, **3**, 1120 (1986).
21. V.E. Chernov and B.A. Zon, *JOSA-B*, **10**, 210 (1993).
22. Y.P. Svirko and N.I. Zheludev, *Phys. Rev. A*, **50**, 709 (1994).

6 Self-Action Polarization Effects in Reflected Light

In this chapter we will study the change of the polarization state of light reflected from a nonlinear medium. The constitutive equations will be adapted to include the boundary between two media. We will find the boundary conditions and derive universal formulae for the polarization state of light normally reflected from the interface between a vacuum and media with cubic optical nonlinearity.

6.1 CONSTITUTIVE EQUATION FOR INHOMOGENEOUS NONLINEAR MEDIUM

Nonlinear polarization effects do not just happen to light passing through a nonlinear medium. Similar polarization effects may also be seen in light reflected from the surface of a nonlinear medium. This is no surprise; if intense light changes the medium refraction, absorption, anisotropy and nonlocality, it will also affect the reflectivity. Generally the effects in transmitted light accumulate with propagation and so grow with the sample thickness. However, only a very thin layer near the interface, approximately one wavelength thick, determines the properties of the reflected light. As a rough guide, reflective effects can be expected to be smaller by a factor of the ratio between the light wavelength and the crystal thickness.

To develop the theory of reflective nonlinear polarization phenomena we need to address two core problems: an adequate constitutive equation and appropriate boundary conditions must be found. We will consider the reflection of intense light waves from the boundary of a medium with cubic nonlinearity. This requires a modification of the constitutive equation (4.1) because the medium changes its properties with coordinate and so the equation must be able to account for such inhomogeneity. The constitutive equation for inhomogeneous media may be derived in a similar way to the equation for homogeneous media (see Chapter 2) for homogeneous medium. However, in inhomogeneous media we must admit that the optical response is coordinate dependent. Consequently, the linear and nonlinear response functions in equation (2.17), that is $\psi^{(1)}_{ij}(\mathbf{r}-\mathbf{r}_1;\, t-t_1)$, $\psi^{(2)}_{ijk}(\mathbf{r}-\mathbf{r}_1,\, t-t_1;\, \mathbf{r}-\mathbf{r}_2,\, t-t_2)$ and $\psi^{(3)}_{ijkl}(\mathbf{r}-\mathbf{r}_1,\, t-t_1;\, \mathbf{r}-\mathbf{r}_2,\, t-t_2;\, \mathbf{r}-\mathbf{r}_3,\, t-t_3)$, depend not only on the differential coordinates $\rho=\mathbf{r}-\mathbf{r}_1$, $\rho_1=\mathbf{r}-\mathbf{r}_2$, $\rho_2=\mathbf{r}-\mathbf{r}_3$, but also on the coordinate $\mathbf{r}$ itself. As a result, the material tensors (2.12) and (2.13), (2.22) and (2.23) and (2.25) and (2.26) will become functions of this coordinate. With this assumption the linear constitutive equation (2.15) in inhomogeneous media must be rewritten [1, 2]:

$$P_i^{(1)}(\mathbf{r}, t) = \frac{1}{4\pi}\{(\epsilon_{ij}(\omega, \mathbf{r}) - \delta_{ij})E_j(\omega, \mathbf{r}) + \nabla_n[\Gamma_{ijn}^{(1)}(\omega, \mathbf{r})E_j(\omega, \mathbf{r})]\}\, e^{-i\omega t} + \text{c.c.} \tag{6.1}$$

Similarly, we must modify the nonlinear constitutive equations for the quadratic (2.21) and cubic (2.24) optical responses. In this chapter we are concerned with nonlinear reflection at the frequency of the incident wave. Therefore, the quadratic terms in the nonlinear constitutive equation are irrelevant. In inhomogeneous media with cubic nonlinearity the constitutive equation (2.24) becomes

$$\begin{aligned} P_i^{(3)}(\mathbf{r}, t) = \sum_{a,b,c} \{ &\chi_{ijkl}^{(3)}(\omega_a + \omega_b + \omega_c, \omega_a, \omega_b, \omega_c; \mathbf{r})E_j(\omega_a, \mathbf{r})E_k(\omega_b, \mathbf{r})E_l(\omega_c, \mathbf{r}) \\ &+ \nabla_n[\Gamma_{ijkln}^{(3)}(\omega_a + \omega_b + \omega_c, \omega_a, \omega_b, \omega_c; \mathbf{r})E_j(\omega_a, \mathbf{r})]E_k(\omega_a, \mathbf{r}) \\ &\times E_l(\omega_c, \mathbf{r})\}\, e^{-i(\omega_a+\omega_b+\omega_c)t} \end{aligned} \tag{6.2}$$

Correspondingly, the constitutive equation for an electromagnetic wave of frequency ω propagating in an inhomogeneous medium takes the following form [3]:

$$\begin{aligned} P_i(\mathbf{r}, t) &= P_i^{(1)}(\omega, \mathbf{r})\, e^{-i\omega t} + P_i^{(3)}(\omega, \mathbf{r})\, e^{-i\omega t} + \text{c.c.} \\ &= \left\{ \frac{\epsilon_{ij}(\omega; \mathbf{r}) - \delta_{ij}}{4\pi} E_j(\omega, \mathbf{r}) + \frac{1}{4\pi}\nabla_n[\Gamma_{ijn}^{(1)}(\omega, \mathbf{r})E_j(\omega, \mathbf{r})] \right. \\ &\quad + 3\chi_{ijkl}^{(3)}(\omega, \omega, \omega, -\omega; \mathbf{r})E_j(\omega, \mathbf{r})E_k(\omega, \mathbf{r})E_l^*(\omega, \mathbf{r}) \\ &\quad + 2\nabla_n[\Gamma_{ijkln}^{(3)}(\omega, \omega, \omega, -\omega; \mathbf{r})E_j(\omega, \mathbf{r})]E_k(\omega, \mathbf{r})E_l^*(\omega, \mathbf{r}) \\ &\quad \left. + \nabla_n[\Gamma_{ijkln}^{(3)}(\omega, -\omega, \omega, \omega; \mathbf{r})E_j^*(\omega, \mathbf{r})]E_k(\omega, \mathbf{r})E_l(\omega, \mathbf{r}) \right\} e^{-i\omega t} + \text{c.c} \end{aligned} \tag{6.3}$$

(Compare with equations (2.14) and (2.53).)

In equation (6.3) the components of the material tensors $\epsilon_{ij}(\omega; \mathbf{r})$, $\Gamma_{ijn}^{(1)}(\omega; \mathbf{r})$, $\chi_{ijkl}^{(3)}(\omega, \omega, \omega - \omega; \mathbf{r})$ and $\Gamma_{ijkln}^{(3)}(\omega, \omega, \omega, \ -\omega; \mathbf{r})$ depend on the coordinate $\mathbf{r}$. This makes it suitable for the study of nonlinear wave propagation effects in media with material parameters that depend on the coordinate.

The constitutive equations (6.1) and (6.3) follow from nothing but the fundamental principle of causality of physical events and therefore represent the most general forms of constitutive equations accounting for weak spatial dispersion. Other approaches to constitutive equations for inhomogeneous media are known when their derivation *a priori* assume other symmetries such as time reversality or reciprocity of optical interactions [16].

6.2 THE BOUNDARY CONDITIONS

We will consider below the propagation of a light wave in the vicinity of a plane boundary between a vacuum and a homogeneous nonlinear medium. The boundary may be treated as a specific case of inhomogeneous media where the medium's properties (i.e. optical susceptibilities) change sharply at the interface. To study wave propagation in the vicinity of the interface we will need equations that relate the electromagnetic fields on both sides of the boundary. These are known as boundary conditions and must follow exclusively from Maxwell's equations.

We assume that the interface between the vacuum and the nonlinear medium coincides with the *YX* plane of the laboratory Cartesian frame. To derive the boundary conditions we introduce, following Ref. [4], a small rectangular loop ABCD which crosses the interface and lies in a plane parallel to the *XZ* plane (see Fig. 6.1). The loop is only to help us derive the boundary conditions; they will not be dependent on the shape of the loop. To simplify the calculations we take a rectangular loop with two sides parallel to the interface plane. To derive of the boundary conditions we will use the Stokes theorem of vector analysis. This theorem relates an integral over the contour $\mathscr{L}$ with an integral over the corresponding surface S:

$$\oint_{\mathscr{L}} (\mathbf{A}\cdot \mathrm{d}\mathbf{l}) = \int_{S} (\mathrm{curl}\,\mathbf{A}\cdot\mathbf{n})\,\mathrm{d}s \tag{6.4}$$

Here $\mathbf{n}$ is a unit vector perpendicular to the loop and pointing in the direction

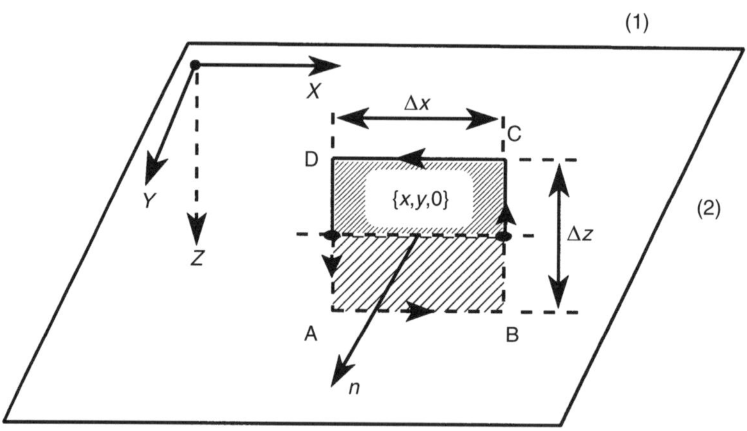

Fig. 6.1 The Cartesian coordinate system and small rectangular loop ABCD used at the boundary between the vacuum (1) and the nonlinear medium (2) for derivation of the boundary conditions

that a right-hand cork screw would move when rotated according to the path followed around the loop. When applied to the first and third Maxwell equations (2.2), the Stokes theorem gives

$$\oint_{\mathscr{L}} (\mathbf{E}\cdot \mathrm{d}\boldsymbol{l}) = -\frac{1}{c}\frac{\partial}{\partial t}\int_s (\mathbf{B}\cdot\mathbf{n})\,\mathrm{d}s$$
$$\oint_{\mathscr{L}} (\mathbf{B}\cdot \mathrm{d}\boldsymbol{l}) = \frac{1}{c}\frac{\partial}{\partial t}\int_s (\mathbf{D}\cdot\mathbf{n})\,\mathrm{d}s \tag{6.5}$$

For the loop presented in Fig. 6.1, the first equation in (6.5) may be rearranged by splitting the contour integrals into sections parallel and perpendicular to the border:

$$\int_{-\Delta z/2}^{\Delta z/2}\left[E_z\left(x-\frac{\Delta x}{2},\, y,\, z',\, t\right) - E_z\left(x+\frac{\Delta x}{2},\, y,\, z',\, t\right)\right]\mathrm{d}z'$$
$$+\int_{x-\Delta x/2}^{x+\Delta x/2}\left[E_x\left(x',\, y,\, -\frac{\Delta z}{2},\, t\right) - E_x\left(x',\, y,\, \frac{\Delta z}{2},\, t\right)\right]\mathrm{d}x'$$
$$= -\frac{1}{c}\frac{\partial}{\partial t}\int_{-\Delta z/2}^{\Delta z/2}\mathrm{d}z'\int_{x-\Delta x/2}^{x+\Delta x/2} B_y(x',\, y,\, z')\,\mathrm{d}x' \tag{6.6}$$

As $\Delta z \to 0$ the first integral on the left-hand side of equation (6.6) tends to zero. On the right-hand side the integral will also tend to zero as $\Delta z \to 0$. Therefore at the boundary, i.e. for $\Delta z \to 0$,

$$\int_{x-\Delta x/2}^{x+\Delta x/2}[E_x^{(1)}(x',\, y,\, 0,\, t) - E_x^{(2)}(x',\, y,\, 0,\, t)]\,\mathrm{d}x' = 0 \tag{6.7}$$

The superscripts (1) and (2) denote the electric fields in the vacuum and in the medium. Since equation (6.7) is valid for arbitrary Δx, we conclude that the expression under the integral sign is zero, i.e. the X components of the electric field on the two sides of the boundary are equal:

$$E_x^{(1)}(\mathbf{r},\, t)|_{z=0} = E_x^{(2)}(\mathbf{r},\, t)|_{z=0} \tag{6.8}$$

This is the first boundary condition.

In order to derive the boundary conditions for the magnetic field, we shall use the second equation in (6.5). Here the right-hand side contains a surface integral over $(\mathbf{D}\cdot\mathbf{n})$ with $\mathbf{D} = \mathbf{E} + 4\pi\mathbf{P}$ and $\mathbf{P}$ given by equation (6.3):

$$\int_s (\mathbf{D}\cdot\mathbf{n})\,\mathrm{d}s = \int_{-\Delta z/2}^{\Delta z/2} \mathrm{d}z' \int_{x-\Delta x/2}^{x+\Delta x/2} D_y(x', y, z')\,\mathrm{d}x'$$

$$= \int_{-\Delta z/2}^{\Delta z/2} \mathrm{d}z' \int_{x-\Delta x/2}^{x+\Delta x/2} \mathrm{d}x' \, \{\epsilon_{yj}(\omega; \mathbf{r})E_j(\omega, \mathbf{r}) + \nabla_n[\Gamma^{(1)}_{yjn}(\omega, \mathbf{r})E_j(\omega, \mathbf{r})]$$

$$+ 12\pi\chi^{(3)}_{yjkl}(\omega, \omega, \omega, -\omega; \mathbf{r})E_j(\omega, \mathbf{r})E_k(\omega, \mathbf{r})E^*_l(\omega, \mathbf{r})$$

$$+ 8\pi\nabla_n[\Gamma^{(3)}_{yjkln}(\omega, \omega, \omega, -\omega; \mathbf{r})E_j(\omega, \mathbf{r})]E_k(\omega, \mathbf{r})E^*_l(\omega, \mathbf{r})$$

$$+ 4\pi\nabla_n[\Gamma^{(3)}_{yjkln}(\omega, -\omega, \omega, \omega; \mathbf{r})E^*_j(\omega, \mathbf{r})]E_k(\omega, \mathbf{r})E_l(\omega, \mathbf{r})\}\mathrm{e}^{-\mathrm{i}\omega t}$$

$$+ \text{c.c.} \tag{6.9}$$

As $\Delta z \to 0$ all terms that do not contain Z derivatives $\nabla_Z = \partial/\partial z$ vanish. Therefore, with an infinitesimally small loop the second equation in (6.5) gives the boundary condition for the X component of the magnetic field:

$$B^{(1)}_x(x, y, 0, t) - B^{(2)}_x(x, y, 0, t) = \frac{\mathrm{i}\omega}{c}\Gamma_{yjz}(\omega)E^{(2)}_j(\omega, x, y, 0)\,\mathrm{e}^{-\mathrm{i}\omega t}$$

$$+ \frac{4\pi\mathrm{i}\omega}{c}[2\Gamma^{(3)}_{yjklz}(\omega, \omega, \omega, -\omega) + \Gamma^{(3)}_{ylkjz}(\omega, -\omega, \omega, \omega]$$

$$\times E^{(2)}_j(\omega, x, y, 0)E^{(2)}_k(\omega, x, y, 0)E^{(2)*}_l(\omega, x, y, 0)\,\mathrm{e}^{-\mathrm{i}\omega t} + \text{c.c} \tag{6.10}$$

Using a loop aligned along the Y direction, boundary conditions similar to (6.8) and (6.10) can be obtained for E_y and $B_{y.}$ Therefore we arrive at the conclusion that the tangential component of the electric field at the boundary is continuous:

$$[\mathbf{z} \times (\mathbf{E}^{(2)}(\mathbf{r}, t) - \mathbf{E}^{(1)}(\mathbf{r}, t))]|_{z=0} = 0 \tag{6.11}$$

while for the tangential component of the magnetic field

$$[\mathbf{z} \times (\mathbf{B}^{(2)}(\mathbf{r}, t) - \mathbf{B}^{(1)}(\mathbf{r}, t))]|_{z=0} = \Delta\mathbf{D} \tag{6.12}$$

where

$$\Delta D_i = \left\{-\frac{\mathrm{i}\omega}{c}\Gamma_{ijz}(\omega, \mathbf{r})E^{(2)}_j(\omega, \mathbf{r})\,\mathrm{e}^{-\mathrm{i}\omega t} - \frac{4\pi\mathrm{i}\omega}{c}[2\Gamma^{(3)}_{ijklz}(\omega, \omega, \omega, -\omega; \mathbf{r})\right.$$

$$\left. + \Gamma^{(3)}_{ilkjz}(\omega, -\omega, \omega, \omega; \mathbf{r})]E_j{}^{(2)}(\omega, \mathbf{r})\mathrm{E}_k{}^{(2)}(\omega, \mathbf{r})E_l{}^{(2)*}(\omega, \mathbf{r})\mathrm{e}^{-\mathrm{i}\omega t} + \text{c.c.}\right\}\Big|_{z=0} \tag{6.13}$$

and $\mathbf{z}$ is a unit vector along the Z direction. Apparently in media with local optical response $\Gamma_{ijz} = 0$ and $\Gamma^{(3)}_{ijklz} = 0$ giving $\Delta\mathbf{D} = 0$.

6.3 THE STOKES PARAMETERS OF NORMALLY REFLECTED LIGHT

In this section, using the boundary conditions (6.11) and (6.12), we will examine the reflection of an electromagnetic wave at normal incidence. In the vicinity of the boundary we have the incident, reflected and transmitted waves. We will denote the amplitudes of the reflected and transmitted waves by superscripts R and T respectively (see Fig. 6.2). Let the incident wave approach the boundary along z. The reflected wave thus propagates along $-\mathbf{z}$. The electric field at frequency ω in the vacuum may be written as follows:

$$\mathbf{E}^{(1)}(\mathbf{r},\, t) = \mathbf{A}\,\mathrm{e}^{-\mathrm{i}\omega t+\mathrm{i}(\omega/c)z} + \mathbf{A}^{\mathrm{R}}\,\mathrm{e}^{-\mathrm{i}\omega t-\mathrm{i}(\omega/c)z} + \mathrm{c.c.} \tag{6.14}$$

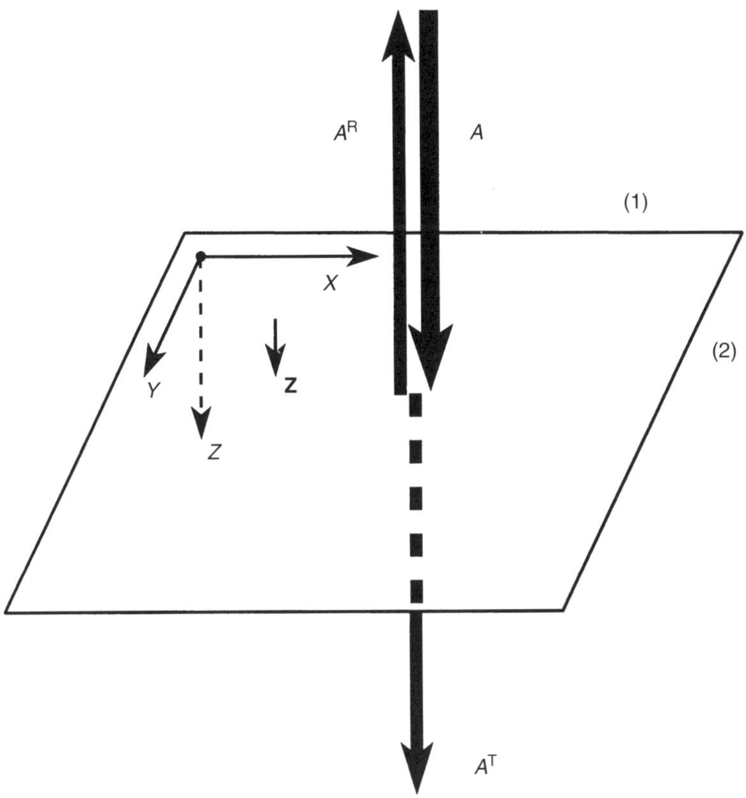

Fig. 6.2 Normal reflection from the boundary between a vacuum (1) and a nonlinear medium (2): incident, reflected and transmitted waves, with amplitudes $\mathbf{A}$, $\mathbf{A}^{\mathrm{R}}$ and $\mathbf{A}^{\mathrm{T}}$ respectively, are shown

Substituting equation (6.14) into the first Maxwell equation (2.2) we obtain the magnetic field on the vacuum side of the interface:

$$\mathbf{B}^{(1)}(\mathbf{r},\ t) = [\mathbf{z} \times \mathbf{A}]\,\mathrm{e}^{-i\omega t + \mathrm{i}(\omega/c)z} - [\mathbf{z} \times \mathbf{A}^{\mathrm{R}}]\,\mathrm{e}^{-i\omega t - \mathrm{i}(\omega/c)z} + \text{c.c.} \qquad (6.15)$$

The amplitude of the electric field inside the nonlinear medium will depend on the coordinate z:

$$\mathbf{E}^{(2)}(\mathbf{r},\ t) = \mathbf{A}^{\mathrm{T}}(z)\,\mathrm{e}^{-\mathrm{i}\omega t + \mathrm{i}kz} + \text{c.c.} \qquad (6.16)$$

where $k = \omega n/c$ and n is the core complex refractive index of the medium. Substituting (6.16) into the first Maxwell equation (2.2) we obtain the magnetic field inside the nonlinear medium:

$$\mathbf{B}^{(2)}(\mathbf{r},\ t) = \left\{ n[\mathbf{z} \times \mathbf{A}^{\mathrm{T}}] - \frac{\mathrm{i}c}{\omega}\left[\mathbf{z} \times \frac{\mathrm{d}\mathbf{A}^{\mathrm{T}}}{\mathrm{d}z}\right] \right\} \mathrm{e}^{-\mathrm{i}\omega t + \mathrm{i}kz} + \text{c.c.} \qquad (6.17)$$

Combining equations (6.14) and (6.16) with the boundary condition (6.11) yields the following relation between the electric field amplitudes of the incident, reflected and transmitted light waves at the boundary:

$$\mathbf{A} + \mathbf{A}^{\mathrm{R}} = \mathbf{A}^{\mathrm{T}}|_{z=0} \qquad (6.18)$$

Similarly, substituting equations (6.15) and (6.17) into (6.12) and (6.13) yields

$$A_i - A_i^{\mathrm{R}} = \mathrm{n}A_i^{\mathrm{T}}|_{z=0} + \frac{c}{\mathrm{i}\omega}\frac{\mathrm{d}A_i^{\mathrm{T}}}{\mathrm{d}z}\bigg|_{z=0} - \frac{\mathrm{i}\omega}{c}\Gamma^{(1)}_{ijz}A_j^{\mathrm{T}}\bigg|_{z=0}$$
$$- \frac{4\pi\mathrm{i}\omega}{c}[2\Gamma^{(3)}_{ijklz}(\omega,\ \omega,\ \omega,\ -\omega) + \Gamma^{(3)}_{ilkjz}(\omega,\ -\omega,\ \omega,\ \omega,\ )]A_j^{\mathrm{T}}A_k^{\mathrm{T}}A_l^{\mathrm{T}*}|_{z=0} \qquad (6.19)$$

Equations (6.18) and (6.19) may be solved in terms of the Cartesian components of the transmitted and reflected waves $A^{\mathrm{T}}_{x,y}|_{z=0}$ and $A^{\mathrm{R}}_{x,y.}$ Equation (6.19) may be rearranged by expressing the derivative on the right-hand side in terms of the field amplitudes from the nonlinear propagation equation (4.3):

$$\frac{\mathrm{d}A_i^{\mathrm{T}}}{\mathrm{d}z}\bigg|_{z=0} = \frac{\mathrm{i}\omega}{2cn}(\epsilon_{ij} + ik\Gamma^{(1)}_{ijz} - n^2\delta_{ij})A_j^{\mathrm{T}}|_{z=0} + \frac{2\pi\mathrm{i}\omega}{cn}(\chi_{ijkl} + ik\Gamma_{ijklz})A_j^{\mathrm{T}}A_k^{\mathrm{T}}A_l^{\mathrm{T}*}|_{z=0} \qquad (6.20)$$

Substituting $\mathrm{d}A_i^{\mathrm{T}}/\mathrm{d}z|_{z=0}$ from equation (6.20) into (6.19) we arrive at

$$A_i - A_i^{\mathrm{R}} = nA_i^{\mathrm{T}}|_{z=0} + \tfrac{1}{2}(\Delta u_{ij}A_j^{\mathrm{T}} + \Delta\kappa_{ijkl}A_j^{\mathrm{T}}A_k^{\mathrm{T}}A_l^{\mathrm{T}*}|_{z=0} \qquad (6.21)$$

Here we have introduced

$$\Delta u_{ij} = \frac{1}{n}(\epsilon_{ij} - ik\Gamma^{(1)}_{ijz} - n^2\delta_{ij}) \qquad (6.22)$$

and

$$\Delta\kappa_{ijkl} = \frac{4\pi}{n}(\chi_{ijkl} + ik\Gamma^{\mathrm{R}}_{ijklz}) \tag{6.23}$$

where

$$\Gamma^{\mathrm{R}}_{ijklz} = -2\Gamma^{(3)}_{ijklz}(\omega, \omega, \omega, -\omega) - 3\Gamma^{(3)}_{ilkjz}(\omega, -\omega, \omega, \omega) \tag{6.24}$$

(Compare with the similar equations (3.19) and (4.4).) If the anisotropy, nonlocality and nonlinearity of the medium are ignored, then $\Delta u_{ij} = 0$ and $\Delta\kappa_{ijkl} = 0$, and we have a simple case of normal reflection from the boundary between a vacuum and an isotropic medium with complex refractive index n. In this case the amplitudes of the reflection and transmission coefficients ($A^{\mathrm{R}}_k = RA_k$ and $A^{\mathrm{T}}_k = TA_k$) will return the standard Fresnel formulae [5]:

$$\begin{aligned} R &= \frac{1-n}{1+n} \\ T &= \frac{2}{1+n} \end{aligned} \tag{6.25}$$

The presence of the nonlinearity, anizotropy and nonlocality will alter the Fresnel formulae for the reflection and transmission coefficients (6.25). We assume that these alterations are small, and therefore may be presented as a power series in Δu_{ij} and $\Delta\kappa_{ijkl}$. Retaining only the first terms of these series, which are proportional to Δu_{ij} and $\Delta\kappa_{ijkl}$, we obtain

$$A^{\mathrm{R}}_k = RA_k - \frac{T^2}{4}(\Delta u_{ij} + |T|^2\Delta\kappa_{ijkl}A_kA^*_l)A_j \tag{6.26}$$

$$A^{\mathrm{T}}_k|_{z=0} = TA_k - \frac{T^2}{4}(\Delta u_{ij} + |T|^2\Delta\kappa_{ijkl}A_kA^*_l)A_j \tag{6.27}$$

The Stokes parameters of a reflected wave can now be calculated. First, we expand Δu_{ij} and $\Delta\kappa_{ijkl}$ over the Pauli matrices as we expanded u_{ij} and κ_{ijkl} in equations (3.20) and (4.5):

$$\begin{aligned} \Delta u_{ij} &= \tfrac{1}{2}v^{\mathrm{R}}_{\alpha}\sigma^{(\alpha)}_{ij} \\ \Delta\kappa_{ijkl} &= \tfrac{1}{4}w^{\mathrm{R}}_{\alpha\beta}\sigma^{(\alpha)}_{ij}\sigma^{(\beta)}_{lk} \end{aligned} \tag{6.28}$$

Summation is presumed over α and $\beta = 0, \ldots, 3$ and the expansion coefficients are

$$\begin{aligned} v^{\mathrm{R}}_{\alpha} &= \sigma^{(\alpha)}_{ji}\Delta u_{ij} \\ w^{\mathrm{R}}_{\alpha\beta} &= \sigma^{(\alpha)}_{ji}\Delta\kappa_{ijkl}\sigma^{(\beta)}_{kl} \end{aligned} \tag{6.29}$$

Resorting equations (6.28) and (6.29), the equations (6.26) and (6.27) for the amplitudes of the transmitted and reflected waves at the boundary may be rewritten as

$$A_k^{\mathrm{R}} = RA_k - \frac{T^2}{8}\Omega_\alpha^{\mathrm{R}}\sigma_{kj}^{(\alpha)}A_j \tag{6.30}$$

$$A_k^{\mathrm{T}}|_{z=0} = TA_k - \frac{T^2}{8}\Omega_\alpha^{\mathrm{R}}\sigma_{kj}^{(\alpha)}A_j \tag{6.31}$$

where $\alpha = 0, 1, 2, 3$ and a summation is presumed over α and j. In equations (6.30) and (6.31) we have introduced

$$\Omega_\alpha^{\mathrm{R}} = v_\alpha^{\mathrm{R}} + \frac{|T|^2}{2}w_{\alpha\beta}^{\mathrm{R}}S_{0\beta} \tag{6.32}$$

The Stokes parameters of the transmitted wave may be defined in a similar manner to those of the incident wave:

$$S_\alpha^{\mathrm{T}} = \langle (A_i^{\mathrm{T}})^*\sigma_{ij}^{(\alpha)}A_j^{\mathrm{T}}\rangle\, \mathrm{e}^{-2\mathrm{Im}\{k\}z} \tag{6.33}$$

Likewise, the Stokes parameters of the reflected wave can be calculated in the laboratory coordinate frame:

$$S_\alpha^{\mathrm{R}} = \langle (A_i^{\mathrm{R}})^*\sigma_{ij}^{(\alpha)}A_j^{\mathrm{R}}\rangle \tag{6.34}$$

It is very important to recall, however, that the reflected wave propagates in the opposite direction to the incident wave. Therefore, to make the calculations of the reflected waves polarization consistent with the definitions of Chapter 3, we need to establish a new system of coordinates X', Y', Z', with the Z' axis pointing in the opposite direction to the Z axis of the laboratory coordinate frame shown in Fig. 6.2. Like the laboratory coordinate frame, the new coordinate frame must be right-handed. It is convenient to set X' to coincide with X since the polarization azimuth is measured from the X direction. Then, to retain the right-handedness of the new coordinate frame, Y' must be directed opposite to Y'. In the new coordinate system the sign of the Y projection of the electric field amplitude A_y will be opposite to that in the laboratory frame. Therefore, in the new coordinate system $X'Y'Z'$ the Stokes parameters S_0 and S_1 will have the same signs as in the XYZ frame, while those of S_2 and S_3 will be reversed (see definitions (3.7)). It follows from here that the polarization azimuth $\Theta = \frac{1}{2}\tan^{-1}(S_2/S_1)$ and the ellipticity angle $\eta = \frac{1}{2}\sin^{-1}(S_3/S_0)$ will have opposite signs in the two frames. For the sake of simplicity, the polarization state of reflected waves will be presented in the same frame as the incident wave, i.e. in the laboratory coordinate frame. If, however, one needs to present the polarization of the reflected wave in the proper coordinate frame, the signs of S_2, S_3, Θ and η must be reversed.

Using equations (6.30) and (6.31) and resorting definitions (6.34) and (6.33) we find that:

$$\mathbf{S}^{\mathrm{R}} = |R|^2\left(\mathbf{S} - \mathrm{Re}\left\{\frac{\Omega_0^{\mathrm{R}}\mathbf{S} + \mathbf{\Omega}^{\mathrm{R}}S_0 + i[\mathbf{\Omega}^{\mathrm{R}} \times \mathbf{S}]}{1 - n^2}\right\}\right) \tag{6.35}$$

$$S_0^{\mathrm{R}} = |R|^2\left(S_0 - \mathrm{Re}\left\{\frac{\Omega_0^{\mathrm{R}}S_0 + (\mathbf{\Omega}^{\mathrm{R}}\cdot\mathbf{S})}{1 - n^2}\right\}\right) \tag{6.36}$$

and

$$\mathbf{S}^{\mathrm{T}}|_{z=0} = |T|^2\left(\mathbf{S} - \tfrac{1}{2}\mathrm{Re}\left\{\frac{\Omega_0^{\mathrm{R}}S + \mathbf{\Omega}^{\mathrm{R}}\mathrm{S}_0 + i[\mathbf{\Omega}^{\mathrm{R}} \times \mathbf{S}]}{1 + n}\right\}\right) \tag{6.37}$$

$$S_0^{\mathrm{T}}|_{z=0} = |T|^2\left(S_0 - \tfrac{1}{2}\mathrm{Re}\left\{\frac{\Omega_0^{\mathrm{R}}S_0 + (\mathbf{\Omega}^{\mathrm{R}}\cdot\mathbf{S})}{1 + n}\right\}\right) \tag{6.38}$$

Formulae (6.35) to (6.38) bring us close to the still ongoing discussion on how to calculate the work which an elliptically polarized wave must do in order to pass through a sharp interface with a chiral (nonlocal) media. Opposing opinions exist on the relevant questions on how to settle the wave energy balance at the chiral interface, and how to calculate correctly the light intensity in a chiral medium in terms of its fields. However, this interesting topic being fundamental for linear optics lays outside of the mainstream of this book which is devoted to nonlinear polarization effects.

6.4 ELEMENTARY NONLINEAR REFLECTIVE POLARIZATION EFFECTS

Polarization changes in the reflected light due to nonlinearity, nonlocality and anisotropy are usually very small:

$$\begin{aligned} \Delta\eta^{\mathrm{R}} &= \eta - \eta^{\mathrm{R}} \ll \pi \\ \Delta\Theta^{\mathrm{R}} &= \Theta - \Theta^{\mathrm{R}} \ll \pi \end{aligned} \tag{6.39}$$

It follows from equations (6.35) and (6.36) that these small changes in the polarization azimuth and ellipticity of the reflected wave may be presented in terms of the polarization azimuth Θ and ellipticity η of the incident wave:

$$\begin{pmatrix} \Delta\Theta^{\mathrm{R}}\cos 2\eta \\ \Delta\eta^{\mathrm{R}} \end{pmatrix} = \begin{pmatrix} -\mathrm{Im} \\ \mathrm{Re} \end{pmatrix} \times \left\{\frac{(\Omega_1^{\mathrm{R}}\cos 2\Theta + \Omega_2^{\mathrm{R}}\sin 2\Theta)\sin 2\eta - \Omega_3^{\mathrm{R}}\cos 2\eta + i(\Omega_2^{\mathrm{R}}\cos 2\Theta - \Omega_1^{\mathrm{R}}\sin 2\Theta)}{2(1 - n^2)}\right\} \tag{6.40}$$

Equation (6.40) allows the calculation of the polarization state of the wave reflected from the boundary between a vacuum and an anisotropic, nonlocal nonlinear medium.

Here we can make an intriguing observation: formula (6.40) may be obtained from formula (5.9) for the polarization state of transmitted light by making the simple replacements: $\Omega \Rightarrow \Omega^{\mathrm{R}}$, $\mathrm{Im}\{\ldots\} \Rightarrow \mathrm{Re}\{\ldots\}$, $\mathrm{Re}\{\ldots\} \Rightarrow -\mathrm{Im}\{\ldots\}$, $\Theta_0 \Rightarrow \Theta$, $\eta_0 \Rightarrow \eta$ and $L \Rightarrow \lambda/\pi(1-n^2)$. The interchange of real and imaginary parts in formulae (5.9) and (6.40) illustrates a very interesting feature of reflective polarization effects: when absorption is moderate, i.e. the refractive coefficient n is predominantly real, the real and imaginary parts of the same susceptibilities control reflective and transmissive polarization effects. For example, as we will see below, $\mathrm{Im}\{\Gamma^{\mathrm{R}}_{yxxxz}\}$ gives nonlinear polarization azimuth rotation on reflection from isotropic gyrotropic media. However, the real part i.e. $\mathrm{Re}\{\Gamma_{yxxxz}\}$ gives nonlinear rotation in transmitted light (see formula (4.23)). Comparison of formulae (5.9) and (6.40) also shows that transmission effects are roughly $\pi n^2 \mathrm{L}/\lambda$ times stronger than reflective effects.

Let us now illustrate the use of formula (6.40). First, we will examine reflection from a linear birefringent medium with local optical response. Transmission through linear birefringent media has been considered in Chapter 3 (see formula (3.40)). In that chapter we calculated the material parameters u_{ij} (formula (3.34)) and $\Omega^{\mathrm{L}}_{\alpha}$ (formula (3.35)). If the birefringent medium shows no nonlocality, that is $\Gamma^{(1)f}_{ijn} = 0$, Δu_{ij} given by formula (6.22) coincides with u_{ij} given by formula (3.43). Using formula (6.40) with $\Omega^{\mathrm{R}}_{\alpha} = \Omega^{\mathrm{L}}_{\alpha}$ we can readily obtain the polarization azimuth and ellipticity of an initially linear polarized wave, reflected from a birefringent medium (see Fig. 3.4 and compare the results with those given by formula (3.41)):

$$\begin{pmatrix} \Delta\Theta^{\mathrm{R}} \\ \Delta\eta^{\mathrm{R}} \end{pmatrix} = \begin{pmatrix} \mathrm{Re} \\ \mathrm{Im} \end{pmatrix} \left\{ \frac{\epsilon_{11} - \epsilon_{33}}{2n(1-n^2)} \right\} \sin(2\Theta - 2\phi) \tag{6.41}$$

Similarly, one can derive a ‘mirror’ analog of formulae (3.48) and (3.49) for an optical active medium. Such derivation may be found in Refs [1] and [11]. Here we point out that the theory of reflective nonlocal effects, in particular the theory of reflective linear nonlocal effects, is very sensitive to the constitutive equation used. If the constitutive equation implies, apart from the causality, some additional symmetries, then some of the reflective effects, such as reflective optical activity may become forbidden (see, for example [16]), and the discussed above correspondence between formulae (5.9) and (6.40) does not hold. Indeed, reflective optical activity implies broken reversality and non-reciprocity of the light-matter interaction process on the border, see appendix 5. Thus, if the theory *a priori* is developed using the constitutive equation of the inhomogeneous media or boundary conditions that implies reversality, the effect of reflective optical activity will be forbidden ‘at root’. Therefore formula (6.40) is the most unrestricted rezsult of the Maxwell theory which is dependent only on the causality principle.

Let us now apply formula (6.40) to nonlinear optics when an intense light wave reflects from a nonlinear medium. We will examine the simple case of an isotropic gyrotopic medium. In a previous chapter we saw that there are three reasons for a change in the polarization of a light wave propagating in an isotropic medium: natural optical activity, nonlinear optical activity and the polarization ellipse self-rotation effect. We will see below that these effects—although usually much smaller—can also be observed in reflected light. The material tensors of isotropic media were introduced in equations (4.11) and (4.12); thus, in isotropic media, equation (6.32) becomes

$$\begin{aligned}
\Omega_1^{\mathrm{R}} &= \frac{4\pi|T|^2}{n}[(\chi_{xxyy} + \chi_{xyyx})S_1 + ik(\Gamma^{\mathrm{R}}_{xxxyz} + \Gamma^{\mathrm{R}}_{xxyxz})S_2] \\
\Omega_2^{\mathrm{R}} &= \frac{4\pi|T|^2}{n}[(\chi_{xxyy} + \chi_{xyyx})S_2 + ik(\Gamma^{\mathrm{R}}_{yxxxz} + \Gamma^{\mathrm{R}}_{xxyxz})S_1] \\
\Omega_3^{\mathrm{R}} &= \frac{4\pi}{n}\{(\chi_{xxyy} - \chi_{xyyx})|T|^2 S_3 + k[2\Gamma^{a} + (\Gamma^{\mathrm{R}}_{yxxxz} - \Gamma^{\mathrm{R}}_{xxyxz})|T|^2 S_0]\} \\
\Omega_0^{\mathrm{R}} &= \frac{4\pi|T|^2}{n}[2(\chi_{xxxx} + \chi_{xxyy})S_0 + k(\Gamma^{\mathrm{R}}_{xxxyz} - \Gamma^{\mathrm{R}}_{xxyxz})S_3]
\end{aligned} \tag{6.42}$$

Substituting equations (6.42) into (6.40) we obtain the following formula for the change in the polarization state of light reflected from an isotropic gyrotropic medium:

$$\begin{aligned}
\Delta\Theta^{\mathrm{R}} &= -\mathrm{Im}\left\{\frac{\chi_{xyyx}}{n(1-n^2)}\right\}|T|^2\frac{8\pi I}{c}\sin 2\eta + \frac{\omega}{c}\,\mathrm{Im}\left\{\frac{\Gamma^{a} + 8\pi\Gamma^{\mathrm{R}}_{yxxxz}|T|^2 I/c}{1-n^2}\right\} \\
\Delta\eta^{\mathrm{R}} &= \\
&\mathrm{Re}\left\{\frac{X_{xyyx}}{n(1-n^2)}\right\}|T|^2\frac{8\pi I}{c}\cos 2\eta\sin 2\eta - \frac{\omega}{c}\,\mathrm{Re}\left\{\frac{\Gamma^{a} + 8\pi\Gamma^{\mathrm{R}}_{yxxxz}|T|^2 I/c}{1-n^2}\right\}\cos 2
\end{aligned} \tag{6.43}$$

where I is the intensity of the incident wave in vacuum.

One can conclude from equations (6.43) that no polarization azimuth change is produced by normal reflection from dissipation-less isotropic nonlinear media (where $\mathrm{Im}\{\chi_{xyyx}\} = 0$, $\mathrm{Im}\{\Gamma^{\mathrm{R}}_{yxxxz}\} = 0$ and $\mathrm{Im}\{\Gamma^{\mathrm{a}}\} = 0$), irrespective of whether it shows natural and nonlinear optical activity or polarization azimuth self-rotation for transmitted light. In dissipation-less isotropic nonlinear media all optical susceptibilities, as well as the refractive coefficient n, are real. Therefore the right-hand side of the first equation of (6.43) is zero. Polarization changes occur only in the ellipticity: for example, incident linearly polarized light will become elliptically polarized. (Compare with formulae (4.21) to (4.24) derived for polarization effects in transmission in an isotropic, gyrotropic, dissipation-less medium). However, natural and intensity-dependent rotation of the polarization azimuth can be seen with absorbing media whose susceptibilities have non zero imaginary parts: in the first equation of (6.43) the first term represents the reflective analogue of the polarization ellipse self-rotation effect (see Fig. 6.3);

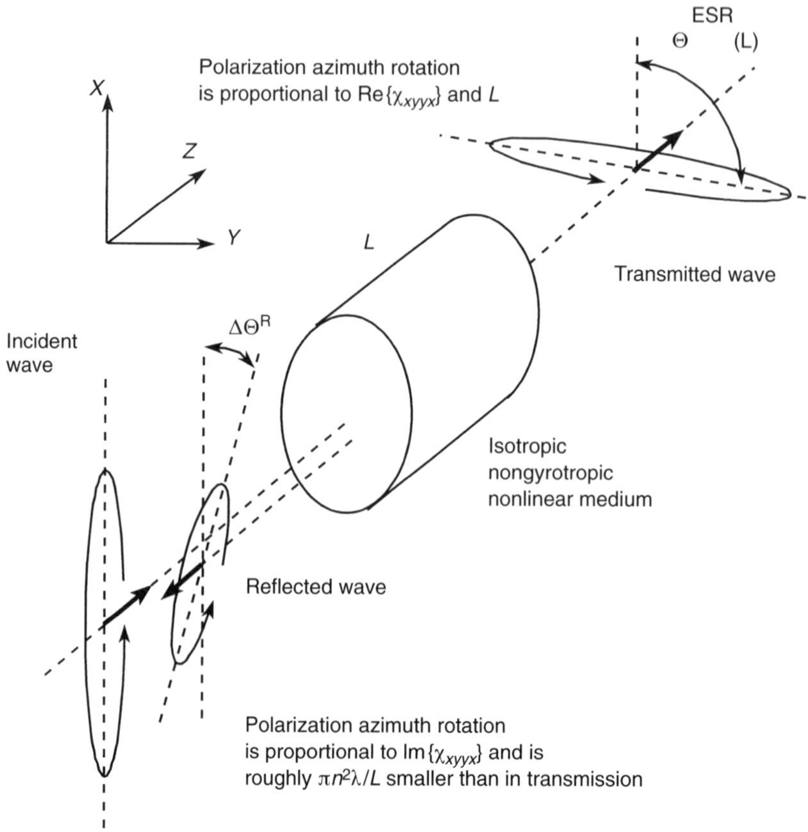

Fig. 6.3 Transmissive and reflective polarization azimuth self-rotation effects in a dissipative, isotropic, nongyrotropic nonlinear medium

the second term shows reflective analogues of linear and nonlinear optical activity. In isotropic non gyrotropic nonlinear media where $\Gamma^{(4)}_{ijk} = 0$ and $\Gamma^{(3)}_{ijken} = 0$ by symmetry, only the polarization ellipse self-rotation effect is present (see Fig. 6.3).

Finally, we will examine linear and nonlinear reflection, along the cubic axis [001], from crystals of the $\bar{4}3$m point group symmetry. The material tensors for this crystal class are listed in Appendix 4 and Chapters 4 and 3, see formulae (4.25) and (4.26). Therefore $\Omega^{\mathrm{R}}_{\alpha}$ may be calculated analogously to $\mathbf{\Omega}$ and Ω_0 as in equations (4.29):

$$\begin{aligned}
\Omega_1^{\mathrm{R}} &= \frac{4\pi|T|^2}{n}[(\chi_{xxxx} - \chi_{xxyy})S_1 + k(\Gamma^{\mathrm{R}}_{xxxyz} - \Gamma^{\mathrm{R}}_{xxyxz})S_3] \\
\Omega_2^{\mathrm{R}} &= \frac{4\pi}{n}[-ik\frac{\Gamma^{\mathrm{s}}}{2\pi} + (\chi_{xxyy} + \chi_{xyyx})|T|^2 S_2 + ik(\Gamma^{\mathrm{R}}_{yxxxz} + \Gamma^{\mathrm{R}}_{xxyxz})|T^2|S_0] \\
\Omega_3^{\mathrm{R}} &= \frac{4\pi|T|^2}{n}[(\chi_{xxyy} - \chi_{xyyx})S_3 + k(\Gamma^{\mathrm{R}}_{yxxxz} - \Gamma^{\mathrm{R}}_{xxyxz})S_1] \\
\Omega_0^{\mathrm{R}} &= \frac{4\pi|T|^2}{n}[(\chi_{xxxx} + \chi_{xxyy})S_0 + ik(\Gamma^{\mathrm{R}}_{xxxyz} + \Gamma^{\mathrm{R}}_{xxyxz})S_2]
\end{aligned} \tag{6.44}$$

Presuming that the light wave is initially linearly polarized,

$$S_1 = S_0 \cos 2\Theta, \qquad S_2 = S_0 \sin 2\Theta, \qquad S_3 = 0$$

we obtain, from equation (6.40), formulae for the polarization azimuth and ellipticity of the reflected wave (compare with equation (4.35)):

$$\begin{aligned}
\Delta\Theta^{\mathrm{R}} &= \mathrm{Re}\left\{\frac{\Delta\chi}{4n(1-n^2)}\right\}|T|^2\frac{8\pi^2 I}{c}\sin 4\Theta \\
&\quad - \frac{2\pi}{\lambda}\,\mathrm{Im}\left\{\frac{\Gamma^{\mathrm{s}} - 8\pi^2\Gamma^{\mathrm{R}}_{yxxxz}|T|^2 I/c}{1-n^2}\right\}\cos 2\Theta \\
\eta^{\mathrm{R}} &= \mathrm{Im}\left\{\frac{\Delta\chi}{4n(1-n^2)}\right\}|T|^2\frac{8\pi^2 I}{c}\sin 4\Theta \\
&\quad + \frac{2\pi}{\lambda}\,\mathrm{Re}\left\{\frac{\Gamma^{\mathrm{s}} - 8\pi^2\Gamma^{\mathrm{R}}_{yxxxz}|T|^2 I/c}{1-n^2}\right\}\cos 2\Theta
\end{aligned} \tag{6.46}$$

where I is the intensity of the incident wave in vacuum. Formulae (6.46) predict the reflective analogues of the nonlinear anisotropic effect and of natural and self-induced gyrotropic linear dichroism also see Appendix 4.

In developing the above treatment of reflective nonlinear polarization effects we tended to follow the same Stokes vector formalism that was used in Chapters 3 to 5. Other approaches to the theory of self-action polarization-sensitive reflective nonlinear effects, including those based on the direct calculations of the electric fields of the propagating waves, rather than of the Stokes parameters, are covered in papers [6] to [8].

As mentioned above, reflective polarization effects are much smaller than their transmissive counterparts. The observation of reflective natural optical activity at glazing incidence is relatively easy [9, 10]. However, the observation of reflective natural optical activity at normal incidence produced several contradictory results [11–13], with positive observation of reflective polarization azimuth rotation being performed in a hugely optical active crystal of α-HgS in the spectral range of strong absorption [11]. Reflective natural gyrotropic linear dichroism is also a small effect whose observation was experimental documented in Ref. [14] and discussed in Refs [17] and [18]. Observations of the effect in the vicinity of the exciton resonance were reported in [19]. To date the study of nonlinear reflective polarization self-action effects has not attracted much attention: there is only one paper known to us on the study of the reflective polarization azimuth rotation [15]. This lack of attention is due to the very considerable difficulties involved in measuring the effects experimentally. However, pump–probe nonlinear polarization effects have received much more attention due to their spectroscopic potential and use in time-resolved measurements. Moreover, pump–probe measurements can be done more accurately and easily than single beam measurements: modulation techniques, which are readily applicable to pump–probe measurements, greatly improve the polarization sensitivity and reliability of the experiments. Polarization-sensitive pump–probe reflective effects will be discussed in the next chapter.

REFERENCES

1. A.R. Bungay, Y.P. Svirko and N.I. Zheludev, *Phys. Rev. B*, **47**, 16141 (1993).
2. A.R. Bungay, Y.P. Svirko and N.I. Zheludev, *Phys. Rev. B*, **47**, 11730 (1993).
3. S.V. Popov, Yu.P. Svirko and N.I. Zheludev. *J. Opt. Soc. Am. B*, **13**, 2729 (1996).
4. J.A. Stratton, *Electromagnetic Theory*, McGraw-Hill, New York, 1991.
5. R.W. Ditchburn, *Light*, Academic Press, London, 1976.
6. A.D. Pertrenko and G.I. Trush, *Kristallografiya*, **37**, 159 (1992).
7. N.I. Zheludev, A.D. Pertrenko and G.I. Trush, *Kristallografiya*, **32**, 399 (1987).
8. A.A. Golubkov and V.A. Makarov, *J. Mod. Opt.*, **37**, 1531 (1990).
9. G.X. Zhou, J.M. Schmitt and C.E. Elliot, *Rev. Sci. Instrum.*, **64**, 2801 (1993).
10. M.P. Silverman, J. Badoz and B. Briat, *Opt. Lett.*, **17**, 886 (1992).
11. A.R. Bungay, Yu.P. Svirko and N.I. Zheludev, *Phys. Rev. Lett.*, **70**, 3039 (1993).
12. A.Yu. Lukianov and M.A. Novikov, *JETP Lett.*, **51**, 673 (1990).
13. T. Takizawa, *J. Phys. Soc. Japan*, **50**, 3054 (1981).
14. A.R. Bungay, S.V. Popov, Yu.P. Svirko and N.I. Zheludev, *Chem. Phys. Lett.*, **217**, 249 (1994).
15. Z.M. Kostov and N.I. Zheludev, *Opt. Lett.*, **13**, 640 (1988).
16. A.A. Golubkov and V.A. Makarov, *Uspekhi Fiz. Nauk*, **165**, 339 (1995).
17. D.J. Bottomley, *Opt. Lett*, **21**, 749 (1996).
18. D.F. Nelson and A.L. Ivanov, *Opt. Lett*, **23**, 86 (1998).
19. G. Mohs, M. Shirane, R. Shimano, M. Kuwata-Gonokami, S. Dhanjal, Yu. P. Svirko and N.I. Zheludev, *International Quantum Electronics Conference*, San. Francisco, California, May 3–8, 1998, Advanced Program, paper. QFC2, p. 138.

7 Pump–Probe Polarization-Sensitive Phenomena

In this chapter we will develop the theory of pump–probe polarization-sensitive effects in nonlinear media with local and nonlocal optical responses. We will also review the key experiments in this area. Equations for the polarization-plane azimuth rotation and ellipticity of the transmitted and reflected probe waves will be obtained and we will discuss several pump–probe nonlinear polarization phenomena, including the inverse Faraday effect, the optical Kerr effect, the nonlinear anisotropic effect, nonlinear optical activity and nonlinear nonlocal birefringence.

7.1 GENERAL THEORY OF REFLECTIVE AND TRANSMISSIVE PUMP–PROBE POLARIZATION EFFECTS

In previous chapters the scope of our consideration was limited to linear and self-action polarization-sensitive phenomena, i.e. essentially to one-wave effects. Spectroscopic applications, however, favour pump–probe phenomena. Using short optical pulses and pump–probe interactions one may study the time dynamics of nonlinear optical responses. By manipulating the frequencies of the pump and probe waves, the complex frequency dispersions of nonlinear susceptibilities can be measured. Importantly, it is often much easier to detect pump–probe nonlinear interactions than single-beam self-action effects: by modulating the pump intensity, one can unmistakably identify the intensity-dependent contribution to the nonlinear effect by detecting probe modulation at the same frequency. The study of pump–probe nonlinear polarization phenomena has attracted considerable attention. The effects have been widely studied experimentally and a number of key experiments in this field are reviewed below. Much effort has also been concentrated on the development of the theory of polarization-sensitive interactions between two waves in isotropic and crystalline media [1–6].

In this chapter we will examine the simplest, but probably the most important, case of polarization-sensitive interaction between two waves, where both waves are incident almost normally on the boundary between a vacuum and a nonlinear medium. We will find out how the pump wave affects the polarization state of the reflected and transmitted probe waves. This analysis will allow us to explain basic polarization sensitive effects and illustrate their potential in spectroscopy. We will compare the formulae for reflective and transmissive effects and introduce the idea of coherent and incoherent polarization-sensitive interactions. The theory will be developed for pump and probe waves of the same frequency.

However, some of the results may be applicable where the frequencies are different. In particular, the formulae for the incoherent components of the polarization effects can be used when the pump and probe are at different frequencies, but both within a spectral range where the frequency dispersions of the nonlinearity and the refractive index are negligible.

Let us consider co-propagating weak probe and strong pump waves with magnitudes $\mathbf{a}$ and $\mathbf{A}$ respectively, incident on the boundary between a vacuum and a nonlinear medium. The combined electric field of these waves is: $\mathbf{E} = (\mathbf{A} + \mathbf{a})\,\mathrm{e}^{-\mathrm{i}\omega t + \mathrm{i}\omega z/c} + \text{c.c.}$ We will denote the magnitudes of the reflected and transmitted waves with superscripts R and T. The electric field associated with the transmitted and reflected waves may be written as $\mathbf{E}^{\mathrm{R}} = (\mathbf{A}^{\mathrm{R}} + \mathbf{a}^{\mathrm{R}})\,\mathrm{e}^{-\mathrm{i}\omega t - \mathrm{i}\omega z/c} + \text{c.c.}$ and $\mathbf{E}^{\mathrm{T}} = (\mathbf{A}^{\mathrm{T}} + \mathbf{a}^{\mathrm{T}})\,\mathrm{e}^{-\mathrm{i}\omega t + \mathrm{i}kz} + \text{c.c.}$, where $|\mathbf{A}^{\mathrm{R,T}}| \gg |\mathbf{a}^{\mathrm{R,T}}|$ (see Fig 7.1).

The transmitted waves propagate in the nonlinear medium, their amplitudes is governed by equation (4.3) written for the total amplitude of the propagating wave:

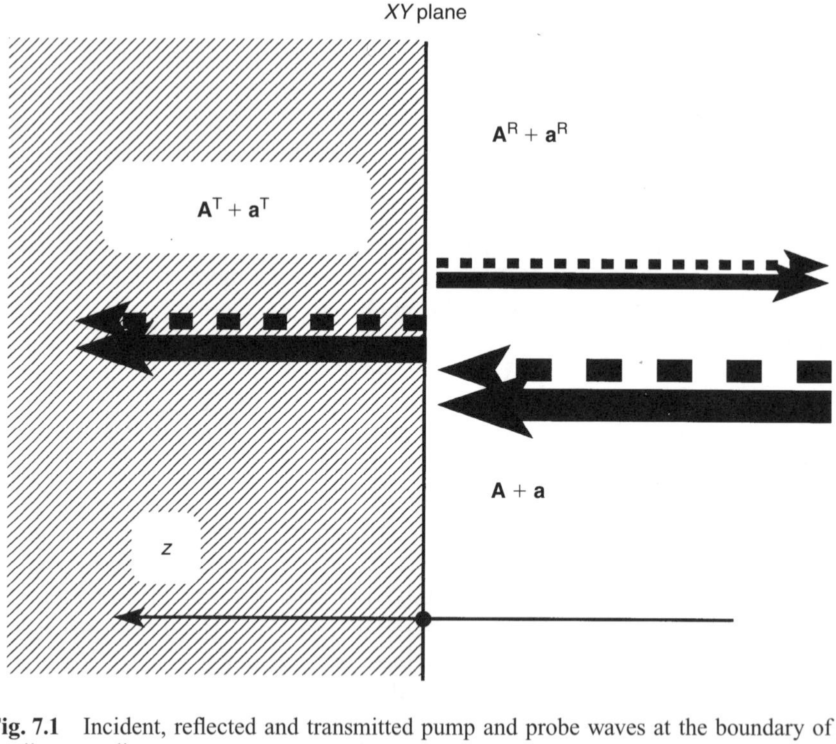

Fig. 7.1 Incident, reflected and transmitted pump and probe waves at the boundary of a nonlinear medium

$$\frac{\mathrm{d}(A_i^{\mathrm{T}} + a_i^{\mathrm{T}})}{\mathrm{d}z} = \frac{\mathrm{i}\omega}{2c}[u_{ij}(A_j^{\mathrm{T}} + a_j^{\mathrm{T}}) + \kappa_{ijkl}(A_j^{\mathrm{T}} + a_j^{\mathrm{T}})(A_k^{\mathrm{T}} + a_k^{\mathrm{T}})(A_l^{\mathrm{T}} + a_l^{\mathrm{T}})^* \,\mathrm{e}^{-2\mathrm{Im}\{k\}z}] \tag{7.1}$$

Since the probe is much weaker than the pump, we can neglect the influence of the probe on the propagation of the pump. As a result, from equation (7.1) we can be split out an equation for the amplitude of the pump wave by dropping the amplitude of the weak transmitted probe wave, $\mathbf{a}^{\mathrm{T}}$:

$$\frac{\mathrm{d}A_i^{\mathrm{T}}}{\mathrm{d}z} = \frac{\mathrm{i}\omega}{2c}[u_{ij}A_j^{\mathrm{T}} + \kappa_{ijkl}A_j^{\mathrm{T}}A_k^{\mathrm{T}}(A_l^{\mathrm{T}})^* \,\mathrm{e}^{-2\mathrm{Im}\{k\}z}] \tag{7.2}$$

Now, subtracting equation (7.2) from (7.1) and neglecting the second- and third-order terms in $\mathbf{a}^{\mathrm{T}}$, we arrive at an equation for the amplitude of the probe wave propagating in the nonlinear medium in the presence of the pump:

$$\frac{\mathrm{d}a_i^{\mathrm{T}}}{\mathrm{d}z} = \frac{\mathrm{i}\omega}{2c}[(u_{ij} + 2\kappa_{ijkl}A_k^{\mathrm{T}}A_l^{\mathrm{T}*} \,\mathrm{e}^{-2\mathrm{Im}\{k\}z})a_j^{\mathrm{T}} + (\kappa_{iklj}A_k^{\mathrm{T}}A_l^{\mathrm{T}} \,\mathrm{e}^{-2\mathrm{Im}\{k\}z})a_j^{\mathrm{T}*}] \tag{7.3}$$

To obtain formulae for pump-induced changes in the polarization state of the reflected and transmitted probe waves we need to derive the boundary condition for the amplitudes of the incident $\mathbf{a}$, transmitted $\mathbf{a}^{\mathrm{T}}$ and reflected $\mathbf{a}^{\mathrm{R}}$ probe waves. It follows from the boundary condition (6.11), for the tangential component of the electric field at the boundary, that

$$\mathbf{A} + \mathbf{a} + \mathbf{A}^{\mathrm{R}} + \mathbf{a}^{\mathrm{R}} = \mathbf{A}^{\mathrm{T}}|_{z=0} + \mathbf{a}^{\mathrm{T}}|_{z=0} \tag{7.4}$$

Similarly, it follows from equation (6.12), for the tangential component of the magnetic field at the boundary, that

$$\begin{aligned}(A_i + a_i) - (A_i^{\mathrm{R}} + a_i^{\mathrm{R}}) = {} & n(A_i^{\mathrm{T}} + a_i^{\mathrm{T}})|_{z=0} \\ & + \tfrac{1}{2}[\Delta u_{ij}(A_j^{\mathrm{T}} + a_j^{\mathrm{T}}) + \Delta\kappa_{ijkl}(A_j^{\mathrm{T}} + a_j^{\mathrm{T}})(A_k^{\mathrm{T}} + a_k^{\mathrm{T}})(A_l^{\mathrm{T}} + a_l^{\mathrm{T}})^*]|_{z=0}\end{aligned} \tag{7.5}$$

The comparative weakness of the probe allows us to neglect the effect of the probe on the boundary conditions for the pump. Therefore, for the pump we can use the boundary conditions already derived in Chapter 6, i.e. formulae (6.18) and (6.21). Subtracting (6.18) from (7.4) and (6.21) from (7.5) we obtain the boundary conditions for the probe wave:

$$\begin{aligned}\mathbf{a}^{\mathrm{T}}|_{z=0} - \mathbf{a}^{\mathrm{R}} - \mathbf{a} &= 0 \\ n\mathbf{a}^{\mathrm{T}}|_{z=0} + \mathbf{a}^{\mathrm{R}} - \mathbf{a} &= \Delta\mathbf{a}\end{aligned} \tag{7.6}$$

where

$$\Delta a_i = -\tfrac{1}{2}[\Delta u_{ij}a_j^{\mathrm{T}} + 2\Delta\kappa_{ijkl}A_k^{\mathrm{T}}(A_l^{\mathrm{T}})^* a_j^{\mathrm{T}} + \Delta\kappa_{iklj}A_k^{\mathrm{T}}A_l^{\mathrm{T}}(a_j^{\mathrm{T}})^*]|_{z=0} \tag{7.7}$$

If the nonlinearity and anisotropy are neglected, that is $\Delta u_{ij} = 0$ and $\Delta\kappa_{ijkl} = 0$, equations (7.6) become $\mathbf{a}^{\mathrm{R}} = R\mathbf{a}$ and $\mathbf{a}^{\mathrm{T}}|_{z=0} = T\mathbf{a}$, where $R = (1-n)/(1+n)$ and $T = 2/(1+n)$ are the Fresnel reflection and transmission coefficients for single-wave reflection (see equations (6.25)).

The presence of nonlinearity and nonlocality modifies the reflectivity and transmission of the interface. These modifications are generally small, and therefore may be presented as power series in Δu_{ij} and $\Delta\kappa_{ijkl}$. Retaining only the first terms of these series, we can write:

$$\begin{aligned} \mathbf{a}^{\mathrm{T}}|_{z=0} &= T\mathbf{a} - T^2\boldsymbol{\Delta}^{\mathrm{R}} \\ \mathbf{a}^{\mathrm{R}} &= R\mathbf{a} - T^2\boldsymbol{\Delta}^{\mathrm{R}} \end{aligned} \tag{7.8}$$

where we have introduced a new vectorial parameter:

$$\boldsymbol{\Delta}_i^{\mathrm{R}} = \Delta u_{ij}a_j + 2|T|^2\Delta\kappa_{ijkl}A_k(A_l)^*a_j + |T|^2\Delta\kappa_{iklj}A_kA_l(a_j)^* \tag{7.9}$$

The amplitude of the probe wave in the bulk of the nonlinear medium may be obtained by solving equation (7.3) with the initial condition for $\mathbf{a}^{\mathrm{T}}|_{z=0}$ given by equations (7.8). This is a rather labourious task; first we need to find $\mathbf{A}^{\mathrm{T}}|_{z=0}$ from equations (6.18) and (6.21). This value of $\mathbf{A}^{\mathrm{T}}|_{z=0}$ must then be used as the initial condition for equation (7.2). This will give $\mathbf{A}^{\mathrm{T}}(z)$, which must then be substituted into equation (7.3). Finally, the resulting set of differential equations for a_x, a_x^*, a_y and a_y^*, with z-dependent coefficients, must be solved. However, since the main use of this theory is in polarization-sensitive pump–probe spectroscopy, it is reasonable to expect that the polarization changes in the propagating probe beam will be small, that is $|\mathbf{a}^{\mathrm{T}}(z) - \mathbf{a}^{\mathrm{T}}(z=0)| \ll |\mathbf{a}^{\mathrm{T}}(z=0)|$. This assumption has practically no implications for the formulae for the reflected wave polarization parameters, which will be obtained later in this chapter (see formulae 25, 31, 34, 42, 47, 49–51 of this chapter). However, it does imply that similar formulae for the polarization parameters of the transmitted wave will be valid only when the changes in the polarization azimuth and ellipticity angle are small in comparison with π, irrespective of whether the changes are due to linear or nonlinear effects (see formulae 26, 32, 35, 43, 48 of this chapter).

If these limitations are accepted, the right-hand side of equation (7.3) can be taken at $z = 0$, and the equation solved by direct integration:

$$\mathbf{a}^{\mathrm{T}}(L) = \mathbf{a}^{\mathrm{T}}|_{z=0} + \left.\frac{\mathrm{d}\mathbf{a}^{\mathrm{T}}}{\mathrm{d}z}\right|_{z=0} L \tag{7.10}$$

The spatial derivatives of the probe wave amplitude at $z = 0$ are determined, here, by the right-hand side of equation (7.3):

$$\left.\frac{\mathrm{d}a_i^{\mathrm{T}}}{\mathrm{d}z}\right|_{z=0} = \frac{\mathrm{i}\omega}{2c}[(u_{ij} + 2\kappa_{ijkl}A_k^{\mathrm{T}}A_l^{\mathrm{T}*})a_j^{\mathrm{T}} + \kappa_{iklj}A_k^{\mathrm{T}}A_l^{\mathrm{T}}a_j^{\mathrm{T}*}]|_{z=0} \tag{7.11}$$

Since we are considering small corrections due to the anisotropy and nonlinearity, we can replace $\mathbf{A}^{\mathrm{T}}|_{z=0}$ with $T\mathbf{A}$ and $\mathbf{a}^{\mathrm{T}}|_{z=0}$ with $T\mathbf{a}$ in the right-hand side of equation (7.11). Here $T = 2/(1+n)$ is the amplitude transmission coefficient at the boundary. Thus, by resorting equations (7.11) and (7.10) we arrive at the formulae for the amplitudes of the reflected and transmitted probe waves:

$$\begin{aligned} \mathbf{a}^{\mathrm{T}}(L) &= T\mathbf{a} + iT\left(\frac{\pi L}{\lambda}\right)\mathbf{\Delta}^{\mathrm{T}} - \frac{T^2}{4}\mathbf{\Delta}^{\mathrm{R}} \\ \mathbf{a}^{\mathrm{R}} &= R\mathbf{a} - \frac{T^2}{4}\mathbf{\Delta}^{\mathrm{R}} \end{aligned} \tag{7.12}$$

where $\mathbf{\Delta}^{\mathrm{R}}$ is as defined in equation (7.9) and we have introduced

$$\mathbf{\Delta}_i^{\mathrm{T}} = u_{ij}a_j + 2|T|^2\kappa_{ijkl}A_k(A_l)^*a_j + |T|^2\kappa_{iklj}A_kA_l(a_j)^* \tag{7.13}$$

where $i = x$ or y. Since $\mathbf{\Delta}^{\mathrm{R}}$ and $\mathbf{\Delta}^{\mathrm{T}}$ are of the same order of magnitude, the last term $T^2\mathbf{\Delta}^{\mathrm{R}}$ in the first equation of (7.12) can be neglected when $L \gg (T/4\pi)\lambda$. In what follows we will calculate the polarization state of transmitted light for relatively thick nonlinear samples, i.e. samples for which $L \gg (T/4\pi)\lambda$.

As the amplitudes of the reflected and transmitted waves are now known, we can find from equations (7.12) the polarization azimuth Θ^{T} and the ellipticity angle η^{T} for the transmitted wave:

$$\begin{aligned} \Theta^{\mathrm{T}}(L) - \Theta &= \frac{\pi L}{\lambda s_0^2 \cos^2 2\eta}\,\mathrm{Im}\left\{J^{\mathrm{T}}\right\} \\ \eta^{\mathrm{T}}(L) - \eta &= -\frac{\pi L}{\lambda s_0^2 \cos 2\eta}\,\mathrm{Re}\left\{J^{\mathrm{T}}\right\} \end{aligned} \tag{7.14}$$

In this chapter the small letters s_0, s_1, s_2 and s_3 denote the Stokes parameters of the probe wave. For the reflected wave we obtain

$$\begin{aligned} \Theta^{\mathrm{R}} - \Theta &= \frac{1}{s_0^2 \cos^2 2\eta}\,\mathrm{Re}\left\{\frac{J^{\mathrm{R}}}{1-n^2}\right\} \\ \eta^{\mathrm{R}} - \eta &= \frac{1}{s_0^2 \cos 2\eta}\,\mathrm{Im}\left\{\frac{J^{\mathrm{R}}}{1-n^2}\right\} \end{aligned} \tag{7.15}$$

In equations (7.14) and (7.15) we have introduced

$$J^{\mathrm{T,R}} = (a_y\Delta_x^{\mathrm{T,R}} - a_x\Delta_y^{\mathrm{T,R}})(a_x^{*2} + a_y^{*2}) \tag{7.16}$$

Formulae (7.14) and (7.15) can be applied to any combination of the incident pump and probe wave polarization states. Equation (7.16) may be rewritten in terms of the material parameters u_{ij}, Δu_{ij}, κ_{ijkl} and $\Delta\kappa_{ijkl}$:

$$
\begin{aligned}
J^{\mathrm{T}} = &-\frac{1}{2}\left(u_{yy} - u_{xx} + \frac{4N}{|1+n|^2}\right)(s_0 s_2 + i s_1 s_3) \\
&-\frac{1}{2}\left(u_{yx} + u_{xy} + \frac{4N_+}{|1+n|^2}\right)(s_0 s_1 - i s_2 s_3) \\
&-\frac{1}{2}\left(u_{yx} - u_{xy} + \frac{4N_-}{|1+n|^2}\right)(s_0^2 - s_3^2)
\end{aligned} \tag{7.17}
$$

In equation (7.17),

$$
\begin{aligned}
N = &(\kappa_{yyxx} - \kappa_{xxxx} + \tfrac{1}{2}\kappa_{yxxy}\,\mathrm{e}^{\mathrm{i}\psi_{xy}} - \tfrac{1}{2}\kappa_{xxxx}\,\mathrm{e}^{\mathrm{i}\psi_{xx}})(S_0 + S_1) \\
&+ (\kappa_{yyyy} - \kappa_{xxyy} + \tfrac{1}{2}\kappa_{yyyy}\,\mathrm{e}^{\mathrm{i}\psi_{yy}} - \tfrac{1}{2}\kappa_{xyyx}\,\mathrm{e}^{\mathrm{i}\psi_{yx}})(S_0 - S_1) \\
&+ (\kappa_{yyxy} - \kappa_{xxxy} + \tfrac{1}{2}\kappa_{yxyy}\,\mathrm{e}^{\mathrm{i}\psi_{yy}} - \tfrac{1}{2}\kappa_{xxyx}\,\mathrm{e}^{\mathrm{i}\psi_{yx}})(S_2 - iS_3) \\
&+ (\kappa_{yyyx} - \kappa_{xxyx} + \tfrac{1}{2}\kappa_{yyxy}\,\mathrm{e}^{\mathrm{i}\psi_{xy}} - \tfrac{1}{2}\kappa_{xyxx}\,\mathrm{e}^{\mathrm{i}\psi_{xx}})(S_2 + iS_3)
\end{aligned} \tag{7.18}
$$

and

$$
\begin{aligned}
N_{\pm} = &(\kappa_{yxxx} \pm \kappa_{xyxx} + \tfrac{1}{2}\kappa_{yxxx}\,\mathrm{e}^{\mathrm{i}\psi_{xx}} \pm \tfrac{1}{2}\kappa_{xxxy}\,\mathrm{e}^{\mathrm{i}\psi_{xy}})(S_0 + S_1) \\
&+ (\kappa_{yxyy} \pm \kappa_{xyyy} + \tfrac{1}{2}\kappa_{yyyx}\,\mathrm{e}^{\mathrm{i}\psi_{yx}} \pm \tfrac{1}{2}\kappa_{xyyy}\,\mathrm{e}^{\mathrm{i}\psi_{yy}})(S_0 - S_1) \\
&+ (\kappa_{yxxy} \pm \kappa_{xyxy} + \tfrac{1}{2}\kappa_{yxyx}\,\mathrm{e}^{\mathrm{i}\psi_{yx}} \pm \tfrac{1}{2}\kappa_{xxyy}\,\mathrm{e}^{\mathrm{i}\psi_{yy}})(S_2 - iS_3) \\
&+ (\kappa_{yxyx} \pm \kappa_{xyyx} + \tfrac{1}{2}\kappa_{yyxx}\,\mathrm{e}^{\mathrm{i}\psi_{xx}} \pm \tfrac{1}{2}\kappa_{xyxy}\,\mathrm{e}^{\mathrm{i}\psi_{xy}})(S_2 + iS_3)
\end{aligned} \tag{7.19}
$$

where S_0, S_1, S_2 and S_3 denote the Stokes parameters of the pump wave. In the above equations $\psi_{ij} = 2[\arg(A_i) - \arg(a_j)]$ is twice the difference between the phases of the complex amplitudes of the pump wave's ith Cartesian component and the probe wave's jth Cartesian component (as usual, i and j denote the X and Yaxes of the laboratory frame). The parameters J^{R}, which are necessary for the calculation of the reflected wave's polarization state, are given by formulae identical to (7.17) to (7.19) but with u_{ij} and κ_{ijkl} replaced by Δu_{ij}, and $\Delta\kappa_{ijkl}$ from equations (6.22) and (6.23).

We will see shortly that polarization-sensitive interactions can be clearly divided into coherent and incoherent effects. Effects that do not depend on the relative phase of the pump and probe waves are incoherent effects. Their observation does not require the pump and probe to be mutually coherent. Formulae for the incoherent contributions to effects can be obtained if, in equations (7.18) and (7.19), the terms containing $\exp(i\psi_{ij})$ are dropped. Those effects that do depend on the phase difference between the pump and probe waves are known as coherent effects. Observation of coherent effects requires that the pump and probe are mutually coherent. Otherwise, phase fluctuations lead to a zero value for the time-averaged coherent term. From an experimental point of view it is always much easier to observe incoherent effects, rather than coherent ones.

To illustrate the difference between coherent and incoherent phenomena, let us consider wave propagation in an absorption-less, nongyrotropic, isotropic nonlinear medium. An intense pump wave $\mathbf{E}_\mathrm{p} = \mathbf{A}\cos(\omega t - kz + \zeta_\mathrm{p})$ changes the refractive index of the medium, by an amount $n^{\mathrm{nonlin}}\mathbf{E}_\mathrm{p}^2 = \frac{1}{2}n^{\mathrm{nonlin}}\mathbf{A}^2 + \frac{1}{2}n^{\mathrm{nonlin}}\mathbf{A}^2\cos(2\omega t - 2kz + 2\zeta_\mathrm{p})$, where n^{nonlin} is the so-called nonlinear refractive coefficient: $n^{\mathrm{nonlin}} = \pi\chi_{xxxx}/n$. The first term in this formula for the refractive index change does not depend on the space–time coordinate, and so it will contribute to a refractive index change for any probe wave, even one of a different frequency. This term is responsible for incoherent pump–probe interactions. Because the second term in the equation for pump-induced refractive index change depends on the space–time coordinate, it only affects probe waves propagating in the same direction as the pump, with the same frequency, that is a wave like $\mathbf{a}\cos(\omega t - kz + \zeta)$. Indeed, for such probe wave in this case that the nonlinear optical response has a spectral component at frequency ω and wave vector $\mathbf{k}$ in the Z direction.

$$\begin{aligned}\mathbf{P}^{\mathrm{NL}}(\omega) &= (\tfrac{1}{4}\pi)[n^{\mathrm{nonlin}}\mathbf{E}_\mathrm{p}^2][\mathbf{a}\cos(\omega t - kz + \zeta)] \\ &= (\tfrac{1}{8}\pi)n^{\mathrm{nonlin}}\mathbf{A}^2\mathbf{a}[\cos(\omega t - kz + \zeta) + \cos(\omega t - kz + \zeta_\mathrm{p})\cos(\zeta_\mathrm{p} - \zeta)]\end{aligned}$$

Note, that the magnitude of this spectral component depends on the phase difference ζ_p-ζ between the pump and probe. This nonlinear modulation adds to the coherent pump–probe interactions.

The main pump–probe polarization effects will be described below, following Ref. [7].

7.2 PUMP–PROBE POLARIZATION PHENOMENA DUE TO $\chi^{(3)}$

In this section we will examine pump–probe polarization-sensitive effects in isotropic media and cubic crystals, neglecting spatial dispersion. As we saw in the last chapter, normal reflection from an isotropic medium or a non-gyrotropic cubic crystal is not accompanied, in a linear regime, by any alteration of the polarization state. However, the presence of an intense pump may induce anisotropy in the medium which can be detected by a linearly polarized, reflected or transmitted probe wave. We presume that both the pump and probe are incident along the Z direction of the laboratory frame and that both the specular and transmissive effects are small: $\Theta^\mathrm{T} - \Theta$; $\Theta^\mathrm{R} - \Theta$; η^T and $\eta^\mathrm{R} \ll \pi$, where Θ denotes the initial orientation of the probe's polarization azimuth.

In isotropic media two major nonlinear pump–probe polarization effects are possible: the optical Kerr effect and the inverse Faraday effect. The optical Kerr effect is the high-frequency analogue of the quadratic electrooptic effect (known as the Kerr effect). In the conventional Kerr effect a static, or relatively slowly varying, electric field is applied to the sample perpendicular to the direction of light propagation, usually via a pair of flat, mutually parallel electrodes. In an isotropic sample this induces birefringence with the optical axis parallel to the direction of the static electric field strength vector (see, for example, the detailed description of the optical Kerr effect, in terms of cubic optical nonlinearities χ_{ijkl}, in Ref. [8]). The induced birefringence is quadratic in the applied field and therefore does not depend on its sign. The static electric field may be replaced with a rapidly oscillating field, for instance the field of an intense electromagnetic wave such as our pump wave. This is how the name 'optical Kerr effect' originated. The effect was first predicted by Buckingham [9]. In the pump–probe optical Kerr effect a weak, linearly polarized light wave acts as the probe, while the pump wave is a linearly polarized wave with a polarization azimuth at some angle $\Delta\Theta$ to the probe polarization (see Fig. 7.2). The effect is permitted in

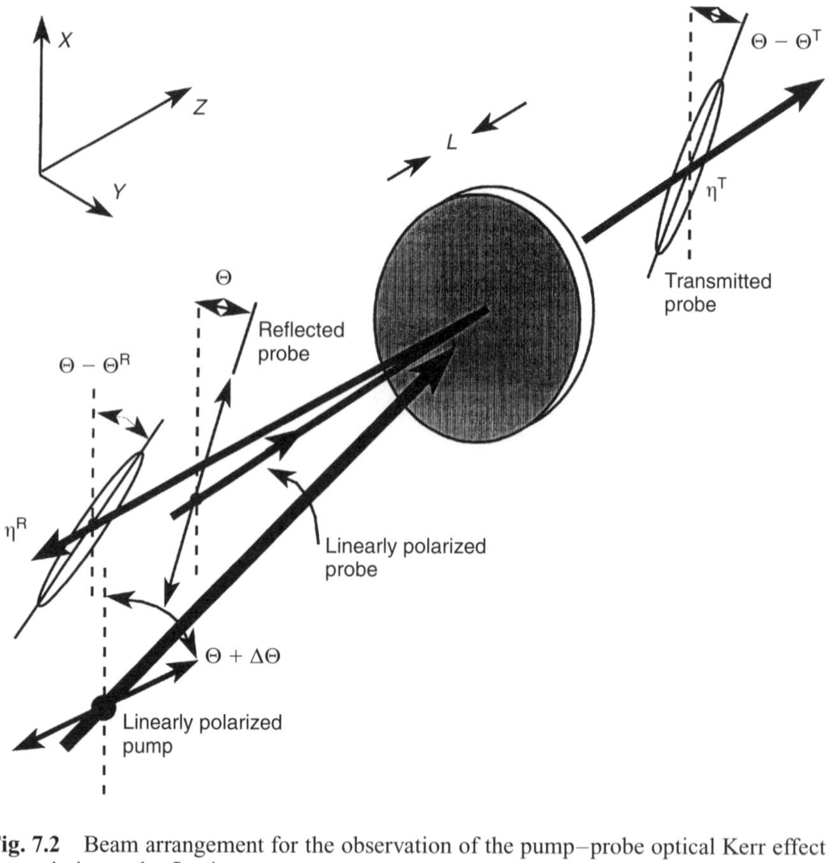

Fig. 7.2 Beam arrangement for the observation of the pump–probe optical Kerr effect in transmission and reflection

media of arbitrary symmetry and, in particular, isotropic nongyrotropic media where we will consider it in more details below. Since we are interested in polarization manifestations of the effect, the polarization changes, $\Theta^R - \Theta$, $\Theta^T - \Theta$, η^R and η^T, induced in the reflected and transmitted probe waves, will be its figures of merit.

In isotropic nongyrotropic media $\Gamma^{(1)}_{ijn} = 0$ and $\Gamma^{(3)}_{ijkln} = 0$; therefore $\Delta u_{ij} = u_{ij} = 0$ and $\kappa_{ijkl} = \Delta\kappa_{ijkl} = 4\pi\chi_{ijkl}/n$. Correspondingly, $J^R = J^T$. The local susceptibility tensor χ_{ijkl} has two independent components, namely χ_{xyyx} and χ_{xxyy}. The following nonzero components are relevant to our analysis:

$$\begin{aligned} \chi_{yxyx} &= \chi_{yyxx} = \chi_{yxyx} = \chi_{xxyy} \\ \chi_{yyyy} &= \chi_{xxxx} = 2\chi_{xxyy} + \chi_{xyyx} \end{aligned} \tag{7.20}$$

To use formulae (7.15) to (7.19) we need to specify the Stokes parameters of the incident probe and pump waves and the doubled phase differences ψ_{ij} between their Cartesian projections. In the Kerr effect the pump and probe are linearly polarized at an angle $\Delta\Theta$ to one another. Therefore for the probe $s = s_0\{1, \cos 2\Theta, \sin 2\Theta, 0\}$ and for the pump $S = S_0\{1, \cos 2(\Theta + \Delta\Theta), \sin 2(\Theta + \Delta\Theta), 0\}$. To find the phase differences between the Cartesian components of the pump and probe we first write down the amplitudes of the waves:

$$\begin{aligned} \mathbf{A} &= \sqrt{S_0}\{\cos\Theta, \sin\Theta\}\, e^{i\Phi} \\ \mathbf{a} &= \sqrt{s_0}\{\cos(\Theta + \Delta\Theta), \sin(\Theta + \Delta\Theta)\}\, e^{i\phi} \end{aligned} \tag{7.21}$$

Here Φ and ϕ are the phases of the pump and probe respectively. By examining equations (7.21) we find that the phase parameters ψ_{ij} appearing in parameters (7.18) and (7.19) are all the same:

$$\psi_{xx} = \psi_{xy} = \psi_{yx} = \psi_{yy} = 2(\Phi - \phi) \tag{7.22}$$

The nonlinear parameters N (7.18) and $N_\pm$ (7.19) can now be calculated as follows:

$$\begin{aligned} N &= -2S_0(\kappa_{xxyy} + \kappa_{xyyx} + \kappa_{xxyy}\, e^{2(\Phi-\phi)}) \cos 2(\Theta + \Delta\Theta) \\ N_+ &= 2S_0(\kappa_{xxyy} + \kappa_{xyyx} + \kappa_{xxyy}\, e^{2(\Phi-\phi)}) \sin 2(\Theta + \Delta\Theta) \\ N_- &= 0 \end{aligned} \tag{7.23}$$

Substituting equations (7.23) into (7.17) we obtain

$$J^R = J^T = -\frac{16\pi(\chi_{xxyy} + \chi_{xyyx} + \chi_{xxyy}\, e^{2i(\Phi-\phi)})}{n|1+n|^2} S_0 s_0^2 \sin 2\Delta\Theta \tag{7.24}$$

and consequently the polarization state of the reflected probe wave is given by

$$\begin{pmatrix} \Theta^{\mathrm{R}} - \Theta \\ \eta^{\mathrm{R}} \end{pmatrix} = -\frac{32\pi^2 I_{\mathrm{p}}}{c|1+n|^2} \begin{pmatrix} \mathrm{Re} \\ \mathrm{Im} \end{pmatrix} \left\{ \frac{\chi_{xxyy} + \chi_{xyyx} + \chi_{xxyy}\, \mathrm{e}^{2\mathrm{i}(\Phi-\phi)}}{n(1-n^2)} \right\} \sin 2\Delta\Theta \tag{7.25}$$

where $I_{\mathrm{p}} = cS_0/(2\pi)$ is the intensity of the pump wave approaching the interface between the vacuum and the nonlinear medium at normal incidence. Similarly, the probe wave polarization change, due to the optical Kerr effect in transmission, is given by

$$\begin{pmatrix} \Theta^{\mathrm{T}} - \Theta \\ \eta^{\mathrm{T}} \end{pmatrix} = -\frac{32\pi^3 L I_{\mathrm{p}}}{\lambda c|1+n|^2} \begin{pmatrix} \mathrm{Im} \\ -\mathrm{Re} \end{pmatrix} \left\{ \frac{\chi_{xxyy} + \chi_{xyyx} + \chi_{xxyy}\, \mathrm{e}^{2\mathrm{i}(\Phi-\phi)}}{n} \right\} \sin 2\Delta\Theta \tag{7.26}$$

The following main features of the optical Kerr effect in isotropic media may be deduced from formulae (7.25) and (7.26):

(a) The magnitudes of both the reflective and transmissive effects reach a maximum when the pump is polarized at $\Delta\Theta = 45°$ or $\Delta\Theta = -45°$ to the probe.
(b) Pump polarizations of $\Delta\Theta$ and $-\Delta\Theta$ to the probe polarization will give effects of opposite sign; the effect disappears when the pump is polarized parallel to the probe.
(c) In an absorption-less medium ($\mathrm{Im}\{n\} = 0$, $\mathrm{Im}\{\chi_{ijkn}\} = 0$) the reflective effect manifest itself as pump-induced polarization azimuth rotation and the transmissive effect as pump-induced ellipticity of the probe wave.
(d) In absorbing media one can expect to see both pump-induced ellipticity and pump-induced azimuth rotation in reflection and transmission.
(e) The transmissive effect grows with the sample thickness L. Assuming equal orders of magnitudes for the real and imaginary parts of the nonlinear susceptibilities, the reflective effect will be approximately $\lambda/(\pi \mathrm{n}^2 \mathrm{L})$ times smaller than the effect in transmission.
(f) In weakly absorbing materials with predominantly real refractive indexes, the real and imaginary parts of the same nonlinear susceptibilities are responsible for the reflective and transmissive effect.
(g) Both the transmissive and reflective effects have coherent components oscillating with Φ-ϕ, and incoherent components which are not sensitive to the phase difference between the pump and the probe.

For example, with nonlinearity $\chi_{xxyy} + \chi_{xyyx} \approx 10^{-11}$ esu, $n \approx 3$, $\lambda = 1\ \mu\mathrm{m} = 5 \times 10^{-5}$ cm, $L = 0.2$ cm and $I_{\mathrm{p}} = 1\ \mathrm{GW/cm^2} = 10^{13}\ \mathrm{erg/(s\,cm^2)}$, one can expect to see a reflective polarization azimuth rotation of approximately 3 μrad and induced ellipticity of the transmitted probe reaching 140 mrad. Since the magnitude of nonlinearities varies hugely from one material to another, from $\chi \sim 10^{-1}$ esu in some organics and excitonic crystals to $\chi \sim 10^{-15}$ in glasses, the optical Kerr effect (OKE) can vary by up to fourteen (!) orders of magnitude. The transmissive effect was first seen by Mayer and Gires in a long cell containing different organic liquids [10] (see Fig. 7.3) and by Bonch-Bruevich *et al.* at resonant conditions, in atomic potassium gas [11]. Kuwata first observed the optical Kerr effect in a reflective configuration [12]. He saw the effect in crystals of HgI_2 exploiting high nonlinearity at excitonic resonance (see Fig. 7.4). The reflective version of the effect is known as SOKE (specular optical Kerr effect) [7].

The second major pump–probe polarization-sensitive effect is the pump–probe inverse Faraday effect (IFE). Here a weak linearly polarized wave acts as the probe, while the pump is circularly polarized (see Fig.7.5). To explain the name we recall the conventional magnetooptical Faraday effect where a static magnetic field, applied along the direction of light propagation, induces circular birefringence and dichroism in isotropic nongyrotropic media. In turn, the medium may be magnetized optically by an intense circularly polarized light wave. This is the inverse Faraday effect [13]. Such magnetization was first demonstrated experimentally by Van der Ziel *et al.* in 1965 [14]. If this magnetization is probed optically, we arrive at the pump–probe inverse Faraday effect in the configuration presented in Fig. 7.4. This pump–probe effect was first discussed by Atkins and Miller [15]. If the pump induces circular birefringence/dichroism in the nonlinear medium a change in the polarization of the reflected and transmitted initially linearly polarized probe waves will result. The pump-induced polarization changes Θ^T-Θ, Θ^R-Θ, η^T and η^R in the transmitted and reflected probe waves are figures of merit of the effect. The inverse Faraday effect is permitted in media of arbitrary symmetry. We will examine, below, its main features in isotropic nongyrotropic media.

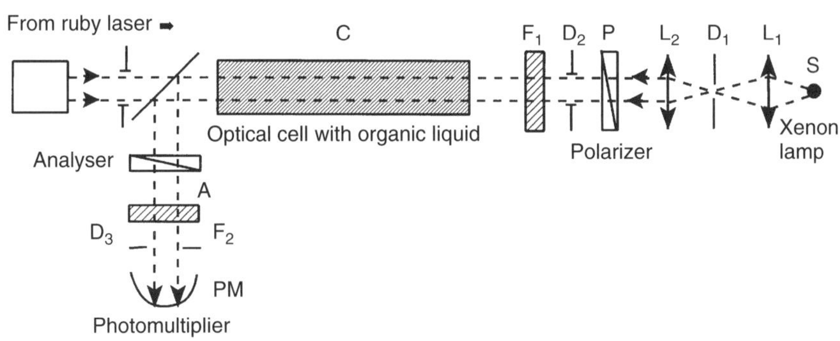

Fig. 7.3 First experiment on the optical Kerr effect. A xenon lamp S was used as the probe light source. A nanosecond pulsed ruby laser was used as a source of pump light contra-propagating with the probe. The effect was seen in several organic liquids including CS_2, $C_6H_5NO_2$ and C_6H_5COCl. The 19 cm long liquid cell C was placed between a polarizer P and an orthogonal analyser A; the optical system also included lenses L_1 and L_2, optical filters F_1 and F_2, photomultiplier PM and diaphragms D_1–D_3 (after Ref. [10])

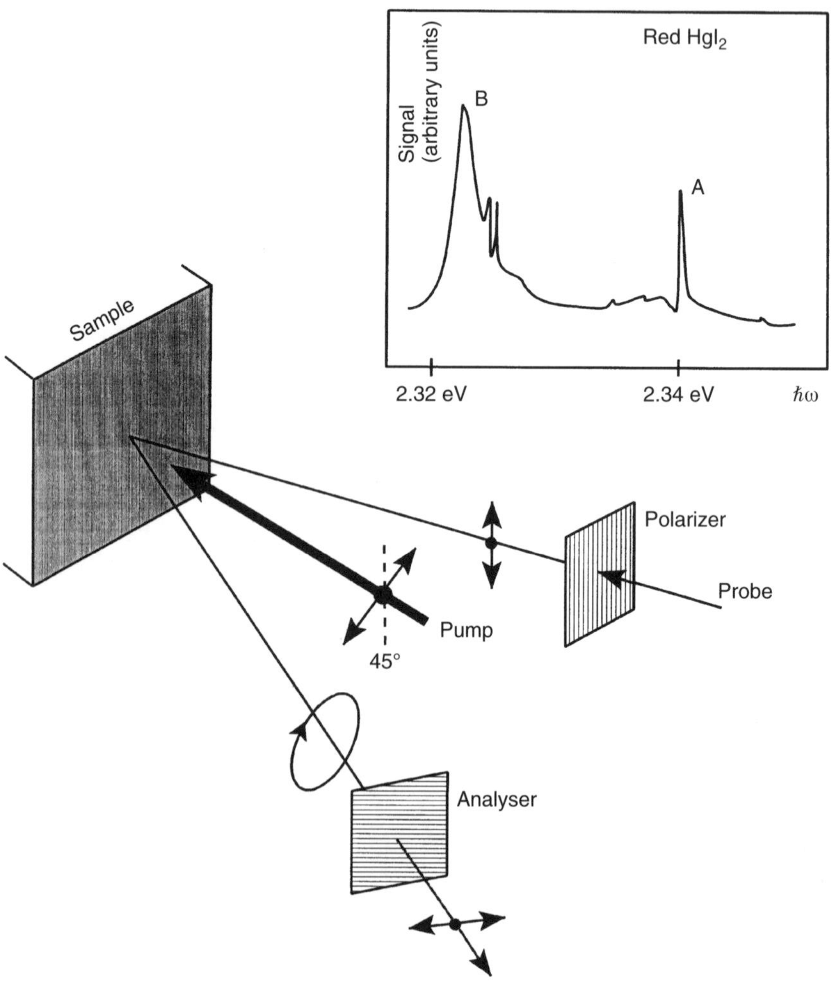

Fig. 7.4 First experimental observation of the specular optical Kerr effect in an HgI_2 crystal. The pump and probe waves were generated by two independent tunable lasers operating at frequencies ω_{pump} and ω_{probe}. The pump and probe beams were linearly polarized at 45° to each other. The intensity of the reflected probe beam was measured after it had passed through a crossed analyser. The spectrum in the inset represents the detected signal as a function of the frequency of the pump: resonance A corresponds to $\omega_{pump} = \omega_{probe}$, with the $\hbar\omega_{probe}$ fixed just above the longitudinal exciton energy in HgI_2; resonance B corresponds to the two-photon absorption resonance of excitonic molecules. (Reprinted from Ref [12] with kind permission of Elsevier Science)

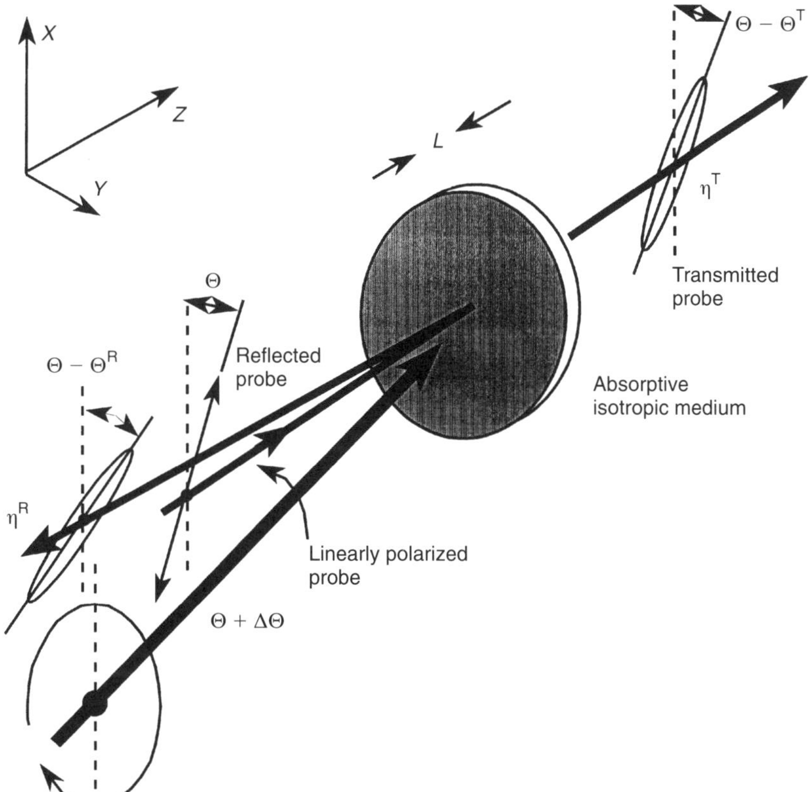

Fig. 7.5 Beam arrangement for the observation of the pump–probe inverse Faraday effect in transmission and reflection

In the inverse Faraday effect the pump wave is circularly polarized, that is $S = S_0\{1, 0, 0, \pm 1\}$ where + or − corresponds to the right- or left-handedness of the polarization. The probe wave is linearly polarized: $s = s_0\{1, \cos 2\Theta, \sin 2\Theta, 0\}$, where Θ refers to the initial orientation of the polarization azimuth with respect to the laboratory Cartesian frame. To define the phase difference between the pump and the probe we write down their amplitudes:

$$\begin{aligned} \mathbf{A} &= \sqrt{\frac{S_0}{2}}(1, \pm i)\,e^{i\Phi} \\ \mathbf{a} &= \sqrt{s_0}(\cos\Theta, \sin\Theta)\,e^{i\phi} \end{aligned} \tag{7.27}$$

where Φ and ϕ are the phases of the pump and probe respectively. From here we find that the phase parameters ψ_{ij} appearing in equations (7.18) and (7.19) are

$$
\begin{aligned}
\psi_{xx} &= \psi_{xy} = 2(\Phi - \phi) \\
\psi_{yx} &= \psi_{yy} = 2(\Phi - \phi) + \pi
\end{aligned}
\tag{7.28}
$$

By substituting equations (7.28) into (7.18) and (7.19) we obtain

$$
\begin{aligned}
N &= -2S_0\kappa_{xxyy}\,\mathrm{e}^{2(\Phi-\phi)} \\
N_+ &= \pm 2iS_0\kappa_{xxyy}\,\mathrm{e}^{2(\Phi-\phi)} \\
N_- &= \pm 2iS_0(\kappa_{xxyy} - \kappa_{xyyx})
\end{aligned}
\tag{7.29}
$$

Now, taking the nonzero components of the nonlinearity tensors (7.20) we obtain

$$
J^{\mathrm{T}} = J^{\mathrm{R}} = \mp i\frac{16\pi(\chi_{xxyy} - \chi_{xyyx} + \chi_{xxyy}\,\mathrm{e}^{2\mathrm{i}(\Phi-\phi\pm\Theta)})}{n|1+n|^2}S_0 s_0^2 \tag{7.30}
$$

Consequently, the polarization azimuth and ellipticity of the reflected probe wave will be given by the following formula:

$$
\begin{pmatrix} \Theta^{\mathrm{R}} - \Theta \\ \eta^{\mathrm{R}} \end{pmatrix} = \pm\frac{32\pi^2 I_{\mathrm{p}}}{c|1+n|^2}\begin{pmatrix} \mathrm{Im} \\ -\mathrm{Re} \end{pmatrix}\left\{\frac{\chi_{xxyy} - \chi_{xyyx} + \chi_{xxyy}\,\mathrm{e}^{2\mathrm{i}(\Phi-\phi\pm\Theta)}}{n(1-n^2)}\right\} \tag{7.31}
$$

where $I_{\mathrm{p}} = cS_0/(2\pi)$ is the intensity of the pump wave approaching the boundary between the nonlinear and the vacuum, and the plus and minus signs refer to right- and left-handed circular polarizations of the pump respectively. Similarly, the inverse Faraday effect in transmission is given by

$$
\begin{pmatrix} \Theta^{\mathrm{T}} - \Theta \\ \eta^{\mathrm{T}} \end{pmatrix} = \pm\frac{32\pi^3 L I_{\mathrm{p}}}{c\lambda|1+n|^2}\begin{pmatrix} \mathrm{Re} \\ \mathrm{Im} \end{pmatrix}\left\{\frac{\chi_{xxyy} - \chi_{xyyx} + \chi_{xxyy}\,\mathrm{e}^{2\mathrm{i}(\Phi-\phi\pm\Theta)}}{n}\right\} \tag{7.32}
$$

The following main features of the inverse Faraday effect in isotropic media may be deduced from formulae (7.31) and (7.32):

(a) The sign of the effect is opposite for right- and left-handed circular polarizations of the pump.
(b) In an absorption-less medium frequency dispersion may be neglected: $\chi_{xxyy} = \chi_{xyyx}$. Therefore, incoherent reflective and transmissive effects vanish. Induced ellipticity in reflected light is seen due to the coherent effect only. The coherent effect gives polarization azimuth rotation for the transmitted light.
(c) In weakly absorbing materials, with predominantly real refractive indices, the real and imaginary parts of the same nonlinear susceptibilities are responsible for the reflective and transmissive effects.
(d) Absorbing media are much more favourable for observation of the effect. In this media $\chi_{xxyy} \neq \chi_{xyyx}$ and one can expect to see pump-induced ellipticity

and pump-induced azimuth rotation due to the incoherent effect, both in reflection and transmission.

(e) The transmissive effects grows with the sample thickness L. Assuming equal orders of magnitude for the real and imaginary parts of the nonlinear susceptibilities, the reflective effect will be approximately $\lambda/(\pi n^2 L)$ times smaller than the transmissive effect;

In a dissipative medium with nonlinearity $\chi_{xxyy} - \chi_{xyyx} \approx 10^{-11}$ esu, $n \approx 3$, $\lambda = 0.5\ \mu m = 5 \times 10^{-5}$ cm and pump intensity $I_p = 1\ GW/cm^2 = 10^{16} erg/(s\ cm^2)$, one can expect to see reflective polarization azimuth rotation of several μrad. Since the magnitude of nonlinearities varies hugely from one material to another, the pump–probe inverse Faraday effect can vary by many orders of magnitude.

The pump–probe inverse Faraday effect was first seen by Arutunyan *et al.* [16]. It was observed in atomic potassium vapour (see Fig. 7.6). With counter-propagating pump and probe beams, the pump–probe inverse Faraday effect forms as a very powerful tool for the Doppler-free spectroscopy of gases [17]. The effect is also useful in the spectroscopy of semiconductors [18,19]. The reflective pump–probe inverse Faraday effect was first observed by Kuwata [12].

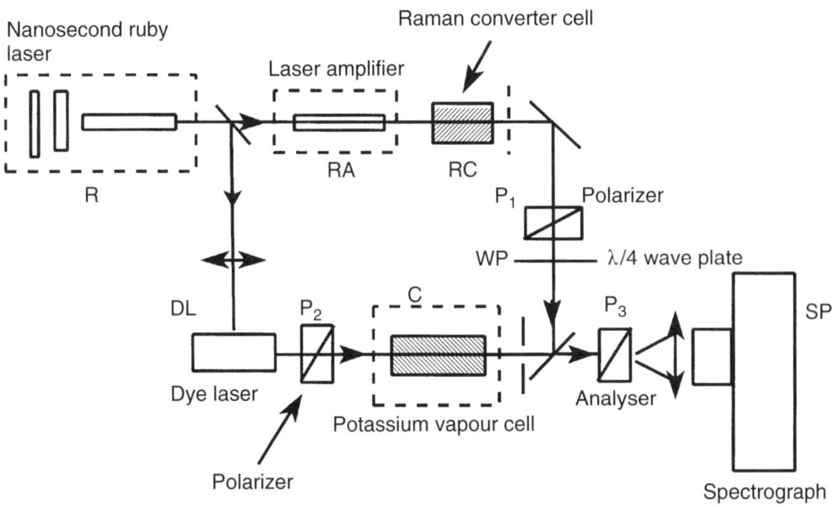

Fig. 7.6 First experimental observation of the inverse optical Faraday effect in potassium vapour. Light from a nanosecond ruby laser R, with amplifier RA and nitrobenzol-stimulated Raman effect converter RC, was circularly polarized by a polarizer P_1 and a quarter-wave plate WP. The probe beam was generated by a dye laser DL and linearly polarized by polarizer P_2. The potassium vapour cell C was placed between polarizer P_2 and analyser P_3. The orthogonal polarization components of the polarimeter signal were detected photographically via a spectrograph SP (Copyright American Institute of Physics, reproduced with permission from Ref. [16])

It is now also known as the specular inverse Faraday effect (SIFE) [7]. Popov *et al.* found that the SIFE has coherent and incoherent contributions [20]. Using modulation techniques and femtosecond lasers, the specular inverse Faraday effect is readily observable in many materials. It is very useful for spectroscopy of the dynamics of cubic nonlinearities in opaque materials such as semiconductors [12,21], superconductors [22] and metals [23] (see Fig. 7.7).

Pump–probe interaction in media with local cubic optical nonlinearities can also lead to the two-beam analogue of the nonlinear anisotropic polarization effect (see Chapter 4). The pump–probe nonlinear anisotropic polarization effect can be observed in cubic crystals and along fourfold symmetry axis in crystals of lower symmetry (see Fig. 7.8). For easy observation of the effect is forbidden in isotropic media or along isotropic directions. To observe the effect the pump and probe waves must be linearly polarized in exactly the same direction: $s = s_0\{1, \cos 2\Theta, \sin 2\Theta, 0\}$ and $S = S_0\{1, \cos 2\Theta, \sin 2\Theta, 0\}$. Here Θ is the angle which the pump and probe polarization azimuths make with the [100] direction of the crystal (see Fig. 7.7). The pump–probe nonlinear anisotropic polarization effect does not interfere with the pump–probe optical Kerr and inverse Faraday effects. Indeed, the inverse Faraday effect disappears with a linearly polarized pump wave and the optical Kerr effect vanishes when the pump is polarized in exactly the same direction as the probe. The pump-induced changes in the reflected probe beam are the figures of merit of the effect. The effect can be seen in reflected and transmitted light and may be treated in the same coordinate system as was used for the analogous self-action phenomenon (see Fig. 4.9).

When a light wave propagates along a fourfold symmetry axis in a nongyrotropic crystal, the tensor of local nonlinear susceptibility has three independent components, χ_{xxxx}, χ_{xxyy} and χ_{xyyx}, relevant to our consideration. From equation (7.17) we obtain

$$J^{\mathrm{T}} = J^{\mathrm{R}} = \frac{8\pi\Delta\chi(1 + \frac{1}{2}\mathrm{e}^{2\mathrm{i}(\Phi-\phi)})}{n|1+n|^2} S_0 s_0^2 \sin 4\Theta \qquad (7.33)$$

where, exactly as in Chapter 4, formula (4.34), we have introduced the anisotropic part of the cubic nonlinear susceptibility $\Delta\chi = \chi_{xxxx} - 2\chi_{xxyy} - \chi_{xyyx}$. Now, by resorting equations (7.14) and (7.15) we obtain the following formulae for the polarization state of the transmitted and reflected probe waves:

$$\begin{pmatrix} \Theta^{\mathrm{R}} - \Theta \\ \eta^{\mathrm{R}} \end{pmatrix} = \frac{16\pi^2 I_{\mathrm{p}}}{c|1+n|^2} \begin{pmatrix} \mathrm{Re} \\ \mathrm{Im} \end{pmatrix} \left\{ \frac{\Delta\chi}{n(1-n^2)} (1 + \tfrac{1}{2}\mathrm{e}^{2\mathrm{i}(\Phi-\phi)}) \right\} \sin 4\Theta \qquad (7.34)$$

$$\begin{pmatrix} \Theta^{\mathrm{T}} - \Theta \\ \eta^{\mathrm{T}} \end{pmatrix} = \frac{16\pi^3 L I_{\mathrm{p}}}{\lambda c|1+n|^2} \begin{pmatrix} \mathrm{Im} \\ -\mathrm{Re} \end{pmatrix} \left\{ \frac{\Delta\chi}{n} (1 + \tfrac{1}{2}\mathrm{e}^{2\mathrm{i}(\Phi-\phi)}) \right\} \sin 4\Theta \qquad (7.35)$$

The following main features of the pump–probe specular nonlinear anisotropic effect may be deduced from formulae (7.34) and (7.35):

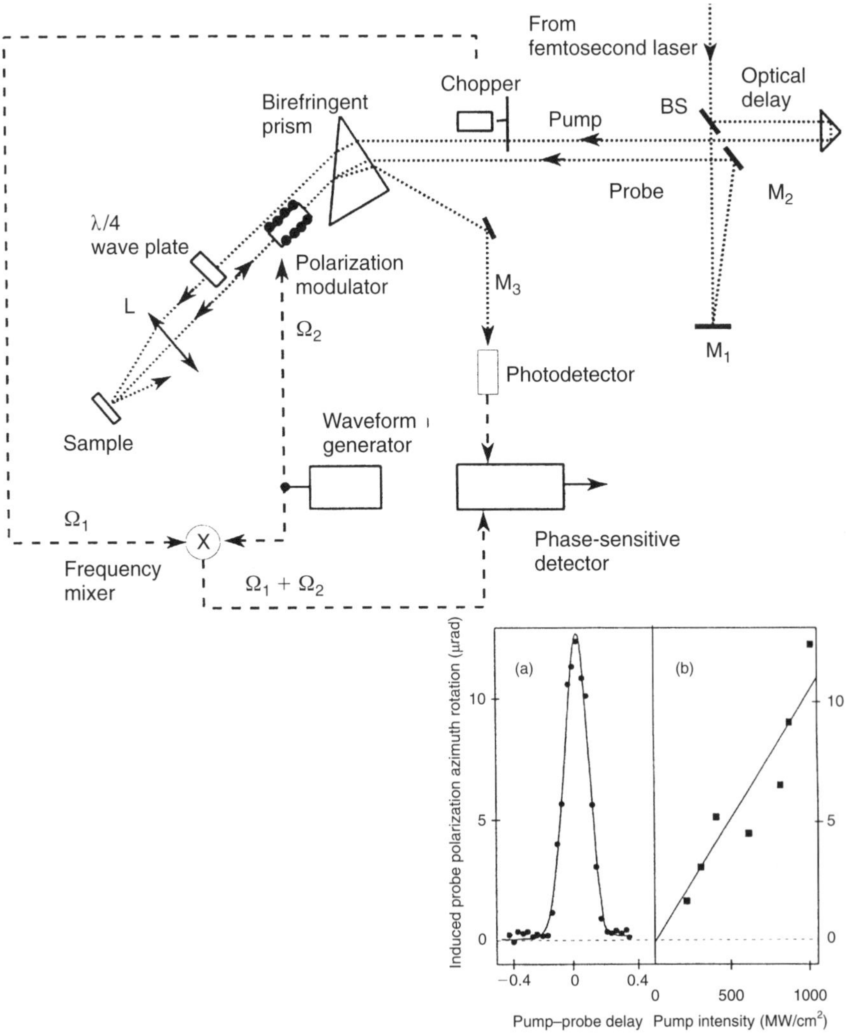

Fig. 7.7 Femtosecond micropolarimeter for time-resolved study of the specular inverse Faraday effect. The polarimeter was based on a birefringent calcite polarization prism. The incident probe light polarization was modulated by the magnetooptical polarization modulator, typically with an amplitude of 10^{-2} rad at a frequency Ω_2 of several kilohertz. The pump beam was modulated in the kilohertz range at frequency Ω_1 and circularly polarized by a $\lambda/4$-wave plate. The reference frequency $\Omega_1 + \Omega_2$ was electronically synthesized. The spectral component of the current of the signal photodetector at frequency $\Omega_1 + \Omega_2$ was proportional to the pump-induced component of the reflected probe polarization azimuth rotation. The inset shows the incoherent SIFE rotation in metallic gold as measured with a femtosecond Cr:forsterite laser at $\lambda = 1260$ nm versus (a) pump–probe delay ($I \sim 1$ GW/cm^2) and (b) pump intensity at zero pump–probe delay (Reproduced by permission of the Optical Society of America from Ref. [23])

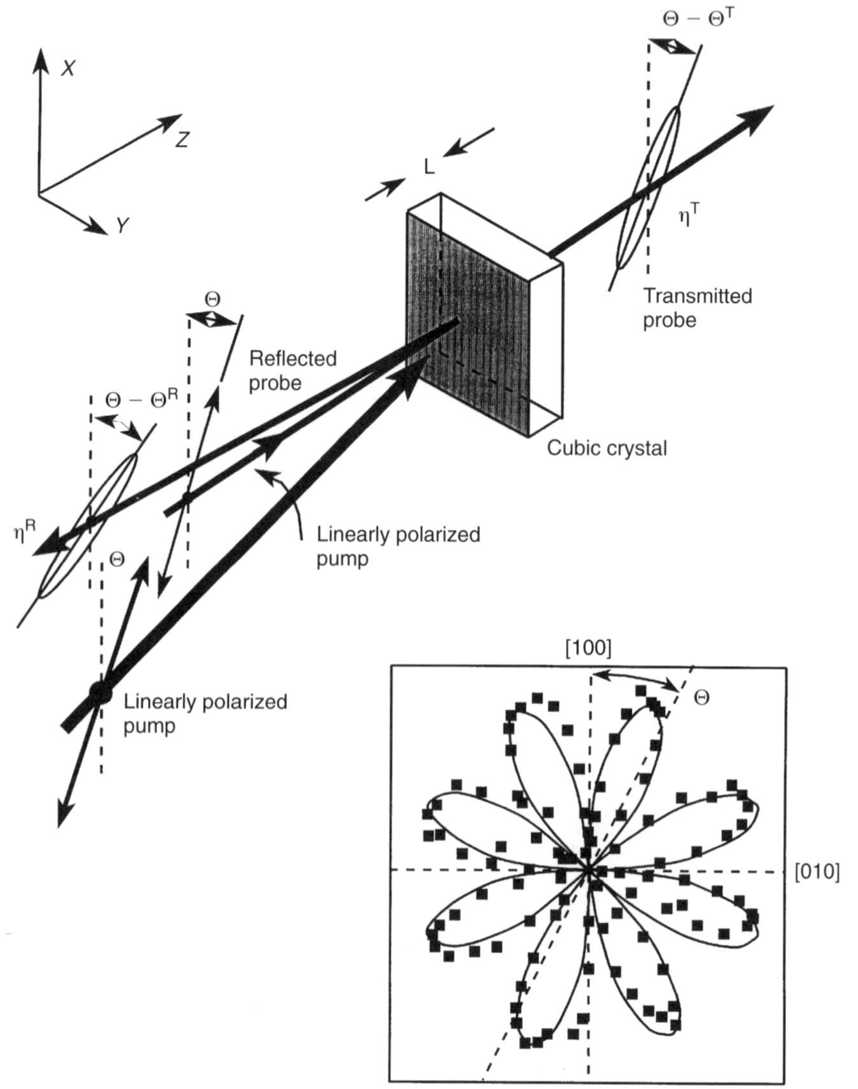

Fig. 7.8 Beam arrangement for the observation of the pump–probe nonlinear anisotropic effect in transmission and reflection. The inset shows the results of the first experimental observation of the specular nonlinear anisotropic effect (SNAPE) in GaAs: the dependence of $\Theta^R - \Theta$ on Θ is presented in polar coordinates. A Ti:sapphire picosecond mode-locked laser was used as a source for both the pump and probe beams (Reproduced by permission of the Optical Society of America from Ref. [25])

(a) The magnitudes of both the reflective and transmissive effects reach a maximum if the pump is initially polarized at $\Theta = \pi/8 + \pi K/4$ to the [100] direction of the cubic crystal, where K is an integer.
(b) The sign of the effect will be opposite for incident waves making angles of Θ and $-\Theta$ to the [100] direction. The effect disappears when the incident waves are polarized along the crystal cubic symmetry axes [100] and [010] or along the bisectors between them.
(c) In an absorption-less crystal ($\mathrm{Im}\{n\} = 0$, $\mathrm{Im}\{\chi_{ijkn}\} = 0$) the reflective effect manifests itself as pump-induced polarization azimuth rotation and the transmissive effect as pump-induced ellipticity of the probe wave.
(d) In weakly absorbing materials, with predominantly real refractive indices, the real and imaginary parts of the same nonlinear susceptibilities are responsible for the reflective and transmissive effects.
(e) In absorbing media one can expect to see both pump-induced ellipticity and pump-induced azimuth rotation in reflection and transmission.
(f) The transmissive effect grows with the sample thickness L. Assuming an equal order of magnitude for the real and imaginary parts of the nonlinear susceptibilities, the reflective effect will be approximately $\lambda/(\pi \mathrm{n}^2 \mathrm{L})$ times smaller than the transmissive effect.
(g) Both transmissive and reflective effects have coherent components oscillating with Φ-ϕ and incoherent components that are not sensitive to the phase difference between the pump and the probe. The coherent and incoherent contributions depend on the same combination of components of the cubic optical nonlinearity tensor.
(h) The effect vanishes completely in isotropic media where $\Delta\chi = 0$.

Fig. 7.9(*overleaf*) Modulation spectrometer for measuring the pump–probe nonlinear anisotropic effect in a cubic crystal of GaAs. A CW mode-locked laser was used as an optical source for both the pump and probe beams. The crystal sample C was rapidly rotated by motor AM with angular frequency f. The rotating sample was placed between polarizer P and analyser A. The pump and the probe beam were identically linearly polarized by the same polarizing prism P and focused into the sample by a lens L. The rotation of the crystal causes the polarization azimuth of the probe beam to oscillate with frequency $4f$, as predicted by formula (7.35). The oscillation of the probe beam intensity was detected by a signal photodetector PD_1 and a phase-sensitive amplifier SD locked at frequency $4f$ with the signal from the reference photodetector $\mathrm{PD}_{\mathrm{ref}}$. The reference signal was generated by interrupting the light from a bulb with a four-slot rotating sample holder. The optical scheme included a computer-controlled pump beam chopper CH and variable pump–probe delay lines. The signal was measured as a function of the angle β between the polarizer P and analyser A. The slope of this dependence is proportional to the amplitude of the nonlinear anisotropic effect. The inset shows this dependence (curve a) for the self-action azimuth rotation as measured only with a probe beam of intensity 4.2 $\mathrm{MW/cm^2}$. Curve b shows a ninefold increased rotation in the presence of a pump beam of intensity approximately 40 $\mathrm{MW/cm^2}$ (after Ref. [24])

The transmissive pump–probe nonlinear anisotropic polarization effect was first seen by Apanasevich *et al.* [24] employing an original modulation technique which used a rotating crystal to exploit the $\sin 4\Theta$ dependence of the effect (see Fig. 7.9). Bungay *et al.* were the first to observe the pump–probe nonlinear anisotropic polarization effect in the reflective configuration [25] (see the inset in Fig. 7.8). The reflective version of the effect is also known as the pump–probe specular nonlinear anisotropic effect (SNAPE) (see Ref. [7]).

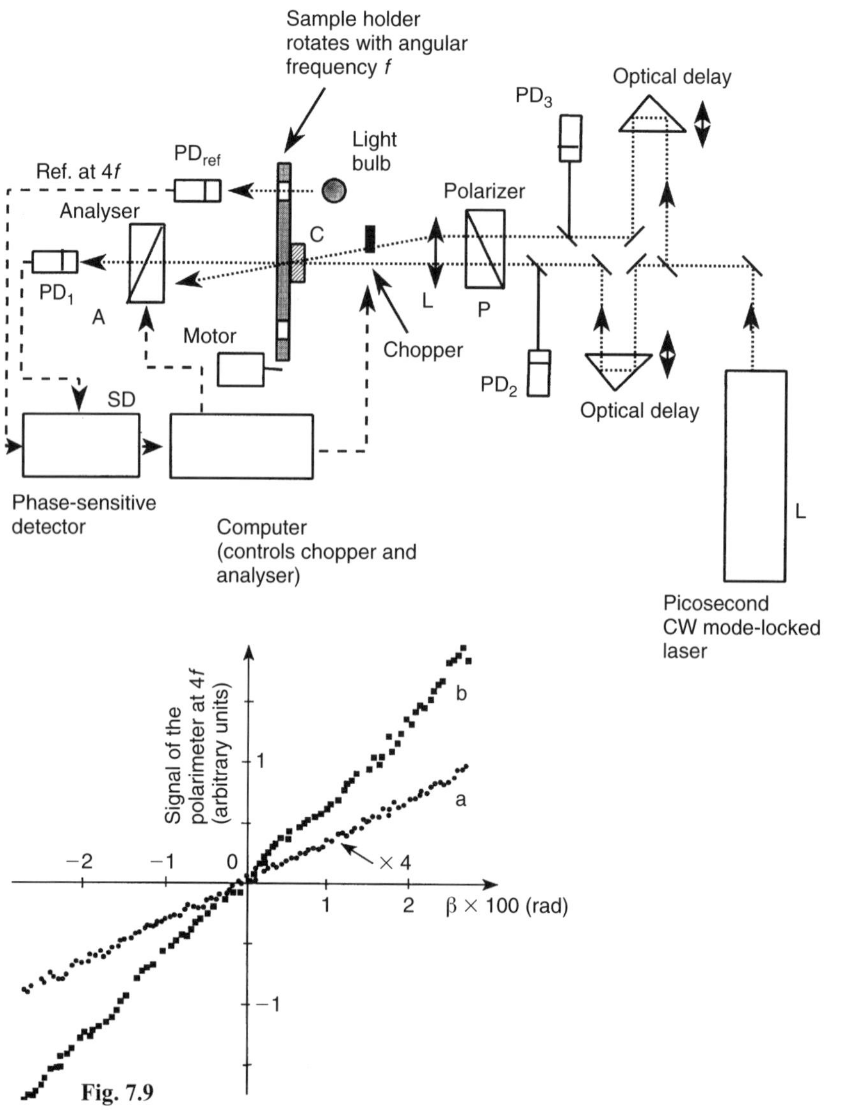

Fig. 7.9

7.3 NONLINEAR PUMP–PROBE POLARIZATION PHENOMENA DUE TO NONLOCALITY

In this section we will analyse pump-induced polarization effects in transmission and reflection which are due to nonlocal nonlinearity. In media with a nonlocal response they appear against a background of natural polarization effects. We will find additional pump-induced contributions to such effects polarization-sensitive effects in nonlinear nonlocal media. We will consider both reflective and transmissive configurations and examine the case of a linearly polarized pump wave in detail.

Pump–probe nonlinear optical activity (NLOA) can be seen in liquids and along isotropic directions in gyrotropic crystals. In this case a strong pump wave changes the circular birefringence and dichroism of a naturally optically active medium while a weak linearly polarized light wave acts as a probe. Let us consider the case when the polarization azimuth of the pump coincides with that of the incident probe, $s = s_0\{1, \cos 2\Theta, \sin 2\Theta, 0\}$ and $S = S_0\{1, \cos 2\Theta, \sin 2\Theta, 0\}$, where Θ is the angle with respect to the laboratory frame. Since the medium is isotropic, the pump–probe nonlinear anisotropic polarization effect considered in the previous section does not occur. The contributions of the inverse Faraday and Kerr effects also vanish (the pump–probe inverse Faraday effect requires a circularly polarized pump while the pump–probe optical Kerr effect is only seen when the probe is polarized at an angle to the pump). This leaves natural and nonlinear optical activity to be the only polarization-sensitive effects observable in isotropic gyrotropic media with co-propagating pump and probe beams. For isotropic media $\Gamma^{(1)}_{ijk} = \Gamma^{\mathrm{a}} \mathrm{e}_{ijk}$. Thus, by resorting equations (3.43) and (6.22) we obtain

$$\begin{aligned} u_{xy} &= -u_{yx} = \frac{\mathrm{i}\omega}{c}\Gamma^{\mathrm{a}} \\ \Delta u_{xy} &= -\Delta u_{yx} = -\frac{\mathrm{i}\omega}{c}\Gamma^{\mathrm{a}} \end{aligned} \tag{7.36}$$

The nonzero components of the nonlinear nonlocal response tensor Γ_{ijkln} which are essential for our consideration, are given by (4.12), while those of the tensor $\Gamma^{\mathrm{R}}_{ijkln}$ are

$$\begin{aligned} \Gamma^{\mathrm{R}}_{xyxxz} &= \Gamma^{\mathrm{R}}_{xxyxz} = -\Gamma^{\mathrm{R}}_{yyxyz} = -\Gamma^{\mathrm{R}}_{yxyyz} \\ \Gamma^{\mathrm{R}}_{yyyxz} &= -\Gamma^{\mathrm{R}}_{xxxyz} = \Gamma^{\mathrm{R}}_{yxxxz} + \Gamma^{\mathrm{R}}_{xyxxz} + \Gamma^{\mathrm{R}}_{xxyxz} \\ \Gamma^{\mathrm{R}}_{xyyyz} &= -\Gamma^{\mathrm{R}}_{yxxxz} = \Gamma^{\mathrm{R}}_{xyxxz} + \Gamma^{\mathrm{R}}_{xxyxz} + \Gamma^{\mathrm{R}}_{xxxyz} \end{aligned} \tag{7.37}$$

With these tensor components the parameters κ_{ijkl} and $\Delta\kappa_{ijkl}$ responsible for the reflective and transmissive nonlinear interactions are calculated as follows (see formulae (4.4) and (6.23)):

$$\begin{aligned}\kappa_{ijkl} &= 4\pi\left(\frac{\chi_{ijkl}}{n} + \frac{\mathrm{i}\omega}{c}\Gamma_{ijklz}\right)\\ \Delta\kappa_{ijkl} &= 4\pi\left(\frac{\chi_{ijkl}}{n} + \frac{\mathrm{i}\omega}{c}\Gamma^{\mathrm{R}}_{ijklz}\right)\end{aligned} \tag{7.38}$$

Here coefficients (7.18) and (7.19) will be

$$\begin{aligned} N &= -\frac{8\pi\mathrm{i}\omega}{c}(\Gamma_{xxxyz} + \Gamma_{xyxxz} + \Gamma_{xyxxz}\,\mathrm{e}^{2\mathrm{i}(\Phi-\phi)})\sin 2\Theta \\ N_+ &= \frac{8\pi\mathrm{i}\omega}{c}[\Gamma_{yxxxz} + \Gamma_{xyxxz} + \tfrac{1}{2}(\Gamma_{yxxxz} + \Gamma_{xxxyz})\,\mathrm{e}^{2\mathrm{i}(\Phi-\phi)}]\cos 2\Theta \\ N_- &= \frac{8\pi\mathrm{i}\omega}{c}[\Gamma_{yxxxz} - \Gamma_{xyxxz} + \tfrac{1}{2}(\Gamma_{yxxxz} - \Gamma_{xxxyz})\,\mathrm{e}^{2\mathrm{i}(\Phi-\phi)}]\end{aligned} \tag{7.39}$$

Finally, by resorting equation (7.17) we derive, for a nonlinear isotropic gyrotropic liquid,

$$J^{\mathrm{T}} = \frac{\mathrm{i}\omega}{c}\left[\Gamma^{\mathrm{a}} - \frac{16\pi S_0}{|1+n|^2}\Gamma_{yxxxz}(1 + \tfrac{1}{2}\mathrm{e}^{2\mathrm{i}(\Phi-\phi)})\right]s_0^2 \tag{7.40}$$

and

$$J^{\mathrm{R}} = \frac{\mathrm{i}\omega}{c}\left[-\Gamma^{\mathrm{a}} - \frac{16\pi S_0}{|1+n|^2}\Gamma^{\mathrm{R}}_{yxxxz}(1 + \tfrac{1}{2}\mathrm{e}^{2\mathrm{i}(\Phi-\phi)})\right]s_0^2 \tag{7.41}$$

Using equations (7.40) and (7.41) and formulae (7.14) and (7.15) we obtain the expressions for the pump-induced polarization azimuth and ellipticity of the transmitted and reflected probe waves:

$$\begin{pmatrix}\Theta^{\mathrm{R}}_{\mathrm{induced}}\\ \eta^{\mathrm{R}}_{\mathrm{induced}}\end{pmatrix} = \frac{128\pi^3 I_{\mathrm{p}}}{c\lambda|1+n|^2}\begin{pmatrix}\mathrm{Im}\\ -\mathrm{Re}\end{pmatrix}\left\{\frac{\Gamma^{\mathrm{R}}_{yxxxz}(1 + \tfrac{1}{2}\mathrm{e}^{2\mathrm{i}(\Phi-\phi)})}{1-n^2}\right\} \tag{7.42}$$

$$\begin{pmatrix}\Theta^{\mathrm{T}}_{\mathrm{induced}}\\ \eta^{\mathrm{T}}_{\mathrm{induced}}\end{pmatrix} = -\frac{128\pi^4 I_{\mathrm{p}} L}{c\lambda^2|1+n|^2}\begin{pmatrix}\mathrm{Re}\\ \mathrm{Im}\end{pmatrix}\{\Gamma_{yxxxz}(1 + \tfrac{1}{2}\mathrm{e}^{2\mathrm{i}(\Phi-\phi)})\} \tag{7.43}$$

Here $I_{\mathrm{p}} = cS_0/(2\pi)$ is the intensity of the pump wave approaching the nonlinear medium from vacuum at normal incidence.

Looking at formulae (7.42) and (7.43) we can see that certain features of the pump–probe nonlinear optical activity effect are similar to those of the other pump–probe effects:

(a) In weakly absorbing materials, with predominantly real refractive indices, the real and imaginary parts of the same nonlinear susceptibilities are responsible for the reflective and transmissive effects.

(b) The transmissive effect grows with the sample thickness L. Assuming an equal order of magnitude for the real and imaginary parts of the nonlinear susceptibilities, the reflective effect will be $\lambda/(\pi n^2 L)$ times smaller than the transmissive effect.

(c) Both transmissive and reflective effects have coherent components oscillating with Φ-ϕ, and incoherent components that are not sensitive to the phase difference between the pump and the probe. The coherent and incoherent contributions depend on the same combination of components of the nonlinearity tensor.

To the best of our knowledge no observation of pump–probe nonlinear optical activity due to the fast, electronic mechanism of nonlinearity has been reported so far. We expect, in particular, the reflective effect to be very small. However, the transmissive effect due to the thermal mechanism has been seen in coloured optically active crystals of quartz in Ref. [26] (about this experiment see Chapter 4) and cholesteric liquid crystals [27]. Nonlinear pump–probe polarization-sensitive experiments have also been performed in optically active liquids [28].

The second pump–probe effect due to nonlocality of nonlinear response is the pump–probe analogue of induced nonlocal birefringence. Like its self-action analogue, this effect can be seen in cubic crystals lacking an inversion centre, such as zinc-blende crystals (see Chapter 4). In this effect, a strong pump wave induces linear birefringence and dichroism via the nonlocality of nonlinear response. To observe pump–probe induced nonlocal birefringence clearly the pump and probe waves must be linearly polarized in exactly the same direction (the Beam arrangement is similar to that in Fig. 7.8). The nonlinear polarization changes in the transmitted and reflected probe beams are figures of merit for the effect. We will examine, below, pump–probe nonlinear nonlocal birefringence in a cubic crystal of the zinc-blende class, $\bar{4}3m$, where the light waves propagate along the cubic axis [0 0 1]. The same Cartesian frame installation as that presented in Fig. 4.11 will be used. For probe and pump waves with identical linear polarization, the optical Kerr and inverse Faraday effects do not appear. However, the nonlinear anisotropic polarization effect interferes with pump–probe-induced nonlocal birefringence. Nevertheless, as we will see below, it has a different dependence on crystal orientation and may be distinguished on this basis.

Since self-action reflective and transmissive effects have already been examined in Chapters 6 and 4, and most of necessary intermediate calculations have already been performed, we can proceed straight to the formulae for the pump–probe effect's nonlinear parameters (7.18) and (7.19):

$$N = 0$$

$$N_{+} = \frac{8\pi i\omega}{c}[\Gamma_{yxxxz} + \Gamma_{xyxxz} + \tfrac{1}{2}(\Gamma_{yxxxz} + \Gamma_{xxxyz})\, e^{2i(\Phi-\phi)}] \tag{7.44}$$

$$N_{-} = \frac{8\pi i\omega}{c}[\Gamma_{yxxxz} - \Gamma_{xyxxz} + \tfrac{1}{2}(\Gamma_{yxxxz} - \Gamma_{xxxyz})\, e^{2i(\Phi-\phi)}] \cos 2\Theta$$

Here only the nonlocal contributions to N and $N_{\pm}$ are presented, i.e. we have dropped χ_{ijkl} from the formulae. By resorting formulae (7.14) and (7.15) we obtain

$$J^{\mathrm{T}} = -\frac{i\omega}{c}\left[\frac{32\pi S_0}{|1+n|^2}\Gamma_{yxxxz}(1 + \tfrac{1}{2}e^{2i(\Phi-\phi)})\right] s_0^2 \cos 2\Theta \tag{7.45}$$

and

$$J^{\mathrm{R}} = -\frac{i\omega}{c}\left[\frac{32\pi S_0}{|1+n|^2}\Gamma^{\mathrm{R}}_{yxxxz}(1 + \tfrac{1}{2}e^{2i(\Phi-\phi)})\right] s_0^2 \cos 2\Theta \tag{7.46}$$

From here we derive equations for the polarization azimuth and ellipticity of the transmitted and reflected probe waves:

$$\begin{pmatrix} \Theta^{\mathrm{R}}_{\mathrm{induced}} \\ \eta^{\mathrm{R}}_{\mathrm{induced}} \end{pmatrix} = \frac{128\pi^3 I_{\mathrm{p}}}{c\lambda|1+n|^2}\begin{pmatrix} \mathrm{Im} \\ -\mathrm{Re} \end{pmatrix}\left\{\frac{\Gamma^{\mathrm{R}}_{yxxxz}(1 + \frac{1}{2}e^{2i(\Phi-\phi)})}{1-n^2}\right\} \cos 2\Theta \tag{7.47}$$

$$\begin{pmatrix} \Theta^{\mathrm{T}}_{\mathrm{induced}} \\ \eta^{\mathrm{T}}_{\mathrm{induced}} \end{pmatrix} = -\frac{128\pi^4 I_{\mathrm{p}}}{c\lambda^2|1+n|^2}\begin{pmatrix} \mathrm{Re} \\ \mathrm{Im} \end{pmatrix}\{\Gamma_{yxxxz}(1 + \tfrac{1}{2}e^{2i(\Phi-\phi)})\} \cos 2\Theta \tag{7.48}$$

Formulae (7.47) and (7.48) give the pump induced polarization change due to the presence of optical nonlocality. In cubic crystals of the $\bar{4}3m$ point group the induced nonlocal birefringence will be observed simultaneously with the nonlinear anisotropic effect described in the previous section, see formulae (7.34) and (7.35), and this has been seen experimentally [23]. However, these two effects may be distinguished because they depend differently on the crystal orientation. Indeed, the pump–probe anisotropic polarization effect depends on the crystal orientation as $\sin 4\Theta$ (see formulae (7.34) and (7.35)), while pump–probe-induced nonlocal birefringence goes as $\cos 2\Theta$ (see formulae (7.47) and (7.48)). Pump–probe-induced nonlocal birefringence was first seen in reflection by Zheludev and Paraschuk [30].

7.4 PUMP–PROBE POLARIZATION SPECTROSCOPY: AN EXAMPLE†

The general approach developed in this chapter is suitable for the description of frequency-degenerate, pump–probe, polarization-sensitive effects in transmission and on reflection. Formulae (7.14)–(7.19) may be used for nonlinear media of arbitrary symmetry with local and nonlocal responses and for arbitrarily polarized pump and probe waves. Below we will show an example how such analysis may be applied to situations more complex than elementary effects discussed in Sections 7.2 and 7.4.

Reflective pump–probe polarization effects have a very considerable spectroscopic potential. Their main advantages are very high sensitivity, accuracy and a dynamic range of pump–probe polarization measurement techniques. Such measurements can be made with femtosecond lasers providing unique opportunities for time-resolved, transient spectroscopy and the measurement of noninstantaneous nonlinearities. Moreover, reflective pump–probe polarization-sensitive phenomena have the advantage of being observable with completely opaque samples, such as narrow-gap semiconductors, superconductors and metals. The formulae derived in this chapter assume instantaneous nonlinearities and do not cover transient spectroscopy. Strictly speaking, they are only applicable for when the duration of the optical pulses is longer than the transient time of the nonlinear response. Nevertheless, even in the case of noninstantaneous nonlinearities these formulae may be used as a guide since they will correctly predict the symmetrical properties of the effects. They will also show which particular components of the tensors of nonlinear susceptibilities shall be involved in the analysis.

The elementary examples of pump–probe polarization effects considered above, such as the optical Kerr and inverse Faraday effects in isotropic media, can, of course, be seen in materials of lower symmetry. In such media a clear separation of the effects may not always be possible. In this case instead of referring to the particular effects, it makes sense to speak about spectroscopy in the 'Kerr' configuration (the pump and probe are linearly polarized at $\Delta\Theta = \pi/4$ to one another) or in the 'Faraday' configuration (the pump is circularly polarized and the probe is linearly polarized). Similarly, one can speak of the 'cubic anisotropy sensitive' configuration, or simply the 'anisotropic' configuration, where the pump and probe are identically linearly polarized.

We will give examples below of experimental results of pump–probe transient spectroscopy in the 'Faraday', 'Kerr' and 'anisotropic' configuations for an InSb crystal belonging to the $\bar{4}3m$ symmetry point group [31]. We will compare them with the results of the phenomenological theory of pump–probe

†This section is written in collaboration with Dr S.V. Popov, University of Southampton, UK.

interactions and see how well the experiments confirm the theory. In the experiments the pump and probe propagated at small angle to the [001] direction of the crystal, which coincided with the Z direction of the coordinate frame. The X and Y axes of the frame coincided with the [100] and [010] crystal directions respectively. We will skip the details of the calculations, as they are similar to those we have already seen in this chapter, and proceed straight to the results. The experiments with InSb were performed using a femtosecond Cr:Forsterite laser generating 90 fs pulses at a wavelength of 1.26 μm. This laser was used as both the pump and probe source. The polarimeter, similar to those in Fig. 7.6, detected only the pump-induced polarization azimuth change and only the incoherent component of the effect was measured. At this wavelength the crystal of InSb is totally opaque and only reflective measurements are possible. Therefore, we will only produce the formulae for the incoherent reflective polarization azimuth rotation.

(a) In the Faraday configuration the theory predicts

$$\Theta^{R} - \Theta = \frac{32\pi^2 I_p}{c|1+n|^2}\left[\mathrm{Im}\left\{\pm\frac{\chi_{xxyy}-\chi_{xyyx}}{n(1-n^2)} + \frac{2\pi}{\lambda}\frac{\Gamma_{yxxxz}+\Gamma_{xyxxz}}{1-n^2}\cos 2\Theta\right\}\right.$$
$$\left.\pm\frac{2\pi}{\lambda}\mathrm{Re}\left\{\frac{\Gamma_{xxxyz}-\Gamma_{xyxxz}}{1-n^2}\right\}\sin 2\Theta\right] \tag{7.49}$$

Here the local term due to χ_{ijkl} is identical to that in an isotropic medium. It represents the 'inverse Faraday effect contribution' (compare with formula (7.31)). The first nonlocal term, which contains Γ_{ijklm}, is similar to the pump–probe nonlocal birefringence term with a linearly polarized pump (see formula (7.47)). It oscillates as $\cos 2\Theta$. However, here, with the circularly polarized pump another term, oscillating as $\sin 2\Theta$ may be generally expected.

This formulae should be compared with the experimental graph presented in Fig. 7.11. Here, the 'inverse Faraday effect' contribution (first term in formula (7.49)) to the reflective pump-induced probe polarization azimuth rotation is clearly seen as a deep trough in the front part of the pump–probe dependence of the induced rotation. It is instantaneous within the time resolution available (see cross-section ABC). Its magnitude, as predicted, does not depend on the crystal orientation. A much slower effect due to nonlocality then follows: it oscillates with the crystal orientation as predicted by (7.49) (see cross-section DEF) and is clearly seen several hundreds femtosecond after pump when the fast 'Faraday effect' rotation disappears.

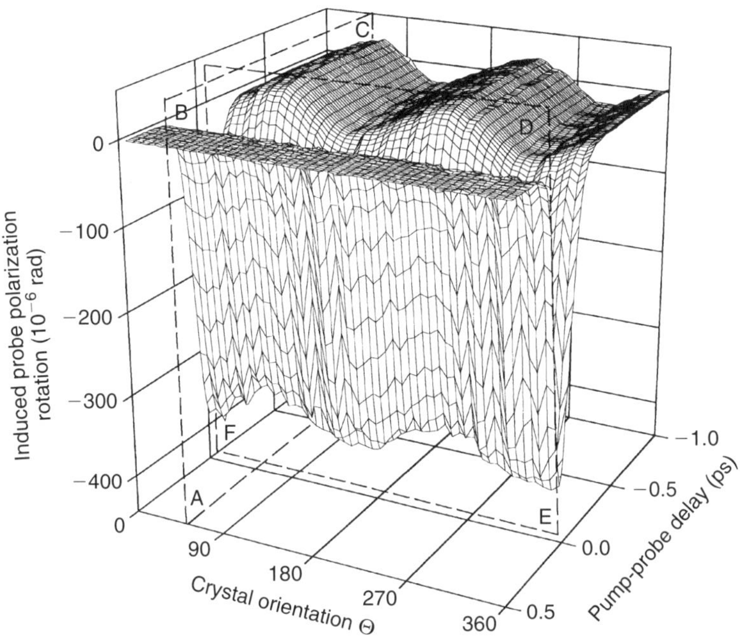

Fig. 7.10 Experimental results from the measurement of the reflective pump–probe polarization effect in InSb using the 'Faraday configuration'. The pump intensity was approximately 0.4 GW/cm^2

(b) In the 'Kerr' configuration the theory predicts

$$\Theta^{\mathrm{R}} - \Theta = \frac{32\pi^2 I_{\mathrm{p}}}{c|1+n|^2}\left[\mp \mathrm{Re}\left\{\frac{\chi_{xxyy} + \chi_{xyyx} + \Delta\chi \sin^2 2\Theta}{n(1-n^2)}\right.\right.$$
$$\left.\left. - \frac{2\pi}{\lambda} \mathrm{Im}\left\{\frac{(\Gamma^{\mathrm{R}}_{yxxxz} + \Gamma^{\mathrm{R}}_{xyxxz}) \cos 2\Theta \mp (\Gamma^{\mathrm{R}}_{yxxxz} - \Gamma^{\mathrm{R}}_{xyxxz}) \sin 2\Theta}{1-n^2}\right\}\right]\right. \quad (7.50)$$

where + or − corresponds to $\Delta\Theta = +45^\circ$ or $\Delta\Theta = -45^\circ$. In formula (7.50) we can clearly see: (i) the anisotropic effect proportional to $\Delta\chi \sin^2 2\Theta$ (compare with formula (7.34)), (ii) the 'optical Kerr effect' contribution, given by the second term, which does not depend on the initial polarization azimuth of the pump and probe, and (iii) the nonlocal contribution oscillating as a harmonic function of 2Θ.

The prediction of this formula is very well confirmed by experiment (see Fig. 7.12): the term oscillating as $\sin^2 2\Theta$ is clearly seen on the experimental

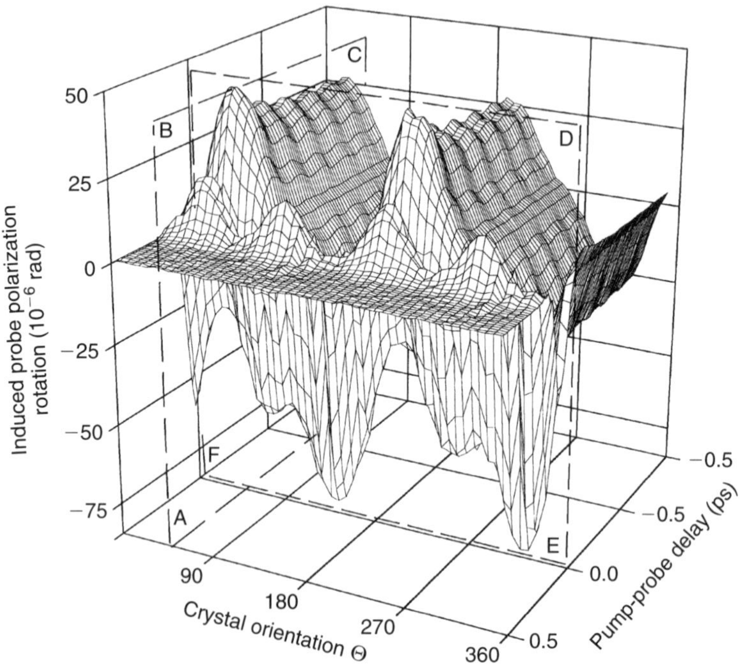

Fig. 7.11 Experimental results from the measurement of the reflective pump–probe polarization effect in InSb using the 'Kerr configuration'. The pump intensity was approximately 0.4 GW/cm^2

graph at zero pump–probe delay (cross-section DEF). It is superimposed on the orientation-independent background due to the optical Kerr effect. A much slower effect due to nonlocality then follows: it oscillates with the crystal orientation as predicted by formula (7.50). It also oscillates in time which may be compared with the results of [33].

(c) In the 'anisotropic' configuration the theory gives

$$\Theta^{\mathrm{R}} - \Theta = \frac{16\pi^2 I_{\mathrm{p}}}{c|1+n|^2} \times \left[\mathrm{Re}\left\{ \frac{\Delta\chi}{n(1-n^2)} \right\} \sin 4\Theta + \frac{8\pi}{\lambda} \,\mathrm{Im}\left\{ \frac{\Gamma^{\mathrm{R}}_{xxxyz}}{1-n^2} \right\} \cos 2\Theta \right] \tag{7.51}$$

Here one can recognize the nonlinear anisotropic component given by formula (7.34) and the contribution from the pump–probe nonlocal birefringence given by formula (7.42).

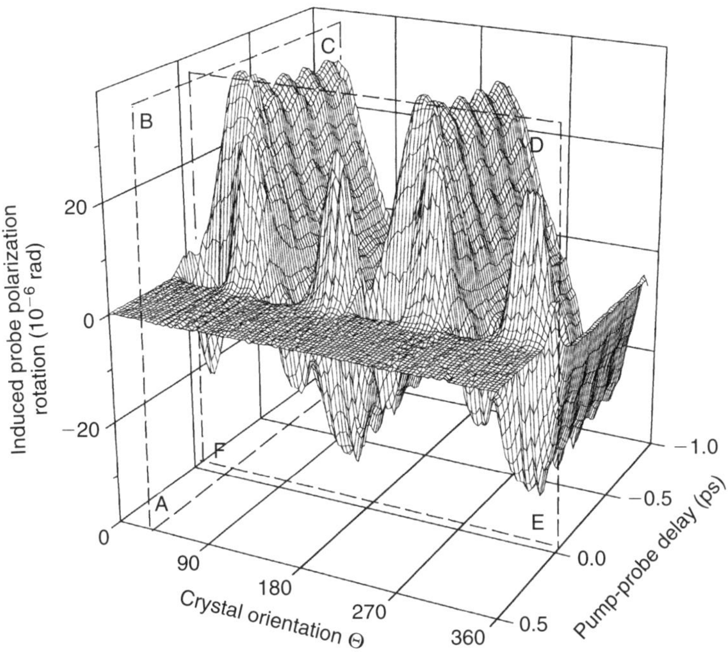

Fig. 7.12 Experimental results from the measurement of the reflective pump–probe polarization effect in InSb using the 'anisotropic configuration'. The pump intensity was approximately 0.3 GW/cm^2

In the experimental graph of Fig. 7.12 these contributions are spectacularly separated both in the time domain, where the anisotropic term is much faster than the nonlocal term (see cross-section ABC), and as functions of the crystal orientation, where the anisotropic term gives a rotation that oscillates with Θ twice as fast as the nonlocal term.

When comparing the experimental results (Figs. 7.11 to 7.13) with the theoretical formulae (7.49) to (7.51) we did not discuss the microscopic nature of the nonlinear response. Such analysis is complex and lies outside the mainstream of this book. We do not address it here. Here we only wanted to show the power of pump–probe polarization-sensitive measurements. Even with high-opacity materials, very accurate reflective experiments, with angular resolutions of 1 μrad and temporal resolutions of 100 fs, are now routinely possible, providing precise information on different components of tensors of local and nonlocal nonlinear optical susceptibilities.

For other numerous examples of the use of pump–probe polarization sensitive spectroscopy see Refs. [32] to [65].

REFERENCES

1. I. Tinoko, *J. Chem. Phys.*, **62**, 1006 (1975).
2. R.A. Harris, *J. Chem. Phys.*, **65**, 5294 (1976).
3. P.W. Atkins and L.D. Barron, *Prog. R. Soc.*, **119A**, 306 (1968).
4. N.I. Zheludev and Yu.P. Svirko, Nonlinear polarization spectroscopy of semiconductors, in S.A. Akhmanov and N.I. Zheludev (eds.), *Nonlinear-Optical Diagnostics of Semiconductors*, Vol. 2 VINITI, Moscow, 1990.
5. A.A. Golubkov and V.A. Makarov, *Optica i Spectroscopiya*, **69**, 622 (1990).
6. A.A. Golubkov and V.A. Makarov, *Optica i Spectroscopiya*, **67**, 1134 (1989).
7. S.V. Popov, Y.P. Svirko and N.I. Zheludev, *JOSA-B*, **13**, 2729 (1996).
8. E.G. Sauter, *Nonlinear Optics*, John Wiley & Sons, New York, Ch. 7, 1996.
9. A.D. Buckingham, *Proc. Phys. Soc. B*, **69**, 344 (1956).
10. G. Mayer and F. Gires, *C. R. Acad. Sci. Paris*, **258**, 2039 (1964).
11. A.M. Bonch-Bruevich, N.N. Kostin and V.A. Khodovoi, *JETP Lett.*, **3**, 279 (1966).
12. M. Kuwata, *J. Luminescence*, **38**, 247 (1987).
13. L.D. Landau, E.M. Lifshitz and L.P. Pitaevskiĭ, *Electrodynamics of Continuous Media*, Pergamon Press, Oxford, 1984, p. 353.
14. J.P. Van der Ziel, P.S. Pershan and L.D. Malmstrom, *Phys. Rev. Lett.*, **15**, 190 (1965).
15. P.W. Atkins and M.H. Miller, *Molec. Phys.*, **15**, 503 (1968).
16. V.M. Arutunyan, T.A. Papazyan, G.G. Adonts, A.V. Karmenyan, S.P. Ishkhanyan and L. Khol'ts, *Soviet Phys. JETP*, **41**, 22 (1975).
17. C. Wieman and T.W. Hänsch, *Phys. Rev. Lett.*, **36**, 1170 (1976).
18. M. Kuwata, *J. Phys. Soc. Japan*, **53**, 4456 (1984).
19. A.M. Danishevski, E.L. Ivchenko, S.F. Kochegarov and V.K. Subashiev, *Soviet Phys. Solid State*, **27**, 439 (1985).
20. S.V. Popov, Yu.P. Svirko and N.I. Zheludev. *Opt. Lett.*, **19**, 13 (1994).
21. N.I. Zheludev, S.V. Popov, A. Malinowski, Yu.P. Svirko, W.Y. Liang and C.T. Lin, *Solid State Commun.*, **90**, 287 (1994).
22. N.I. Zheludev, M.A. Brummel, R.T. Harley, A. Malinowski, S.V. Popov, D.E. Ashenford and B. Lunn. *Solid State Commun.*, **89**, 823 (1994).
23. N.I. Zheludev, P.J. Bennet, H. Loh, S.V. Popov, I.R. Shatwell, Y.P. Svirko, V.E. Gusev, V.F. Kamalov and E.V. Slobodchikov, *Opt. Lett.*, **20**, 1368–70 (1995).
24. S.P. Apanasevich, D.N. Dovchenko and N.I. Zheludev, *Opt. Spectrosc. (USSR)*, **62**, 287 (1987).
25. A.R. Bungay, S.V. Popov, N.I. Zheludev and Yu.P. Svirko, *Opt. Lett.*, **20**, 356 (1995).
26. D.V. Vlasov and V.P. Zaitseva, *JETP Lett.*, **14**, 112 (1971).
27. O.V. Garibian, B.V. Zhdanov, N.I. Zheludev, A.I. Kovrigin and V.I. Kuznetsov, *Sov. Phys.- Crystallogr.*, **26**, 446 (1981).
28. T. Yakozawa, Y. Matsumoto, H. Ashitaka; Y. Yokoh and M. Kuwata-Gonokami, *Jap. J. Appl. Phys.*, **34**, 3522 (1995).
29. N.I. Zheludev and D.Yu. Paraschuk, *JETP Lett.*, **52**, 34 (1990).
30. N.I. Zheludev, D.Yu. Paraschuk, A. Malinowskii and Y.P. Svirko, *Phys. Rev. B*, **50**, 11508 (1994).
31. The reported experimental results were obtained by S.V. Popov, Yu.P. Svirko and N.I. Zheludev at the University of Southampton, UK, in 1996–7.
32. A. Tackeuchi, S. Muto, T. Inata and T. Fuji, *Appl. Phys. Lett.*, **56**, 2213 (1990).
33. G.C. Cho, W. Kütt and H. Kurz, *Phys. Rev. Lett.*, **65**, 764 (1990).
34. L. Min and R.J.D. Miller, *Appl. Phys. Lett.*, **56**, 524 (1990).
35. J.J. Baumberg, S.A. Crooker, D.D. Awshalom, N. Samarth, H. Luo and J.K. Furdyna, *Phys. Rev. B*, **50**, 7689 (1994).

36. V.M. Lyubin and V.K. Tikhomirov, *JETP Lett.*, **52**, 78 (1990).
37. T. Kawazoe, Y. Masumoto and T. Mishina, *Phys. Rev. B*, **47**, 10452 (1993).
38. J.B. Stark, W.H. Khox and D.S. Chemla, *Phys. Rev. Lett.*, **68**, 3080 (1992).
39. R.E. Worsley, N.J. Traynor, T. Grevatt and R.T. Harley, *Phys. Rev. Lett.*, **76**, 3224 (1996).
40. E.Y. Perlin, *Optika I Spektroskopiya*, **50**, 706 (1981).
41. G.E. Jellison and D.H. Lowndes, *Applied Physics Letters*, **47**, 718 (1985).
42. G.E. Jellison and D.H. Lowndes, *Applied Optics*, **24**, 2948 (1985).
43. A.M. Danishevskii and E.Y. Perlin, *JETP Letters*, **41**, 524 (1985).
44. A.M. Danishevskii, E.L. Ivchenko and V.K. Subashiev, *Fizika Tverdogo Tela*, **28**, 291 (1986).
45. N.I. Zheludev, V.Y. Karasev, Z.M. Kostov and M.S. Nunuparov, *JETP Letters*, **43**, 747 (1986).
46. G.E. Jellison and D.H. Lowndes, *Applied Physics Letters*, **51**, 352 (1987).
47. Z.M. Kostov and N.I. Zheludev, *Optical and Quantum Electronics*, **20**, 30 (1988).
48. S.A. Akhmanov, N.I. Zheludev, C.N. Ironside, Z.M. Kostov and S.V. Popov, *JETP Letters*, **49**, 195 (1989).
49. N.I. Zheludev, I.S. Ruddock and R. Illingworth, *Applied Physics B—Photophysics and Laser Chemistry*, **49**, 65 (1989).
50. N.I. Zheludev and D.Y. Parashchuk, *JETP Letters*, **52**, 34 (1990).
51. E.Y. Perlin and A.M. Danishevskii, *Zhurnal Eksperimentalnoi I Teoreticheskoi Fiziki*, **106**, 503 (1994).
52. W.A. Schroeder, D.S. Mccallum, D.R. Harken, M.D. Dvorak, D.R. Andersen, A.L. Smirl and B.S. Wherrett, *Journal of the Optical Society of America B—Optical Physics*, **12**, 401 (1995).
53. H.K. Tsang, R.V. Penty, I.H. White, R.S. Grant, W. Sibbett, J.B.D. Soole, H.P. Leblanc, N.C. Andreadakis, E. Colas and M.S. Kim, *Applied Physics Letters*, **59**, 3440 (1991).
54. S. Patkar, A.E. Paul, W. Sha, J.A. Bolger and A.L. Smirl, *Physical Review B—Condensed Matter*, **51**, 10789 (1995).
55. D.S. Mccallum, X.R. Huang, A.L. Smirl, D. Sun and E. Towe, *Applied Physics Letters*, **66**, 2885 (1995).
56. A.E. Paul, I.A. Bolger, A.L. Smirl and J.G. Pellegrino, *Journal of the Optical Society of America B—Optical Physics*, **13**, 1016 (1996).
57. J.A. Bolger, A.A. Paul and A.L. Smirl, *Physical Review B—Condensed Matter*, **54**, 11666 (1996).
58. W.J. Walecki, D.N. Fittinghoff, A.L. Smir and R. Trebino, *Optics Letters*, **22**, 81 (1997).
59. P. Unsbo and C. Flytzanis, *Journal of the Optical Society of America B—Optical Physics*, **14**, 560 (1997).
60. R. Pankoke, P. Leisching, C. Buss, R. Frey, J. Cibert and C. Flytzanis, *Physica Status Solidi B—Basic Research*, **204**, 49 (1997).
61. A.R. Bungay, S.V. Popov, I.R. Shatwell and N.I. Zheludev, *Physics Letters A*, **234**, 379 (1997).
62. S. Dhanjal, S.V. Popov, I.R. Shatwell, Y.P. Svirko, N.I. Zheludev and V.E. Gusev, *Optics Letters*, **22**, 1879 (1997).
63. B.J. Bennett, S. Dhanjal, Yu.P. Svirko and N.I. Zheludev, *Journal of Moderm Optics*, **45**, 1009 (1998).
64. P. Leisching, C. Buss, J. Cibert, R. Frey and C. Flytzanis, *Journal of Crystal Growth*, **185**, 984 (1998).
65. G.P. Ju, A. Vertikov, A.V. Nurmikko, C. Candy, G. Xiao, R.F.C. Farrow and A. Cebollada, *Phys. Rev. B*, **57**, R700 (1998)

8 The Oscillator Model of Nonlinear Optical Response

In this chapter we will show how local and nonlocal optical susceptibilities can be calculated from a classical microscopic model. We will begin with the Lorentz-type model which represents a molecule as an ensemble of charged anharmonic oscillators. Light-induced vibrations of these oscillators cause electric current at the optical frequency. Expansion of the current density as a series in the magnitude of the light field will give us the optical susceptibilities of the medium. We will then develop two- and multi-electron oscillator models of the optical properties of gyrotropic media.

8.1 LORENTZ'S MODEL

At the end of the nineteenth century H.A. Lorentz developed a simple approach for describing the dispersion of the dielectric properties of a medium in the vicinity of an absorption resonance [1]. This model became very popular in linear [2] and nonlinear [3–8] optics. In this chapter we will use the Lorentzian approach to calculate nonlinear optical susceptibilities and we will show how it may be modified to include medium response nonlocality. To begin with, we will briefly describe the Lorentz model itself.

According to Lorentz, the optical properties of a medium are determined by the motion of bound electrons in the fields of a propagating light wave. The electrons involved in light–matter interactions are often referred to as optical electrons. We will first examine the simplest example of a molecular medium where: (a) each molecule of the medium has only one optical electron, (b) molecules do not interact with one another and (c) the motion of each optical electron is constrained to one dimension, say along the X axis of the laboratory coordinate frame (see Fig. 8.1).

Let the electron have potential energy U and an equilibrium position $\mathbf{r}_0 = \{x_0, y_0, z_0\}$, where $\mathbf{r}_0$ is the solution to the equation $\partial U/\partial \mathbf{r} = 0$. On being displaced from the equilibrium, the electron moves under the restoring force $F_x = -\partial U/\partial x$. Therefore, its displacement x from the equilibrium position satisfies the following equation:

$$\frac{\mathrm{d}^2 x}{\mathrm{d}t^2} + \gamma \frac{\mathrm{d}x}{\mathrm{d}t} + \frac{1}{m}\frac{\partial U}{\partial x} = 0 \tag{8.1}$$

where m is the mass of the electron and γ is a damping parameter. U may be expanded as a power series in x near the equilibrium position of the electron:

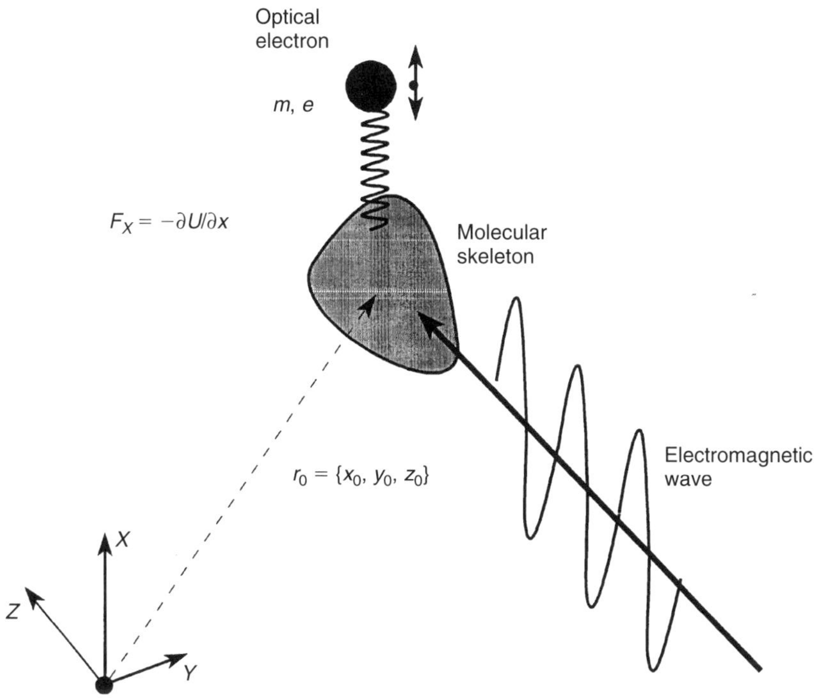

Fig. 8.1 One-dimensional Lorentzian molecular model. An optical electron with charge e and mass m is attached by an elastic spring to the molecular skeleton. When driven by an electromagnetic wave, it oscillates near the equilibrium position $\mathbf{r}_0$

$$U(x_0 + x,\ y_0,\ z_0) - U(x_0,\ y_0,\ z_0) = \left.\frac{\partial^2 U}{\partial x^2}\right|_{x=0} \frac{x^2}{2!} + \left.\frac{\partial^3 U}{\partial x^3}\right|_{x=0} \frac{x^3}{3!} + \left.\frac{\partial^4 U}{\partial x^4}\right|_{x=0} \frac{x^4}{4!} + \dots \tag{8.2}$$

The first term on the right-hand side of expansion (8.2), quadratic in x, gives the potential energy of a harmonic oscillator. At this stage we will retain only the first term of equation (8.2) and ignore all higher order terms. An electron interacting with the plane light wave at frequency ω,

$$\mathbf{E}(r_0,\ t) = \mathbf{A}\,\mathrm{e}^{-\mathrm{i}\omega t + \mathrm{i}(\mathbf{k}\cdot\mathbf{r}_0)} + \mathbf{A}^*\,\mathrm{e}^{\mathrm{i}\omega t - \mathrm{i}(\mathbf{k}\cdot\mathbf{r}_0)} \tag{8.3}$$

moves according to equation (8.1) with the Lorentz force $-e\mathbf{E} - (e/c)[\mathbf{v} \times \mathbf{B}]$ on the right-hand side. Here $-e$ is the charge of the electron and $\mathbf{v}$ is its velocity. The magnetic part of the Lorentz force is perpendicular to the velocity of the electron. Since the motion is restricted to X direction, the velocity is directed along X and, therefore, the magnetic part of the Lorentz force has no component in the X direction and does not affect the electron's motion. Accordingly, only the component of the electrostatic part of the Lorentz force, $-e\mathbf{E}(\mathbf{r}_0, t)$, in the direction of electron motion should be included in the equation of motion:

$$\frac{d^2x}{dt^2} + \gamma \frac{dx}{dt} + \omega_0^2 x = -\frac{e}{m} E_x(\mathbf{r}_0, t) \tag{8.4}$$

Here we have introduced the natural frequency of the oscillator:

$$\omega_0^2 = \frac{1}{m} \frac{d^2U}{dx^2}\bigg|_{x=0} \tag{8.5}$$

Substituting equation (8.3) into (8.4), we readily obtain the following formula for the displacement of the electron from its equilibrium position $\mathbf{r}_0$ in the presence of the light wave:

$$x(\mathbf{r}_0, t) = -\frac{e}{m} A_x \frac{e^{-i\omega t + i(\mathbf{k}\cdot\mathbf{r}_0)}}{\omega_0^2 - i\gamma\omega - \omega^2} + \text{c.c.} \tag{8.6}$$

We will now show how this result can be used to obtain the constitutive equation of a medium consisting of uniformly distributed molecules.

To derive the constitutive equation we have to calculate the electric field polarization response $\mathbf{P}(\mathbf{r}, t) = \int \mathbf{J}(\mathbf{r}, t')\,dt'$, where $\mathbf{J}(\mathbf{r}, t)$ is the induced current density at point $\mathbf{r}$ in the medium (see definition (2.3). To obtain $\mathbf{J}(\mathbf{r}, t)$ as a *macroscopic* quantity the current density involved in Maxwell's equations (2.2), we have to smooth individual *microscopic* molecular currents: all of the molecular currents induced by the light wave in individual molecules located within a small volume δV, which contains point $\mathbf{r}$, must be added together and then divided by δV. A small cube of side Δ may be chosen as this volume. The cube's sides should be much less than the distance over which the electric field of the light wave varies appreciably, but, on the other hand, large enough such that the cube contains many molecules and the principles of statistical mechanics are applicable [1] (see Fig. 8.2).

The current associated with an individual optical electron is proportional to its velocity $dx(\mathbf{r}_0, t)/dt$, and therefore the total current density induced in a volume δV is given by

$$J_x(\mathbf{r}, t) = \frac{1}{\delta V} \sum_{\mathbf{r}_0 \in \delta V} \left[-e \frac{dx(\mathbf{r}_0, t)}{dt}\right] \tag{8.7}$$

where $x(\mathbf{r}_0, t)$ is given by equation (8.6) and summation takes place over all molecular electrons located within δV. If the molecular volume number density $N_0 \gg 1$, summation (8.7) may be rewritten as follows:

$$J_x(\mathbf{r}, t) = \frac{N_0}{\Delta^3} \int_{x-\Delta/2}^{x+\Delta/2} dx_0 \int_{y-\Delta/2}^{y+\Delta/2} dy_0 \int_{z-\Delta/2}^{z+\Delta/2} dz_0 \left[-e \frac{dx(\mathbf{r}_0, t)}{dt}\right] \tag{8.8}$$

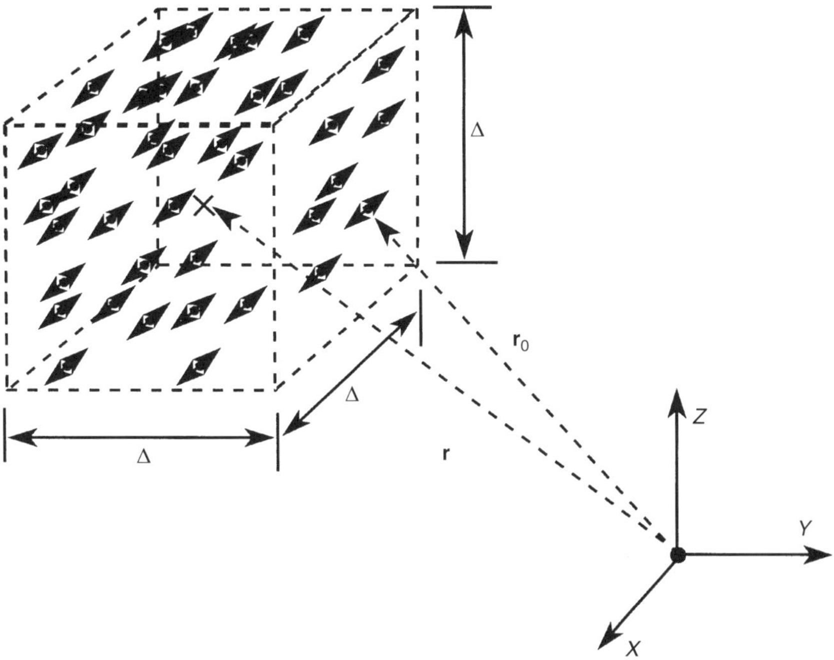

Fig. 8.2 An ensemble of identically oriented Lorentz molecules surrounding point with coordinate **r**. The integration volume contains $N_0\Delta^3 \gg 1$ molecules, but the side of the cube is much smaller that the wavelength of the light wave

By substituting solution (8.6) into (8.8) and integrating over x_0, y_0 and z_0 we obtain

$$J_x(\mathbf{r},\, t) = \left[-i\omega \frac{N_0 e^2}{m(\omega_0^2 - \mathrm{i}\gamma\omega - \omega^2)} A_x e^{-\mathrm{i}\omega t + \mathrm{i}(\mathbf{k}\cdot\mathbf{r})} + \mathrm{c.c.}\right] \times \frac{\sin\frac{1}{2}k_x\Delta}{\frac{1}{2}k_x\Delta}\, \frac{\sin\frac{1}{2}k_y\Delta}{\frac{1}{2}k_y\Delta}\, \frac{\sin\frac{1}{2}k_z\Delta}{\frac{1}{2}k_z\Delta} \tag{8.9}$$

Since $\Delta \ll \lambda$, $\sin(k_{x,y,z}\Delta/2)/(k_{x,y,z}\Delta/2) \simeq 1$ and the Δ-dependent factors in formula (8.9) can be dropped. Now, recalling that $\mathbf{P}(\mathbf{r},\, t) = \int J(\mathbf{r},\, t')\,\mathrm{d}t'$, we can obtain the constitutive equation for the X projection of the electric field displacement:

$$P_x(\mathbf{r},\, t) = \frac{N_0 e^2}{m(\omega_0^2 - \mathrm{i}\gamma\omega - \omega^2)} A_x\, \mathrm{e}^{-\mathrm{i}\omega t + \mathrm{i}(\mathbf{k}\cdot\mathbf{r})} + \mathrm{c.c.} \tag{8.10}$$

or, in the spectral representation,

$$P_x(\omega, \mathbf{r}) = \frac{1}{4\pi}[\epsilon_{xx}(\omega) - 1]E_x(\omega, \mathbf{r}) = \frac{N_0 e^2}{m(\omega_0^2 - \mathrm{i}\gamma\omega - \omega^2)} E_x(\omega, \mathbf{r}) \quad (8.11)$$

where $\mathbf{E}(\omega, \mathbf{r}) = \mathbf{A}\exp\{i\mathbf{k}\cdot\mathbf{r}\}$ (compare equation (8.11) with local term in the phenomenologically introduced constitutive equation (2.14)). Therefore, the dielectric tensor $\epsilon_{ij}(\omega)$ of a medium consisting of equally oriented molecules has the following non-zero components:

$$\begin{aligned} \epsilon_{xx}(\omega) &= 1 + \frac{4\pi N_0 e^2}{m} \frac{1}{\omega_0^2 - \mathrm{i}\gamma\omega - \omega^2} \\ \epsilon_{yy}(\omega) &= \epsilon_{zz}(\omega) = 1 \end{aligned} \quad (8.12)$$

Note that with identically oriented molecules, the medium shows birefringence with an optical axis along the X direction.

This model can also be modified to describe isotropic media. Isotropy can be achieved if the molecules are randomly oriented. Their optical electrons will therefore oscillate in different directions. If an electron with equilibrium position $\mathbf{r}_0$ oscillate along a unit vector $\mathbf{n}$, its amplitude of its forced oscillations, $u(\mathbf{r}_0, t)$, depends on the component of the electric field of the light wave in the direction of $\mathbf{n}$:

$$u(\mathbf{r}_0, t) = -\frac{e}{m}(\mathbf{n}\cdot\mathbf{A})\frac{\mathrm{e}^{-\mathrm{i}\omega t + \mathrm{i}(\mathbf{k}\cdot\mathbf{r}_0)}}{\omega_0^2 - \mathrm{i}\gamma\omega - \omega^2} + \text{c.c.} \quad (8.13)$$

In this case formula (8.7) becomes

$$\mathbf{J}(\mathbf{r}, t) = \frac{1}{\delta V}\sum_{\mathbf{r}_0 \in \delta N}\left[-e\mathbf{n}\frac{\mathrm{d}u(\mathbf{r}_0, t)}{\mathrm{d}t}\right] \quad (8.14)$$

We need to account for the random orientation of molecular oscillators and so we rewrite formula (8.14) as follows:

$$\mathbf{J}(\mathbf{r}, t) = \frac{N_0}{\Delta^3}\int_{x-\Delta/2}^{x+\Delta/2}\mathrm{d}x_0\int_{y-\Delta/2}^{y+\Delta/2}\mathrm{d}y_0\int_{z-\Delta/2}^{z+\Delta/2}\mathrm{d}z_0\left\langle -e\mathbf{n}\frac{\mathrm{d}u(\mathbf{r}_0, t)}{\mathrm{d}t}\right\rangle \quad (8.15)$$

where angular brackets $\langle\ldots\rangle$ denote an orientation average:

$$\left\langle -e\mathbf{n}\frac{\mathrm{d}u(\mathbf{r}_0, t)}{\mathrm{d}t}\right\rangle = \frac{-\mathrm{i}\omega e^2 e^{-\mathrm{i}\omega t + \mathrm{i}(\mathbf{k}.\mathbf{r}_0)}}{m(\omega_0^2 - \mathrm{i}\gamma\omega - \omega^2)}\langle\mathbf{n}(\mathbf{n}\cdot\mathbf{A})\rangle + \text{c.c.} \quad (8.16)$$

For a randomly oriented unit vector $\mathbf{n}$ and an arbitrary vector $\mathbf{A}$ the following identity holds: $\langle\mathbf{n}(\mathbf{n}\cdot\mathbf{A})\rangle = \mathbf{A}/3$. Thus, for $\Delta \ll \lambda$, we arrive at the constitutive equation for isotropic media:

$$\mathbf{P}(\mathbf{r}, t) = \frac{N_0 e^2}{3m(\omega_0^2 - \mathrm{i}\gamma\omega - \omega^2)}\mathbf{A}\,\mathrm{e}^{-\mathrm{i}\omega t + \mathrm{i}(\mathbf{k}\cdot\mathbf{r})} + \text{c.c.} \quad (8.17)$$

The dielectric tensor of such medium consisting of randomly oriented molecules is isotropic, $\epsilon_{ij}(\omega) = \epsilon(\omega)\delta_{ij}$ and has the real and imaginary parts given by

$$\begin{aligned} \mathrm{Re}\,\epsilon(\omega) &= 1 + \frac{4\pi N_0 e^2}{3m}\frac{\omega_0^2 - \omega^2}{(\omega_0^2 - \omega^2)^2 + \gamma^2\omega^2} \\ \mathrm{Im}\,\epsilon(\omega) &= \frac{4\pi N_0 e^2}{3m}\frac{\gamma\omega}{(\omega_0^2 - \omega^2)^2 + \gamma^2\omega^2} \end{aligned} \tag{8.18}$$

In the next section we will show how the Lorentz model may be generalized to describe optical nonlinearity.

8.2 NONLINEARITY IN THE LORENTZ MODEL

In order to make the Lorentz model suitable for the description of nonlinear optical effects, we need to consider the anharmonic terms in expansion (8.2). These higher order terms will ensure nonlinearity in the restoring force that acts on the electric charge of the optical electrons. When the second and third anharmonic corrections to the potential are taken into account, we have to deal with an extended, nonlinear version of the equation of motion for the electron at $\mathbf{r}_0$:

$$\frac{d^2x}{dt^2} + \gamma\frac{dx}{dt} + \omega_0^2 x + ax^2 + bx^3 = -\frac{e}{m}A_x\,\mathrm{e}^{-\mathrm{i}\omega t + \mathrm{i}(\mathbf{k}\cdot\mathbf{r}_0)} + \mathrm{c.c.} \tag{8.19}$$

where the anharmonic parameters a and b are

$$a = \frac{1}{2m}\frac{\partial^3 U}{\partial x^3}\bigg|_{x=0}, \qquad b = \frac{1}{6m}\frac{\partial^4 U}{\partial x^4}\bigg|_{x=0} \tag{8.20}$$

The displacement of the electron can now be found as a series:

$$x(t) = \sum_{n=1}^{\infty} x_n(t) \tag{8.21}$$

where $x_n(t)$ is proportional to the nth power of the electric field strength. If the electron is driven at a frequency ω, with a restoring force containing nonlinear terms which are quadratic and cubic in x, its displacement from the equilibrium position has spectral components, not only of frequency ω, but also of its harmonics 2ω, 3ω and so on.

By substituting expansion (8.21) into (8.19) we can obtain a series of consecutive equations each of which involves terms of the same order with respect to A_x. The first order equation is

$$\frac{\mathrm{d}^2 x_1}{\mathrm{d}t^2} + \gamma \frac{\mathrm{d}x_1}{\mathrm{d}t} + \omega_0^2 x_1 = -\frac{e}{m}(A_x \, \mathrm{e}^{-\mathrm{i}\omega t + \mathrm{i}(\mathbf{k}\cdot\mathbf{r}_0)} + A_x^* \, \mathrm{e}^{\mathrm{i}\omega t - \mathrm{i}(\mathbf{k}.\mathbf{r}_0)}) \tag{8.22}$$

The second- and third-order equations are

$$\frac{\mathrm{d}^2 x_2}{\mathrm{d}t^2} + \gamma \frac{\mathrm{d}x_2}{\mathrm{d}t} + \omega_0^2 x_2 = -a x_1^2 \tag{8.23}$$

and

$$\frac{\mathrm{d}^2 x_3}{\mathrm{d}t^2} + \gamma \frac{\mathrm{d}x_3}{\mathrm{d}t} + \omega_0^2 x_3 = -2a x_1 x_2 - b x_1^3 \tag{8.24}$$

These equations can be solved by consecutive approximations. The solution of the first one, i.e. equation (8.22), has already been found (see formula (8.6)). By substituting solution (8.6) into equation (8.23) we observe that the electron motion has a spectral component at 2ω, the magnitude of which is proportional to A_x^2. It also shows a static displacement proportional to $|A_x|^2$:

$$\begin{aligned} x_2(\mathbf{r}_0, t) = & -\frac{e^2}{m^2} \frac{a A_x^2 \, \mathrm{e}^{-2\mathrm{i}\omega t + 2\mathrm{i}(\mathbf{k}\cdot\mathbf{r}_0)}}{(\omega_0^2 - 2\mathrm{i}\gamma\omega - 4\omega^2)(\omega_0^2 - \mathrm{i}\gamma\omega - \omega^2)^2} \\ & - \frac{e^2}{m^2} \frac{a|A_x|^2}{\omega_0^2 |\omega_0^2 - \mathrm{i}\gamma\omega - \omega^2|^2} + \text{c.c.} \end{aligned} \tag{8.25}$$

By resorting equation (8.6) and substituting (8.25) into (8.24) we see that, in the next approximation, the electron motion also has a spectral component at ω which is proportional to $|A_x|^2 A_x$ and a spectral component at 3ω which is proportional to ${A_x}^3$:

$$\begin{aligned} x_3(\mathbf{r}_0, t) = & \frac{e^3}{m^3} \frac{+3b - 4a^2/\omega_0^2 - 2a^2/(\omega_0^2 - 2\mathrm{i}\gamma\omega - 4\omega^2)}{(\omega_0^2 - \mathrm{i}\gamma\omega - \omega^2)^2 |\omega_0^2 - \mathrm{i}\gamma\omega - \omega^2|^2} |A_x|^2 \, A_x \, \mathrm{e}^{-\mathrm{i}\omega t + \mathrm{i}(\mathbf{k}\cdot\mathbf{r}_0)} \\ & + \frac{e^3}{m^3} \frac{b - 2a^2/(\omega_0^2 - 2\mathrm{i}\gamma\omega - 4\omega^2)}{(\omega_0^2 - \mathrm{i}\gamma\omega - \omega^2)^3 (\omega_0^2 - 3\mathrm{i}\gamma\omega - 9\omega^2)} A_x^3 \, \mathrm{e}^{-3\mathrm{i}\omega t + 3\mathrm{i}(\mathbf{k}\cdot\mathbf{r}_0)} + \text{c.c.} \end{aligned} \tag{8.26}$$

Thus, the-light induced electron vibrations produce an electric current density $\mathbf{J}(\mathbf{r}, t)$ with spectral components at ω, 2ω and 3ω. Applying a similar procedure of integration over a small volume as for the linear optical response, we obtain the quadratic and cubic terms of the medium's nonlinear constitutive equations (compare with local terms in phenomenologically introduced constitutive equations (2.47) and (2.53)):

$$
\begin{aligned}
P_x^{(2)}(2\omega, \mathbf{r}) &= \chi_{xxx}^{(2)}(2\omega, \omega, \omega)E_x^2(\omega, \mathbf{r}) \\
P_x^{(2)}(0, \mathbf{r}) &= 2\chi_{xxx}^{(2)}(0, \omega, -\omega)|E_x(\omega, \mathbf{r})|^2 \\
P_x^{(3)}(\omega, \mathbf{r}) &= 3\chi_{xxxx}^{(3)}(\omega, \omega, \omega, -\omega)|E_x(\omega, \mathbf{r})|^2 E_x(\omega, \mathbf{r}) \\
P_x^{(3)}(3\omega, \mathbf{r}) &= \chi_{xxxx}^{(3)}(3\omega, \omega, \omega, \omega)E_x^3(\omega, \mathbf{r})
\end{aligned}
\tag{8.27}
$$

Within this microscopic model the Y and Z Cartesian components of the nonlinear parts, namely $\mathbf{P}^{(2)}(2\omega, \mathbf{r})$, $\mathbf{P}^{(2)}(0, \mathbf{r})$, $\mathbf{P}^{(3)}(\omega, \mathbf{r})$ and $\mathbf{P}^{(3)}(3\omega, \mathbf{r})$ of the polarization response are zero. The only nonzero components of the second-order local susceptibilities are

$$
\begin{aligned}
\chi_{xxx}^{(2)}(2\omega, \omega, \omega) &= \frac{aN_0e^3}{m^2(\omega_0^2 - 2\mathrm{i}\gamma\omega - 4\omega^2)(\omega_0^2 - \mathrm{i}\gamma\omega - \omega^2)^2} \\
\chi_{xxx}^{(2)}(0, \omega, -\omega) &= \frac{aN_0e^3}{m^2\omega_0^2|\omega_0^2 - \mathrm{i}\gamma\omega - \omega^2|^2}
\end{aligned}
\tag{8.28}
$$

and the only nonzero components of the third-order local susceptibilities are

$$
\chi_{xxxx}^{(3)}(\omega, \omega, \omega, -\omega) = \frac{N_0e^4}{m^3}\frac{-b + 4a^2/3\omega_0^2 + 2a^2/[3(\omega_0^2 - 2\mathrm{i}\gamma\omega - 4\omega^2)]}{(\omega_0^2 - \mathrm{i}\gamma\omega - \omega^2)^2|\omega_0^2 - \mathrm{i}\gamma\omega - \omega^2|^2}
$$

$$
\chi_{xxxx}^{(3)}(3\omega, \omega, \omega, \omega) = \frac{N_0e^4}{m^3}\frac{-b + 2a^2/(\omega_0^2 - 2\mathrm{i}\gamma\omega - 4\omega^2)}{(\omega_0^2 - \mathrm{i}\gamma\omega - \omega^2)^3(\omega_0^2 - 3\mathrm{i}\gamma\omega - 9\omega^2)}
\tag{8.29}
$$

Hence, the model of oriented nonlinear oscillators shows anisotropy. In order to make the response of an isotropic medium one must admit random orientation of the molecules and go through a procedure similar to that used in the calculation of isotropic linear response. This will give:

$$
\mathbf{P}^{(2)}(2\omega, \mathbf{r}) = 0,
$$

$$
\begin{aligned}
\mathbf{P}^{(3)}(\omega, \mathbf{r}) = \tfrac{2}{15}\chi_{xxxx}^{(3)}(\omega, \omega, \omega, -\omega)\,&[2(\mathbf{E}(\omega, \mathbf{r})\cdot\mathbf{E}^*(\omega, \mathbf{r}))\mathbf{E}(\omega, \mathbf{r}) \\
&+ (\mathbf{E}(\omega, \mathbf{r})\cdot\mathbf{E}(\omega, \mathbf{r}))\mathbf{E}^*(\omega, \mathbf{r})]
\end{aligned}
$$

and

$$
\mathbf{P}^{(3)}(3\omega, \mathbf{r}) = \tfrac{1}{5}\chi_{xxxx}^{(3)}(3\omega, \omega, \omega, \omega)\,((\mathbf{E}(\omega, \mathbf{r})\cdot\mathbf{E}(\omega, \mathbf{r}))\mathbf{E}(\omega, \mathbf{r}).
$$

The linear and nonlinear models examined above did not include the nonlocality of optical response. The description of nonlocality requires a more complicated molecular model and in the next section we will show that such a model can be constructed using two interacting oscillators moving in different directions.

8.3 TWO-ELECTRON MODEL OF CHIRAL MOLECULES

To describe the nonlocality of optical response and phenomena such as optical activity, we need a molecular model consisting of more than one optical electron. In 1915 Born suggested a model consisting of two identical optical electrons [9–11], similar to those used by Lorentz. The electrons in the molecule are separated by a distance D and move in perpendicular directions (see Fig. 8.3). We choose the Cartesian coordinate frame in such a way that the molecular oscillators vibrate along its X and Y axes. As shown in Fig. 8.3, the electron equilibrium positions are slightly off the Z direction, making a chiral structure that does not coincide with its mirror image. The electrons are bound to each other by an elastic spring, and on being displaced from their equilibrium positions they begin to oscillate. The centre of the molecule is located at $\mathbf{r}_0$.

The potential energy of the chiral molecule presented in Fig. 8.3 depends both on the displacement u_1 of the first electron in the X direction and on the displacement u_2 of the second electron in the Y direction. Therefore, the potential energy of the electrons may be written as follows:

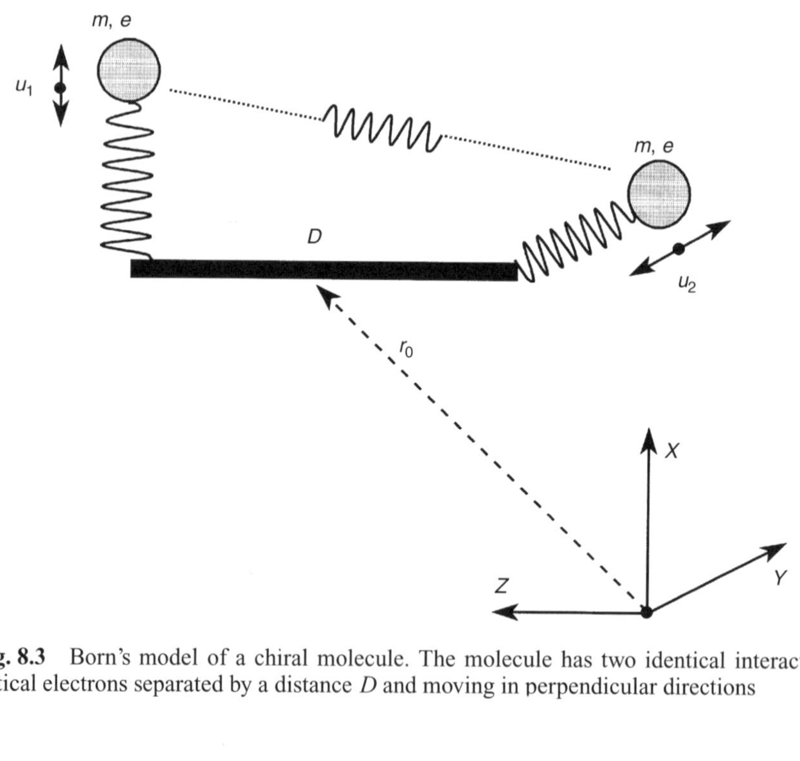

Fig. 8.3 Born's model of a chiral molecule. The molecule has two identical interacting optical electrons separated by a distance D and moving in perpendicular directions

$$U(u_1, u_2) = U\left(x_0 + u_1, y_0, z_0 + \frac{D}{2}; x_0, y_0 + u_2, z_0 - \frac{D}{2}\right)$$
$$- U\left(x_0, y_0, z_0 + \frac{D}{2}; x_0, y_0, z_0 - \frac{D}{2}\right)$$
$$= \left.\frac{\partial^2 U}{\partial u_1^2}\right|_{u_{1,2}=0} \frac{u_1^2}{2!} + \left.\frac{\partial^2 U}{\partial u_2^2}\right|_{u_{1,2}=0} \frac{u_2^2}{2!} + \left.\frac{\partial^2 U}{\partial u_1 \partial u_2}\right|_{u_{1,2}=0} u_1 u_2 + \dots \tag{8.30}$$

or

$$U - U_0 = \tfrac{1}{2} m\omega_0^2 (u_1^2 + u_2^2) + m\xi u_1 u_2 \tag{8.31}$$

where $\omega_0^2 = (1/m)\partial^2 U/\partial u_1^2 = (1/m)\partial^2 U/\partial u_2^2$ is the square of the natural frequency of the oscillators with no bound between them. The second term, containing $m\xi = \partial^2 U/(\partial u_1 \partial u_2)$, describes the elastic interaction between the electrons.

We will consider the interaction of this molecule with a plane light wave propagating in the Z direction. In the electric field of the light wave the electrons are forced to move according to the following equations:

$$\frac{d^2 u_1}{dt^2} + \gamma \frac{du_1}{dt} + \omega_0^2 u_1^2 + \xi u_2 = -\frac{e}{m} A_x \, e^{-i\omega t + ik(z_0 + D/2)} + \text{c.c.}$$
$$\frac{d^2 u_2}{dt^2} + \gamma \frac{du_2}{dt} + \omega_0^2 u_2^2 + \xi u_1 = -\frac{e}{m} A_y \, e^{-i\omega t + ik(z_0 - D/2)} + \text{c.c} \tag{8.32}$$

Here, the centre of the molecule is at $\{x_0; y_0; z_0\}$ while the X and Y vibrating oscillators are at $\{x_0; y_0; z_0 + D/2\}$ and $\{x_0; y_0; z_0 - D/2\}$ respectively. The difference between the phase factors of the forces driving the X and Yoscillators is due to the wave travel time between the oscillators. The set of linear differential equations (8.32) may readily be solved, to give

$$u_1(\mathbf{r}_0, t) = -\frac{e}{m\omega_0^2}[F_+(\omega) A_x \, e^{ikD/2} + F_-(\omega) A_y \, e^{-ikD/2}] \, e^{-i\omega t + ikz_0} + \text{c.c.}$$
$$u_2(\mathbf{r}_0, t) = -\frac{e}{m\omega_0^2}[F_+(\omega) A_y \, e^{-ikD/2} + F_-(\omega) A_x \, e^{ikD/2}] \, e^{-i\omega t + ikz_0} + \text{c.c.} \tag{8.33}$$

where

$$
\begin{aligned}
F_+(\omega) &= \frac{\omega_0^2(\omega_0^2 - \mathrm{i}\gamma\omega - \omega^2)}{(\omega_0^2 - \mathrm{i}\gamma\omega - \omega^2)^2 - \xi^2} \\
F_-(\omega) &= -\frac{\xi\omega_0^2}{(\omega_0^2 - \mathrm{i}\gamma\omega - \omega^2)^2 - \xi^2}
\end{aligned}
\tag{8.34}
$$

Now the current density induced in the medium can be calculated. We will consider a medium consisting of equally oriented 'Born molecules' with N_0 molecules per unit volume. At this stage in the one-dimensional single-electron Lorentz model, we were able to write down the constitutive equation by simply summing the contributions from the individual electrons within a small volume (see formula (8.7)). The treatment of optical nonlocality requires a more sophisticated approach: we need to account for the fact that the moving electrons are spatially separated inside the molecule, i.e. we must account for the fact that, due to the finite size of the molecule, the light wave reaches different electrons of the molecule at different moments in time. Moreover, the electric fields produced by the motions of the spatially separated charges reach an observer with a certain time lag.

To account for the finite size of the molecule, we start from the general formula for the current density induced at point $\mathbf{r}$ by a moving electron at point $\mathbf{R}$: $\mathbf{j}(\mathbf{r}) = -e\mathbf{v}\delta(\mathbf{r} - \mathbf{R})$. Here $\mathbf{v}$ is the velocity of the electron and the delta function pinpoints the actual position of the electron. Since each molecule contains two electrons, the current density at point $\mathbf{r}$, created by an ensemble of identically oriented Born molecules, may be written as

$$
\mathbf{J}(\mathbf{r},\, t) =
$$

$$
\frac{1}{\delta V}\int_{\delta V}(-e)\left[\mathbf{n}_1\frac{\mathrm{d}u_1(\mathbf{r}_0,\, t)}{\mathrm{d}t}\delta(\mathbf{r} - \mathbf{r}_0 - \mathbf{r}_1^{(0)}) + \mathbf{n}_2\frac{\mathrm{d}u_2(\mathbf{r}_0,\, t)}{\mathrm{d}t}\delta(\mathbf{r} - \mathbf{r}_0 - \mathbf{r}_2^{(0)})\right]\mathrm{d}\mathbf{r}_0
\tag{8.35}
$$

Here $\mathbf{r}_0$ is the coordinate of the molecular centre, $\mathbf{r}_{1,2}^{(0)} = \{0,\, 0,\, \pm D/2\}$ are the positions of the electrons with respect to the molecular centre and $\mathbf{n}_1 = \{1,\, 0,\, 0\}$ and $\mathbf{n}_2 = \{0,\, 1,\, 0\}$ are unit vectors along the directions of the molecular oscillator's vibrations. The delta functions in equation (8.35) pinpoint the positions of the electrons with respect to the molecular centre. The integration over $\mathbf{r}_0$ gives a factor of $N_0\delta V$, and formula (8.35) reduces to

$$
\mathbf{J}(\mathbf{r},\, t) = -eN_0\left[\mathbf{n}_1\left.\frac{\mathrm{d}u_1(\mathbf{r}_0,\, t)}{\mathrm{d}t}\right|_{\mathbf{r}_0 = \mathbf{r} - \mathbf{r}_1^{(0)}} + \mathbf{n}_2\left.\frac{\mathrm{d}u_2(\mathbf{r}_0,\, t)}{\mathrm{d}t}\right|_{\mathbf{r}_0 = \mathbf{r} - \mathbf{r}_2^{(0)}}\right]
\tag{8.36}
$$

Substituting $\mathrm{d}u_{1,2}/\mathrm{d}t$ from equation (8.33) into (8.36) we obtain the following formulae for the Cartesian components of the current density:

$$
\begin{aligned}
J_x(\mathbf{r}, t) &= -\mathrm{i}\omega \frac{N_0 e^2}{m\omega_0^2} \times [F_+(\omega)A_x + F_-(\omega)A_y e^{-ikD}]\, \mathrm{e}^{-\mathrm{i}\omega t + \mathrm{i}kz} + \text{c.c.} \\
J_y(\mathbf{r}, t) &= -\mathrm{i}\omega \times \frac{N_0 e^2}{m\omega_0^2} \times [F_+(\omega)A_y + F_-(\omega)A_x e^{ikD}]\, \mathrm{e}^{-\mathrm{i}\omega t + \mathrm{i}kz} + \text{c.c.}
\end{aligned} \tag{8.37}
$$

Since $D \ll \lambda$, that is $kD \ll 1$, we can expand $\mathrm{e}^{\pm \mathrm{i}kD} \simeq 1 \pm \mathrm{i}kD \ldots$ If in this expansion only the first term is retained, $\mathrm{e}^{\pm \mathrm{i}kD} \simeq 1$, we ignore the size of the molecule in comparison with the wavelength, i.e. assume that the induced current is localized in its centre. Therefore, retaining only this term implies that we neglect all of the retardation effects within the molecule. This is known as the calculation of optical response in the local approximation. We will, however, also retain the second term of this expansion $\mathrm{e}^{\pm \mathrm{i}kD} \simeq 1 \pm \mathrm{i}kD$. In this case the induced current is not necessarily localized at the centre of the molecule. Optical effects which are due to the term linear in k are referred to as nonlocal or first-order spatial dispersion phenomena.

From $\mathbf{P}(\omega, \mathbf{r}) = \mathbf{J}(\omega, \mathbf{r})/(-i\omega)$ using (8.37) we obtain a constitutive equation which accounts for nonlocality in the first-order spatial dispersion approximation:

$$
P_i(\omega, \mathbf{r}) = \frac{1}{4\pi}[\epsilon_{ij}(\omega) - \delta_{ij}]E_j(\omega, \mathbf{r}) + \frac{1}{4\pi}\Gamma^{(1)}_{ijn}(\omega)\nabla_n E_j(\omega, \mathbf{r}) \tag{8.38}
$$

where ϵ_{ij} is the dielectric tensor and $\Gamma^{(1)}_{ijn}$ is the nonlocality tensor (compare with the phenomenological constitutive equation (2.14)). The dielectric tensor has the following nonzero components:

$$
\begin{aligned}
\epsilon_{xx}(\omega) &= \epsilon_{yy}(\omega) = 1 + \frac{4\pi N_0 e^2}{m} \frac{\omega_0^2 - \omega^2 - \mathrm{i}\gamma\omega}{(\omega_0^2 - \omega^2 - \mathrm{i}\gamma\omega)^2 - \xi^2} \\
\epsilon_{xy}(\omega) &= \epsilon_{yx}(\omega) = -\frac{4\pi N_0 e^2}{m} \frac{\xi}{(\omega_0^2 - \omega^2 - \mathrm{i}\gamma\omega)^2 - \xi^2} \\
\epsilon_{zz}(\omega) &= 1
\end{aligned} \tag{8.39}
$$

while the nonzero components of the nonlocality tensor are as follows:

$$
\Gamma^{(1)}_{xyz}(\omega) = -\Gamma^{(1)}_{yxz}(\omega) = -D\epsilon_{xy}(\omega) = \frac{4\pi D N_0 e^2}{m} \frac{\xi}{(\omega_0^2 + \xi - \mathrm{i}\gamma\omega - \omega^2)^2 - \xi^2} \tag{8.40}
$$

If there is no interaction between the electrons in the molecule, that is $\xi = 0$, the nonlocality tensor vanishes in spite of the spatial separation of the electrons within the molecules. Interactions between the electrons ($\xi \neq 0$) and the finite size of the Born molecule ($D \neq 0$) ensure the presence of the nonlocal term in constitutive equation (8.38). The bigger the molecule and the stronger the coupling between optical electrons on opposite sides of the molecule, the stronger the nonlocality.

In isotropic media, after averaging over molecular orientations we obtain: $\epsilon_{ij} = \epsilon\delta_{ij}$, where $\epsilon = (\epsilon_{xx} + \epsilon_{yy} + \epsilon_{zz})/3$ and $\Gamma^{a} = \Gamma^{(1)}_{xyz}/3$. The frequency dependencies of the real and imaginary parts of ϵ and Γ^{a} are presented in Fig. 8.4 for different ratios between $|\omega_2 - \omega_1|$ and the damping parameter γ where $\omega_{1,2}^2 = \omega_0^2 \pm \xi$.

If $|\omega_2 - \omega_1| \ll \gamma$, the frequency dispersions of the real and imaginary parts of the dielectric coefficient ϵ are similar to those for a one-electron molecule (compare with formula (8.18)). In this case the frequency dispersion of Re $\{\Gamma^{a}\}$ is similar to d(Re $\{\epsilon\})/\mathrm{d}\omega$, while that of Im $\{\Gamma^{a}\}$ is similar to d(Im $\{\epsilon\})/\mathrm{d}\omega$ (see Fig. 8.4a). If $|\omega_2 - \omega_1| \gg \gamma$, we observe two separated resonances at frequencies ω_1 and ω_2. Here the frequency dependencies of the real and imaginary parts of the dielectric coefficient ϵ and Γ^{a} are similar (see Fig. 8.4b).

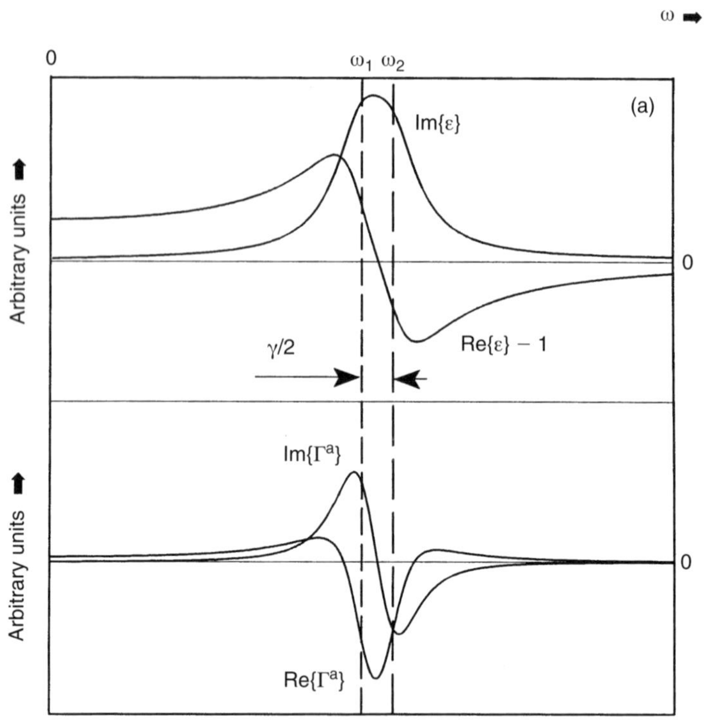

Fig. 8.4 The frequency dispersion of the real and imaginary parts of local and nonlocal optical susceptibility in the two-electron molecular model for: (a) $|\omega_2 - \omega_1| = 0.5\gamma$ and (b) $|\omega_2 - \omega_1| = 4\gamma$

In the next section we will show how the two-electron chiral molecular model can be modified to describe nonlinear nonlocal effects.

8.4 NONLINEARITY IN THE TWO-ELECTRON CHIRAL MOLECULE

To make the two-electron model suitable for the description of nonlinear nonlocal optical effects, the anharmonic terms in the expansion (8.30) of the potential $U(u_1, u_2)$ must be considered [7]. In this section we will analyse the case where (a) the binding between the oscillators is elastic and (b) the first nonvanishing anharmonic terms in $U(u_1, u_2)$ are of the fourth order in u_1 and u_2. The potential energy of the oscillators can then be written as follows (compare with equation (8.31)):

$$U - U_0 = \frac{m}{2}[\omega_0^2(u_1^2 + u_2^2) + 2\xi u_1 u_2] + \frac{m}{4}b(u_1^4 + u_2^4) \qquad (8.41)$$

The equations of motion of the electrons in the field of the light wave are now nonlinear (compare with linear equations (8.32)):

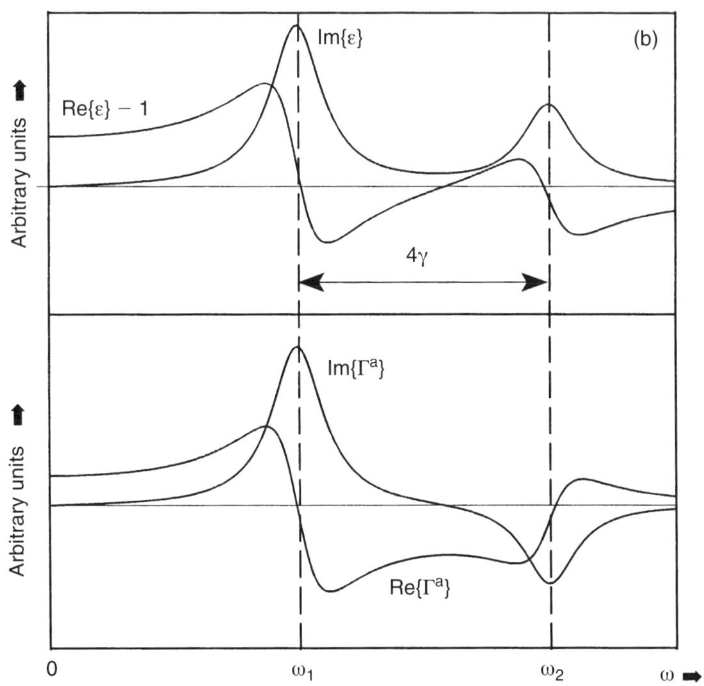

$$\frac{\mathrm{d}^2u_{1,2}}{\mathrm{d}t^2}+\gamma\frac{\mathrm{d}u_{1,2}}{\mathrm{d}t}+\omega_0^2u_{1,2}+\xi u_{2,1}+bu_{1,2}^3=-\frac{e}{m}A_{x,y}\,\mathrm{e}^{\pm\mathrm{i}kD/2}\,\mathrm{e}^{-\mathrm{i}\omega t+\mathrm{i}kz_0}+\mathrm{c.c.} \tag{8.42}$$

To calculate the optical response at frequency ω we first need to find the spectral components of the amplitudes $u_{1,2}$ at frequency ω. They can be found from equations (8.42) as a series of terms of different order in the light wave amplitude $\mathbf{A}$:

$$u_{1,2}(\omega)=u_{1,2}^{(1)}+u_{1,2}^{(2)}+u_{1,2}^{(3)} \tag{8.43}$$

where the superscripts indicate the order of the term in $\mathbf{A}$. The linear terms in the amplitude of the electric field are (compare with equation (8.33)):

$$\begin{aligned}u_1^{(1)}&=-\frac{e}{m\omega_0^2}[F_+(\omega)A_x\,\mathrm{e}^{\mathrm{i}k(z_0+D/2)}+F_-(\omega)A_y\,\mathrm{e}^{\mathrm{i}k(z_0-D/2)}]\\ u_2^{(1)}&=-\frac{e}{m\omega_0^2}[F_+(\omega)A_y\,\mathrm{e}^{\mathrm{i}k(z_0-D/2)}+F_-(\omega)A_x\,\mathrm{e}^{\mathrm{i}k(z_0-D/2)}]\end{aligned} \tag{8.44}$$

and the cubic terms are

$$\begin{aligned}u_1^{(3)}&=-\frac{3b}{\omega_0^2}[F_+(\omega)|u_1^{(1)}|^2u_1^{(1)}+F_-(\omega)|u_2^{(1)}|^2u_2^{(1)}]\\ u_2^{(3)}&=-\frac{3b}{\omega_0^2}[F_+(\omega)|u_2^{(1)}|^2u_2^{(1)}+F_-(\omega)|u_1^{(1)}|^2u_1^{(1)}]\end{aligned} \tag{8.45}$$

where the spectral response functions $F_\pm(\omega)$ are as defined in equatons (8.34). Note that the second-order term in equation (8.43) does not show up in equation (8.42), that is $u_{1,2}^{(2)}=0$.

Using the same procedure as in the previous section, we can now write down the spectral component of the current density of the charges at frequency ω. This will give us the constitutive equation of a medium consisting of two-electron nonlinear molecules of number density N_0 (compare with the phenomenologically introduced constitutive equations (2.16) and (2.56)):

$$\begin{aligned}P_i(\omega,\mathbf{k})&=P_i^{(1)}(\omega,\mathbf{k})+P_i^{(3)}(\omega,\mathbf{k})\\ &=\frac{1}{4\pi}(\epsilon_{ij}-\delta_{ij}+ik\Gamma_{ijz}^{(1)})A_j+[\chi_{ijkl}+ik\Gamma_{ijklz}]A_jA_kA_l^*\end{aligned} \tag{8.46}$$

In constitutive equation (8.46), ϵ_{ij} and $\Gamma_{ijn}^{(1)}$ are given by formulae (8.39) and (8.40) while the components of the local and nonlocal nonlinear susceptibility tensors are

$$
\begin{aligned}
\chi_{xxxx} &= \chi_{yyyy} = -\frac{3N_0 e^4}{m^3\omega_0^8}[|F_+(\omega)|^2 F_+^2(\omega) + |F_-(\omega)|^2 F_-^2(\omega)] \\
\chi_{xxyy} &= \chi_{yyxx} = \chi_{xyyx} = \chi_{yxxy} = -\frac{3N_0 e^4}{m^3\omega_0^8}[|F_+(\omega)|^2 F_-^2(\omega) + |F_-(\omega)|^2 F_+^2(\omega)] \\
\chi_{xxxy} &= \chi_{yyyx} = -\frac{3N_0 e^4}{m^3\omega_0^8}[F_+^3(\omega)F_-^*(\omega) + F_-^3(\omega)F_+^*(\omega)] \\
\chi_{xxyx} &= \chi_{yyxy} = \chi_{xyyy} = \chi_{yxxx} = -\frac{3N_0 e^4}{m^3\omega_0^8}[|F_+(\omega)|^2 + |F_-(\omega)|^2]F_+(\omega)F_-(\omega) \\
\Gamma_{xxyxz} &= \Gamma_{xyyyz} = -\Gamma_{yyxyz} = -\Gamma_{yxxxz} = -D\chi_{xxyx} \\
\Gamma_{xyyxz} &= -\Gamma_{yxxyz} = -D\chi_{xyyx} \\
\Gamma_{xxxyz} &= -\Gamma_{yyyxz} = D\chi_{xxxy}
\end{aligned}
\tag{8.47}
$$

Above, we analysed a medium consisting of identically oriented molecules. This model shows anisotropy as, for example, $\chi_{xxxx} \neq \chi_{zzzz} = 0$. An isotropic medium can be described by an ensemble of randomly oriented two-electron molecules. The corresponding constitutive equation can be obtained by averaging the current density over the molecular orientation (see the orientation averaging procedure described in section 8.1. This gives the constitutive equation for nonlinear nonlocal isotropic media is

$$
\begin{aligned}
\mathbf{P}(\omega, \mathbf{k}) = \mathbf{P}^{(1)}(\omega, \mathbf{k}) + \mathbf{P}^{(3)}(\omega, \mathbf{k}) &= \frac{1}{4\pi}((\epsilon - 1)\mathbf{A} + i\Gamma^{(1)}[\mathbf{k} \times \mathbf{A}]) \\
&+ \chi_1^{(3)}(\mathbf{A}\cdot\mathbf{A}^*)\mathbf{A} + \chi_2^{(3)}(\mathbf{A}\cdot\mathbf{A})\mathbf{A}^* \\
&+ i\Gamma_1^{(3)}(\mathbf{A}\cdot\mathbf{A}^*)[\mathbf{k} \times \mathbf{A}] + i\Gamma_2^{(3)}(\mathbf{k}\cdot[\mathbf{A} \times \mathbf{A}^*])\mathbf{A}
\end{aligned}
\tag{8.48}
$$

where

$$
\epsilon = \tfrac{1}{3}(1 + \epsilon_{xx} + \epsilon_{yy})
$$

$$
\begin{aligned}
\Gamma^{(1)} &= \tfrac{1}{3}\Gamma^{(1)}_{xyz} \\
\chi_1^{(3)} &= \tfrac{2}{15}(2\chi_{xxxx} + 3\chi_{xxyy} - \chi_{xyyx}) \\
\chi_2^{(3)} &= \tfrac{2}{15}(\chi_{xxxx} - \chi_{xxyy} + 2\chi_{xyyx}) \\
\Gamma_1^{(3)} &= -\tfrac{1}{15}(\Gamma_{xxxyz} - 2\Gamma_{xxyxz} + \Gamma_{xyyyz}) \\
\Gamma_2^{(3)} &= \tfrac{1}{15}(3\Gamma_{xxxyz} + 2\Gamma_{xxyxz} + \Gamma_{xyyyz})
\end{aligned}
\tag{8.49}
$$

One can find the following relationship between $\chi_2^{(3)}$ and $\Gamma_1^{(3)}$ and between $\Gamma^{(1)}$ and ϵ:

$$\frac{\Gamma_1^{(3)}}{\chi_2^{(3)}} = \frac{D}{2}\,\frac{F_+^*(\omega)F_-(\omega) + F_-^*(\omega)F_+(\omega)}{|F_+(\omega)|^2 + |F_-(\omega)|^2}$$
$$\frac{\Gamma^{(1)}}{\epsilon - 1} = D\,\frac{F_+ - F_-}{F_+ + F_-} \tag{8.50}$$

or, far from absorption resonances ($\omega \ll \omega_0$),

$$\frac{\Gamma_1^{(3)}}{\chi_2^{(3)}} = \frac{\xi}{2\omega_0^2}D$$
$$\frac{\Gamma^{(1)}}{\epsilon - 1} = \frac{\xi}{\omega_0^2}D \tag{8.51}$$

The two-electron model allows an estimate to be made of the non-resonant nonlocal nonlinearity $\Gamma^{(3)}$ if information on $\Gamma^{(1)}$ and the conventional cubic nonlinearity $\chi^{(3)}$ is available: $\Gamma^{(3)} \simeq \Gamma^{(1)}\chi^{(3)}/2(\epsilon - 1)$, or, assuming $\xi \simeq \omega$, $\Gamma^{(3)} \simeq D\chi^{(3)}/2$ and $\Gamma^{(1)} \simeq D(\epsilon - 1)$.

8.5 CLASSICAL OSCILLATOR MODEL AT LARGE

To describe the optical response of complex molecules, the two-electron model presented in the previous section is often insufficient. In particular, it often does not allow an adequate representation of the symmetrical properties of such molecules. In this section we will develop a more sophisticated model which can be used to calculate the local and non-local optical susceptibilities of polyatomic molecules [12].

Let us examine a molecule which has N optical electrons numbered by subscript p. Let $\mathbf{r}_0$ be the coordinate of the molecular centre and $\mathbf{r}_p^{(0)}$ be the equilibrium position of the pth electron with respect to the centre. Each of these electrons moves along a certain direction represented by unit vectors $\mathbf{n}_p (p = 1, \ldots, N)$ (see Fig. 8.5). If $U(\mathbf{r}_1, \ldots, \mathbf{r}_N)$ is the total potential energy of these optical electrons, their equilibrium positions $\mathbf{r}_p^{(0)}$ may be found from equations $\partial U/\partial \mathbf{r}_p = 0$, $(p = 1, \ldots, N)$. Optical excitation stimulates vibrations of the electrons near $\mathbf{r}_p^{(0)}$. Electron displacements from equilibrium will be denoted $\mathbf{u}_p = u_p \mathbf{n}_p$.

The potential energy of optical electrons $U(\mathbf{r}_1, \ldots, \mathbf{r}_N)$ can be presented as a Taylor series near their equilibrium positions:

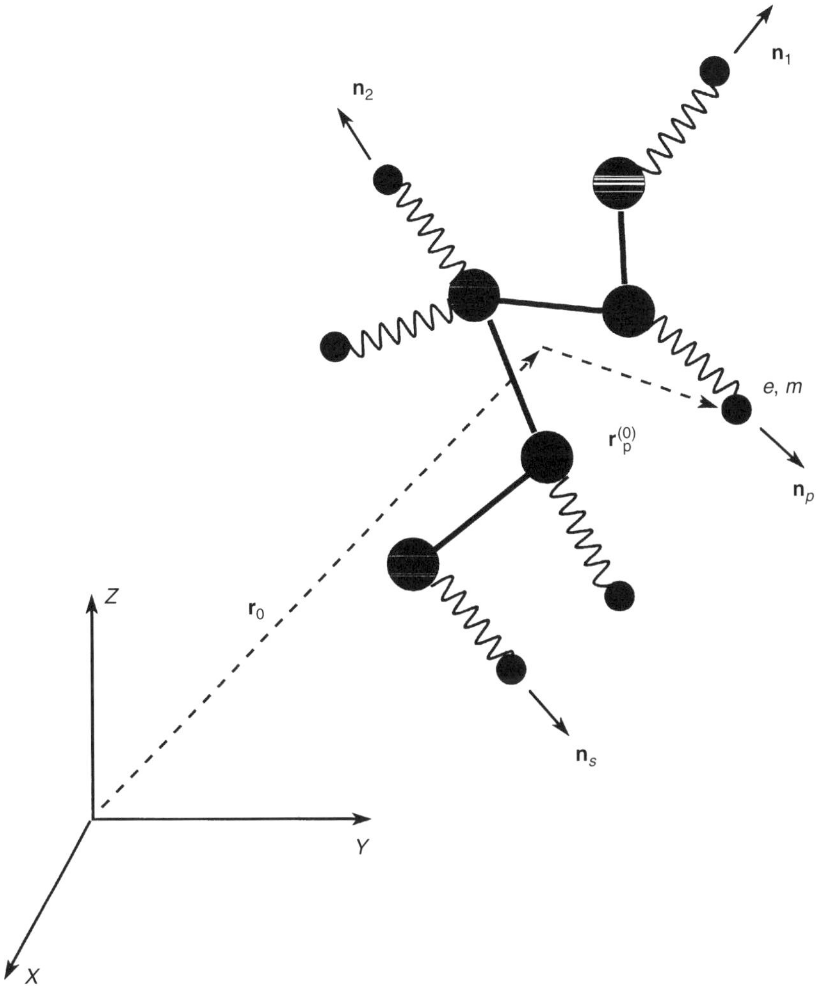

Fig. 8.5 Molecular skeleton of a hypotetical complex polyatomic molecule. Optical electrons are represented as anharmonic oscillators with electric charge e and mass m. The electrons interact with one another and move along certain directions indicated by unit vectors $\mathbf{n}_{\mathrm{p}}$

$$U(\mathbf{r}_1^{(0)} + \mathbf{n}_1 u_1, \ldots, \mathbf{r}_N^{(0)} + \mathbf{n}_N u_N) - U_0(\mathbf{r}_1^{(0)}, \ldots, \mathbf{r}_N^{(0)})$$

$$= \frac{m}{2} \sum_{p,q=1}^{N} \xi_{pq} u_p u_q + \frac{m}{3} \sum_{p,q,r=1}^{N} a_{pqr} u_p u_q u_r + \frac{m}{4} \sum_{p,q,r,s=1}^{N} b_{pqrs} u_p u_q u_r u_s + \ldots \quad (8.52)$$

where

$$\xi_{pq} = n_{pi} n_{qj} \frac{\partial^2 U}{\partial r_{pi} \partial r_{qj}}$$

$$a_{pqr} = \tfrac{1}{2} n_{pi} n_{qj} n_{rl} \frac{\partial^3 U}{\partial r_{pi} \partial r_{qj} \partial r_{rl}} \tag{8.53}$$

$$b_{pqrs} = \tfrac{1}{6} n_{pi} n_{qj} u_{rl} u_{lm} \frac{\partial^4 U}{\partial r_{pi} \partial r_{qj} \partial r_{rl} \partial r_{sm}}$$

are the Taylor series coefficients where we presume summation over indices i, j, k and l, which refer to the Cartesian components, and indices p, q, r and s, which number the oscillators.

The motion in the light wave of the pth optical electron is governed by the following equation:

$$m \frac{\mathrm{d}^2 u_p}{\mathrm{d}t^2} + m\gamma_p \frac{\mathrm{d}u_p}{\mathrm{d}t} + \frac{\partial U}{\partial u_p} = R_p \tag{8.54}$$

where γ_p is the damping parameter for pth electron and R_p is the component of the Lorentz force in the direction of $\mathbf{n}_p$. The Lorentz force is associated with the electric $\mathbf{E}(\mathbf{r}_p, t)$ and magnetic $\mathbf{B}(\mathbf{r}_p, t)$ fields of the light wave at the point $\mathbf{r}_p$:

$$\mathbf{F}_p = -e\mathbf{E}(\mathbf{r}_p, t) - \frac{e}{c} \frac{\mathrm{d}u_p}{\mathrm{d}t} [\mathbf{n}_p \times \mathbf{B}(\mathbf{r}_p, t)] \tag{8.55}$$

Consequently, R_p is given by

$$R_p = (\mathbf{n}_p \cdot \mathbf{F}_p) \tag{8.56}$$

In equation (8.55) the electric and magnetic fields are taken at the position of the pth electron $\mathbf{r}_p = \mathbf{r}_0 + \mathbf{r}_p^{(0)} + \mathbf{n}_p u_p$. The vector of the magnetic part of the Lorentz force is perpendicular to the velocity of the electron, i.e. to its oscillation direction $\mathbf{n}_p$. Consequently, the magnetic part of the Lorentz force has no components in the direction of $\mathbf{n}_p$; it does not drive the electron and it can be dropped from any further consideration.

We can also assume that the light-induced electron displacement is much smaller than the molecular size, $|u_p| \ll |\mathbf{r}_p^{(0)}|$. Indeed, it follows from equation (8.54) that the magnitude of the electron vibration in the electric field, at frequency ω, can be estimated as $u_p \simeq e|E|/(m\omega^2)$. Even when driven by a light wave of high intensity, 1 GW/cm^2, the electron would not oscillate by more than 2×10^{-9} cm, which is much smaller than the typical molecular size $|r_p^{(0)}|$ of $5{-}50 \times 10^{-8}$ cm. Therefore, we can approximate $\mathbf{r}_p = \mathbf{r}_0 + \mathbf{r}_p^{(0)} + \mathbf{n}_p u_p$ to $\mathbf{r}_p = \mathbf{r}_0 + \mathbf{r}_p^{(0)}$ and rearrange formula (8.56):

$$R_p = -e(\mathbf{n}_p \cdot \mathbf{E}(\mathbf{r}_0 + \mathbf{r}_p^{(0)}, t)) \tag{8.57}$$

The electric current density, produced by the motion of all the electrons in the molecule, is

$$\mathbf{J}(\mathbf{r},\ t) = \frac{1}{\delta V}\int_{\delta V} \mathrm{d}\mathbf{r}_0 \sum_{p=1}^{N} (-e)\ \mathbf{n}_p \frac{\mathrm{d}u_p}{\mathrm{d}t}\ \delta(\mathbf{r} - \mathbf{r}_0 - \mathbf{r}_p^{(0)}) \tag{8.58}$$

where the δ functions pinpoint the positions of the optical electrons (compare with formula (8.35) for the two-electron model) and $\mathrm{d}u_p/\mathrm{d}t$ must be obtained from the set of equations (8.54) giving the following expression for the current density $\mathbf{J}(\mathbf{r},\ t)$:

$$\mathbf{J}(\mathbf{r},\ t) = N_0 \times \sum_{p=1}^{N} (-e)\ \mathbf{n}_p \frac{\mathrm{d}u_p}{\mathrm{d}t}\bigg|_{\mathbf{r}_0 = \mathbf{r} - \mathbf{r}_p^{(0)}} \tag{8.59}$$

The above formula leads to the constitutive equation for an ensemble of oriented molecules. If the molecules are randomly oriented (as in a fluid) or partially oriented (as in a liquid crystal), expression (8.59) should be averaged over the orientations of the molecules.

With potential energy (8.52), the equation of motion (8.54) of the pth molecular electron in a light wave $\mathbf{E}(\mathbf{r},\ t) = \mathbf{E}(\omega,\ \mathbf{r})\,\mathrm{e}^{-\mathrm{i}\omega t}$ + c.c. of frequency ω, becomes

$$\frac{\mathrm{d}^2 u_p}{\mathrm{d}t^2} + \gamma_p \frac{\mathrm{d}u_p}{\mathrm{d}t} + \sum_{q=1}^{N} \xi_{pq} u_q + \sum_{q,r=1}^{N} a_{pqr} u_q u_r + \sum_{q,r,s=1}^{N} b_{pqrs} u_q u_r u_s$$
$$= -\frac{e}{m}(\mathbf{n}_p \cdot \mathbf{E}_p) e^{-\mathrm{i}\omega t} + \text{c.c.} \tag{8.60}$$

where $\mathbf{E}_p = \mathbf{E}(\omega,\ \mathbf{r}_0 + \mathbf{r}_p^{(0)})$. We can solve eqaution (8.60) by presenting u_p as a series of terms proportional to the nth power of the electric field strength at different spectral harmonics:

$$u_p(t) = u_p^{(1)}(\omega)\,\mathrm{e}^{-\mathrm{i}\omega t} + u_p^{(2)}(0) + u_p^{(2)}(2\omega)\,\mathrm{e}^{-2\mathrm{i}\omega t}$$
$$+ u_p^{(3)}(\omega)\,\mathrm{e}^{-\mathrm{i}\omega t} + u_p^{(3)}(3\omega)\,\mathrm{e}^{-3\mathrm{i}\omega t} + \text{c.c.} + \ldots \tag{8.61}$$

Here the upper indices denote the order of the term in the electric field strength.

When equaiton (8.61) is substituted in equation (8.60) and terms of the same order are grouped together, we arrive at a series of equations each of which includes terms of the same order in $\mathbf{E}_p$:

$$(-\omega^2 - \mathrm{i}\omega\gamma_p) u_p^{(1)} + \sum_{q=1}^{N} \xi_{pq} u_q^{(1)} = -\frac{e}{m}(\mathbf{n}_p \cdot \mathbf{E}_p) \tag{8.62}$$

$$\sum_{q=1}^{N} \xi_{pq} u_q^{(2)}(0) = -2 \sum_{q,r=1}^{N} a_{pqr} u_q^{(1)} u_r^{(1)*} \tag{8.63}$$

$$(-4\omega^2 - 2\mathrm{i}\omega\gamma_p)\, u_p^{(2)}(2\omega) + \sum_{q=1}^{N} \xi_{pq} u_q^{(2)}(2\omega) = -\sum_{q,r=1}^{N} a_{pqr} u_q^{(1)} u_r^{(1)} \tag{8.64}$$

$$(-\omega^2 - \mathrm{i}\omega\gamma_p)\, u_p^{(3)} + \sum_{q=1}^{N} \xi_{pq} u_q^{(3)} =$$

$$-2 \sum_{q,r=1}^{N} a_{pqr}[u_q^{(2)}(2\omega) u_r^{(1)*} + u_q^{(0)}(0) u_r^{(1)}] - 3 \sum_{q,r,s=1}^{N} b_{pqrs} u_q^{(1)} u_r^{(1)} u_s^{(1)*} \tag{8.65}$$

Linear equation (8.62) has the following solution:

$$u_p^{(1)} = -\sum_{q=1}^{N} \frac{e}{m} f_{pq}(\omega)(\mathbf{n}_q \cdot \mathbf{E}_q) \tag{8.66}$$

where $f_{pq}(\omega) = \|g_{pq}\|^{-1}$ is the reciprocal matrix to matrix $g_{pq}(\omega) = \xi_{pq} - (\omega^2 + \mathrm{i}\omega\gamma_p)\delta_{pq}$. If the oscillators within the molecule interact with one another (that is $\xi_{pq} \neq 0$ for $p \neq q$), then matrices $g_{pq}(\omega)$ and $f_{pq}(\omega)$ are nondiagonal. In this case the displacement of the pth electron depends on the magnitude of the electric field strength at the locations of the qth electrons in the molecule. Such a molecule will show nonlocal response.

The amplitudes of the spectral components of the displacement at different frequencies, 2ω, 3ω and zero frequency follow from equations (8.62) to (8.65):

$$u_p^{(2)}(0) = -2 \sum_{p',q,r=1}^{N} f_{pp'}(0) a_{p'qr} u_q^{(1)} u_r^{(1)*} \tag{8.67}$$

$$u_p^{(2)}(2\omega) = -\sum_{p',q,r=1}^{N} f_{pp'}(2\omega) a_{p'qr} u_q^{(1)} u_r^{(1)} \tag{8.68}$$

and

$$u_p^{(3)}(\omega) = -3 \sum_{p',q,r,s=1}^{N} f_{pp'}(\omega) \overline{b}_{p'qrs} u_q^{(1)} u_r^{(1)} u_s^{(1)} \tag{8.69}$$

where

$$\overline{b}_{p'qrs} = b_{p'qrs}$$

$$-\frac{2}{3}\sum_{q',r'=1}^{N}[a_{qrq'}f_{q'r'}(2\omega)\ a_{r'p's} + a_{p'q'r}f_{q'r'}(0)\ a_{qr's} + a_{p'qr'}f_{r'q'}(0)a_{q'rs}] \tag{8.70}$$

Now the linear component of the current density (8.59) is

$$\mathbf{J}^{(1)}(\omega,\,\mathbf{r}) = -\mathrm{i}\omega N_0\sum_{p=1}^{N}(-e)\mathbf{n}_p u_p^{(1)}|_{\mathbf{r}_0=\mathbf{r}-\mathbf{r}_p^{(0)}}$$

$$= -\mathrm{i}\omega\sum_{p,q=1}^{n}\frac{e^2}{m}f_{pq}(\omega)\ \mathbf{n}_p\ (\mathbf{n}_p\cdot\mathbf{E}(\omega,\ \mathbf{r}+\mathbf{r}_q^{(0)}-\mathbf{r}_p^{(0)})) \tag{8.71}$$

For nonlinear response at frequency 2ω we obtain

$$\mathbf{J}^{(2)}(2\omega,\,\mathbf{r}) = -2\mathrm{i}\omega N_0\sum_{p=1}^{N}(-e)\mathbf{n}_p u_p^{(2)}(2\omega)|_{\mathbf{r}_0=\mathbf{r}-\mathbf{r}_p^{(0)}}$$

$$= 2\mathrm{i}\omega\sum_{p,q,r=1}^{N}\frac{e^3}{m^2}A_{pqr}(\omega)\mathbf{n}_p$$

$$\times\,(\mathbf{n}_q\cdot\mathbf{E}(\omega,\ \mathbf{r}+\mathbf{r}_q^{(0)}-\mathbf{r}_p^{(0)}))(\mathbf{n}_r\cdot\mathbf{E}(\omega,\ \mathbf{r}+\mathbf{r}_r^{(0)}-\mathbf{r}_p^{(0)})) \tag{8.72}$$

where

$$A_{pqr}(\omega) = -\sum_{p',q',r'=1}^{N}f_{pp'}(2\omega)f_{qq'}(\omega)f_{rr'}(\omega)a_{p'q'r'} \tag{8.73}$$

Similarly, for the third-order response at frequency ω we get

$$\mathbf{J}^{(3)}(\omega,\,\mathbf{r}) = -\mathrm{i}\omega N_0\sum_{p=1}^{N}(-e)\mathbf{n}_p u_p^{(3)}|_{\mathbf{r}_0=\mathbf{r}-\mathbf{r}_p^{(0)}}$$

$$= -\mathrm{i}\omega N_0\sum_{p,q,r,s=1}^{N}\frac{e^4}{m^3}B_{pqrs}(\omega)\ \mathbf{n}_p(\mathbf{n}_q\cdot\mathbf{E}(\omega,\ \mathbf{r}+\mathbf{r}_q^{(0)}-\mathbf{r}_p^{(0)}))$$

$$\times\,(\mathbf{n}_r\cdot\mathbf{E}(\mathbf{r}+\mathbf{r}_r^{(0)}-\mathbf{r}_p^{(0)}))(\mathbf{n}_s\cdot\mathbf{E}^*(\mathbf{r}+\mathbf{r}_s^{(0)}-\mathbf{r}_p^{(0)})) \tag{8.74}$$

where

$$B_{pqrs}(\omega) = -3 \sum_{p',q',r',s'=1}^{N} f_{pp'}(\omega) f_{qq'}(\omega) f_{rr'}(\omega) f_{ss'}(-\omega) \bar{b}_{p'q'r's'} \qquad (8.75)$$

We have found that the linear and nonlinear responses of the medium depend on the Taylor coefficients, ξ_{pq}, a_{pqr}, and b_{pqrs}, of the molecular potential energy and also on the actual locations of the optical electrons inside the molecule, i.e. on molecular size. Below, as we did for the two-electron molecular model, we will presume that the size of the molecule $D = \max|\mathbf{r}_q^{(0)} - \mathbf{r}_p^{(0)}|$ is much smaller than the optical wavelength λ. In this case nonlocality can be considered as a first-order effect in D/λ and so in the current density only the terms proportional to D/λ will be retained. To do this we expand the electric field of the light wave in equations (8.71), (8.72) and (8.74) as Taylor series:

$$\mathbf{E}(\omega, \mathbf{r} + \mathbf{r}_q^{(0)} - \mathbf{r}_p^{(0)}) = \mathbf{E}(\omega, \mathbf{r}) + ((\mathbf{r}_q^{(0)} - \mathbf{r}_q^{(0)}) \cdot \nabla)\, \mathbf{E}(\omega, \mathbf{r}) \qquad (8.76)$$

The linear part of the molecular current density can now be written as

$$\mathbf{J}(\omega, \mathbf{r}) = -i\omega N_0 \sum_{p,q=1}^{N} \frac{e^2}{m} f_{pq}(\omega) \mathbf{n}_p (\mathbf{n}_q \cdot \mathbf{E}(\omega, \mathbf{r}))$$
$$- i\omega N_0 \sum_{p,q=1}^{N} \frac{e^2}{m} f_{pq}(\omega) \mathbf{n}_p ((\mathbf{r}_p^{(0)} - \mathbf{r}_p^{(0)}) \cdot \nabla)(\mathbf{n}_q \cdot \mathbf{E}(\omega, \mathbf{r})) \qquad (8.77)$$

This gives the following constitutive equation in terms of optical susceptibilities (compare with (2.14)):

$$P_i(\omega, \mathbf{r}) = \frac{1}{4\pi} [(\epsilon_{ij}(\omega) - \delta_{ij}) E_j(\omega, \mathbf{r}) + \Gamma_{ijn}^{(1)}(\omega) \nabla_n E_j(\omega, \mathbf{r})] \qquad (8.78)$$

Here

$$\epsilon_{ij}(\omega) = \delta_{ij} + 4\pi N_0 \sum_{p,q=1}^{N} \frac{e^2}{m} f_{pq}(\omega) n_{pi} n_{qj} \qquad (8.79)$$

and

$$\Gamma_{ijn}^{(1)}(\omega) = 4\pi N_0 \sum_{p,q=1}^{N} \frac{e^2}{m} f_{pq}(\omega) n_{pi} n_{qj} (r_{qn}^{(0)} - r_{pn}^{(0)}) \qquad (8.80)$$

are the dielectric and nonlocal optical response tensors of a medium consisting of identically oriented polyatomic molecules.

The polarization response at frequency 2ω, quadratic in the electric field of the optical wave, is given by (compare with the phenomenologically introduced constitutive equation (2.47))

$$P_j^{(2)}(2\omega, \mathbf{r}) = \chi_{ijk}^{(2)}(2\omega, \omega, \omega)E_j(\omega, \mathbf{r})E_k(\omega.\mathbf{r})$$
$$+ \Gamma_{ijkn}^{(2)}(2\omega, \omega, \omega)E_k(\omega, \mathbf{r})\nabla_n E_j(\omega, \mathbf{r}) \quad (8.81)$$

where

$$\chi_{ijk}^{(2)}(2\omega, \omega, \omega) = -N_0 \sum_{p,q,r=1}^{N} \frac{e^3}{m^2} A_{pqr}(\omega)n_{pi}n_{qj}n_{rk} \quad (8.82)$$

and

$$\Gamma_{ijkn}^{(2)}(2\omega, \omega, \omega) = -N_0 \sum_{p,q,r=1}^{N} \frac{2e^3}{m^2} A_{pqr}(\omega)n_{pi}n_{qj}n_{rm}(r_{qn}^{(0)} - r_{pn}^{(0)}) \quad (8.83)$$

The polarization response at frequency ω, cubic in the electric field of the optical wave, is given by

$$P_i^{(3)}(\omega, \mathbf{r}) = 3\chi_{ijkl}^{(3)}(\omega, \omega, \omega, -\omega)E_j(\omega, \mathbf{r})E_k(\omega, \mathbf{r})E_l^*(\omega, \mathbf{r})$$
$$+ 2\Gamma_{ijkln}^{(3)}(\omega, \omega, \omega, -\omega)E_k(\omega, \mathbf{r})E_l^*(\omega.\mathbf{r})\nabla_n E_j(\omega, \mathbf{r})$$
$$+ \Gamma_{ijkln}^{(3)}(\omega, -\omega, \omega, \omega)E_k(\omega, \mathbf{r})E_l(\omega, \mathbf{r})\nabla_n E_j^*(\omega, \mathbf{r}) \quad (8.84)$$

(compare with constitutive equation (2.53)). Here, the cubic optical susceptibility is

$$\chi_{ijkl}^{(3)}(\omega, \omega, \omega, -\omega) = N_0 \sum_{p,q,r,s=1}^{N} \frac{e^4}{m^3} B_{pqrs}(\omega)n_{pi}n_{qi}n_{rk}n_{sl} \quad (8.85)$$

and the cubic nonlocal optical susceptibility is

$$\Gamma_{ijkln}^{(3)}(\omega, \omega, \omega, -\omega) = N_0 \sum_{p,q,r,s=1}^{N} \frac{e^4}{m^3} B_{pqrs}(\omega)\; n_{pi}n_{qj}n_{rk}n_{sl}(r_{qn}^{(0)} - r_{pn}^{(0)}) \quad (8.86)$$

$$\Gamma_{ijkln}^{(3)}(\omega, -\omega, \omega, \omega) = N_0 \sum_{p,q,r,s=1}^{N} \frac{e^4}{m^3} B_{psrq}(\omega)\; n_{pi}n_{qj}n_{rk}n_{sl}(r_{qn}^{(0)} - r_{pn}^{(0)}) \quad (8.87)$$

When the medium interacts with the plane light wave $\mathbf{E}(\omega, \mathbf{r}) = \mathbf{A}\exp\{i\mathbf{k}{\cdot}\mathbf{r}\}$, the nonlinear constitutive equation (8.84) reduces to (compare with (2.56))

$$P_i^{(3)}(\omega, \mathbf{k}) = (\chi_{ijkl} + ik_n\Gamma_{ijkln})A_jA_kA_l^* \quad (8.88)$$

where

$$\chi_{ijkl} = 3\chi^{(3)}_{ijkl}(\omega, \omega, \omega, -\omega) = N_0 \sum_{p,q,r,s=1}^{N} \frac{3e^4}{m^3} B_{pqrs}(\omega) n_{pi} n_{qj} n_{rk} n_{sl} \tag{8.89}$$

and the cubic nonlocal optical susceptibility is

$$\begin{aligned}\Gamma_{ijkln} &= \Gamma^{(3)}_{ijkln}(\omega, \omega, \omega, -\omega) + \Gamma^{(3)}_{ikjln}(\omega, \omega, \omega, -\omega) - \Gamma^{(3)}_{ilkjn}(\omega, -\omega, \omega, \omega) \\ &= N_0 \sum_{p,q,r,s=1}^{N} \frac{e^4}{m^3} [B_{pqrs}(\omega) + B_{prqs}(\omega) - B_{psrq}(\omega)] n_{pi} n_{qj} n_{k} n_{sl} (r^{(0)}_{qn} - r^{(0)}_{pn})\end{aligned} \tag{8.90}$$

If there is no interaction between the electrons in the molecule, the components of the nonlinear nonlocal response tensor vanish in spite of the spatial separation of the electrons within the molecules. Again, as in linear optics, the interaction between the electrons and the finite size of the molecule ensure a nonlocal term in the nonlinear part of the constitutive equation.

Formulae (8.89) and (8.90) can be used to calculate the optical response of complex molecules or crystals of low symmetry. Examples of similar calculations of local and nonlocal linear and nonlinear responses for crystals with sixfold symmetry axes and noncentrosymmetric cubic crystals can be found in Refs. [12] and [13]. Similar nonlinear oscillator models have also been used to describe intrinsic polarization multistability [7].

REFERENCES

1. J. Van Kranendonk and J.E. Sipe, *Progress in Optics*, **15**, 245 (1977).
2. M.V. Volkenstein, *Molecular Optics*, Gostekhizdat, Moscow, 1951.
3. S.A. Akhmanov and R.V. Khohlov, *Problems of Nonlinear Optics: Electromagnetic waves in nonlinear dispersive media*, Gordon and Breach, New York, 1972.
4. T.P. McLean, Linear and nonlinear optics of condensed matter, in *Interaction of Radiation with Condensed Matter*, Vol. 1, Lectures Presented at a Winter College, Triest, 14 January–26 March 1976, International Atomic Energy Agency, Vienna, 1977.
5. N. Blombergen, *Nonlinear Optics*, W.A. Benjamin, New York, 1965.
6. S.A. Akhmanov and S.A. Nikitin, *Physical Optics*, Oxford University Press, 1997.
7. S.A. Akhmanov, N.I. Zheludev and Y.P. Svirko, *Izv. Aka. Nauk SSSR Ser. Fiz.*, **46**, 1070 (1982).
8. J.A. Goldstone and E. Garmire, *Phys. Rev. Lett.*, **53**, 910 (1984).
9. M. Born, *Physikal. Z.*, **16**, 251 (1915).
10. M. Born, *Optik*, Springer, Berlin, 1933.

11. W. Kuhn, *Trans. Faraday Soc.*, **26**, 293 (1930).
12. N.I. Zheludev and Yu.P. Svirko, Nonlinear polarization spectroscopy of semiconductors, in S.A. Akhmanov and N.I. Zheludev (eds.), *Nonlinear-Optical Diagnostic of Semiconductors*, Vol. 2, VINITI, Moscow, 1990.
13. N.I. Zheludev and Yu.P. Svirko, *Kristallografiya*, **35**, 279 (1990).

9 Optical Susceptibilities in Quantum Mechanics

In this chapter we will derive the nonlinear local and nonlocal optical susceptibilities of a nonlinear medium using a quantum mechanical model. After a short introduction to the time-dependent perturbation theory, we will define a Hamiltonian which describes the interaction of bound molecular electrons with a light wave. Using time-dependent perturbation theory, we will calculate expectation values for the current density operator and linear and consequently derive the constitutive equation.

9.1 TIME-DEPENDENT PERTURBATION THEORY

The classical oscillator models developed in the previous chapter were a powerful and simple tool for calculating local and nonlocal molecular optical responses and molecular nonlinearities. However, a more accurate and versatile theory of optical susceptibilities can be based on quantum mechanics. The procedures for calculating optical susceptibilities in quantum mechanics and the classical oscillator model are similar. In classical mechanics, however, we calculated the current density directly, while in quantum mechanics it is given by the expectation value of the current density operator. Expanding of this expectation value as a series in the amplitude of the light wave gives the optical susceptibilities.

We will begin with a brief revision of time-dependent perturbation theory [1]. When molecular electrons interact with a light wave, the energy of the molecule varies with time. The wavefunction $\psi(t)$ of the molecule obeys the Schrödinger equation

$$\frac{\partial \psi(t)}{\partial t} = -\frac{i}{\hbar}[\hat{H}_0 + \hat{H}^{\text{int}}(t)]\psi(t) \tag{9.1}$$

where $\hat{H}_0$ is a time-independent, nonperturbed Hamiltonian with eigenvalues E_n (energy levels of the molecule) and corresponding eigenfunctions u_n:

$$\hat{H}_0 u_n = E_n u_n \tag{9.2}$$

The eigenfunctions u_n form a complete orthonormal set, i.e.

$$\int d\mathbf{r}\, u_n^*(\mathbf{r}) u_k(\mathbf{r}) = \delta_{nk} \tag{9.3}$$

In equation (9.1), $\hat{H}^{\text{int}}(t)$ is the Hamiltonian of interaction with the electromagnetic wave. The time-dependent solution of the Schrödinger equation can be expanded in terms of u_n:

$$\psi(t) = \sum_n a_n(t) u_n e^{-iE_n t/\hbar} \tag{9.4}$$

If the molecule is unperturbed, it is in its ground state, which will be denoted by subscript g. In expansion (9.4) the coefficients $a_n(t)$ obey the following set of equations:

$$\frac{da_k}{dt} = -\frac{i}{\hbar} \sum_n a_n(t) H_{kn}^{\text{int}} e^{i\omega_{kn} t} \tag{9.5}$$

which can be obtained by substituting expansion (9.4) into equation (9.1).

It will be shown below that the Hamiltonian of the interaction of the molecule with the light wave at frequency ω can be presented in the following form:

$$\hat{F}^{\text{int}}(t) = \hat{F} e^{-i\omega t} + \hat{G} e^{i\omega t} \tag{9.6}$$

where $\hat{F}$ and $\hat{G}$ are time-independent operators. To ensure that the expectation value of the interaction energy is real, the operator $\hat{H}^{\text{int}}(t)$ must be Hermitian, i.e. its matrix elements

$$H_{nk}^{\text{int}} = \int u_n^* \hat{H}^{\text{int}} u_k \, d\mathbf{r} \tag{9.7}$$

must satisfy $H_{kn}^{\text{int}} = (H_{nk}^{\text{int}})^*$. It follows from equation (9.6), therefore, that $F_{kn} = (G_{nk})^*$.

With the given interaction Hamiltonian, the coefficients $a_n(t)$ can be found as a power series expansion in $|H_{nk}^{\text{int}}|/(\hbar\omega) \ll 1$:

$$a_n(t) = a_n^{(0)}(t) + a_n^{(1)}(t) + a_n^{(2)}(t) + \cdots \tag{9.8}$$

Here $a_n^{(0)}(t) = \delta_{ng}$, where δ_{ng} is the Kronekker symbol and index g denotes the molecular ground state. From now on, in order to simplify the notation, the time argument in the coefficients $a_n^{(i)}$ will be dropped. Substituting equation (9.8) into (9.5) and grouping together terms of the same order in $|H_{nk}^{\text{int}}|/(\hbar\omega)$ produces the following chain of equations:

$$\begin{aligned}
\frac{da_k^{(1)}}{dt} &= -\frac{i}{\hbar} \sum_n a_n^{(0)} H_{kn}^{\text{int}} e^{i\omega_{kn} t} \\
\frac{da_k^{(2)}}{dt} &= -\frac{i}{\hbar} \sum_n a_n^{(1)} H_{kn}^{\text{int}} e^{i\omega_{kn} t} \\
\frac{da_k^{(3)}}{dt} &= -\frac{i}{\hbar} \sum_n a_n^{(2)} H_{kn}^{\text{int}} e^{i\omega_{kn} t} \\
&\cdots
\end{aligned} \tag{9.9}$$

where $\omega_{nk} = (E_n - E_k)/\hbar$. These equations can readily be solved:

$$a_k^{(1)} = -\frac{F_{kg}\,\mathrm{e}^{\mathrm{i}(\omega_{kg}-\omega)t}}{\hbar(\omega_{kg}-\omega)} - \frac{G_{kg}\,\mathrm{e}^{\mathrm{i}(\omega_{kg}+\omega)t}}{\hbar(\omega_{kg}+\omega)} \tag{9.10}$$

$$a_k^{(2)} = \frac{1}{\hbar^2}\sum_n\left[\frac{F_{kn}F_{ng}\,\mathrm{e}^{\mathrm{i}(\omega_{kg}-2\omega)t}}{(\omega_{ng}-\omega)(\omega_{kg}-2\omega)} + \frac{F_{kn}G_{ng}\,\mathrm{e}^{\mathrm{i}\omega_{kg}t}}{\omega_{kg}(\omega_{ng}+\omega)} + \frac{G_{kn}F_{ng}\,\mathrm{e}^{\mathrm{i}\omega_{kg}t}}{\omega_{kg}(\omega_{ng}-\omega)} + \frac{G_{kn}G_{ng}\,\mathrm{e}^{\mathrm{i}(\omega_{kg}+2\omega)t}}{(\omega_{ng}+\omega)(\omega_{kg}+2\omega)}\right] \tag{9.11}$$

$$\begin{aligned} a_k^{(3)} = -\frac{1}{\hbar^3}\sum_{nm} & \left[\frac{F_{kn}F_{nm}F_{mg}\,\mathrm{e}^{\mathrm{i}(\omega_{kg}-3\omega)t}}{(\omega_{mg}-\omega)(\omega_{ng}-2\omega)(\omega_{kg}-3\omega)} + \frac{F_{kn}F_{nm}G_{mg}\,\mathrm{e}^{\mathrm{i}(\omega_{kg}-\omega)t}}{(\omega_{mg}+\omega)\omega_{ng}(\omega_{kg}-\omega)}\right. \\ & + \frac{F_{kn}G_{nm}F_{mg}\,\mathrm{e}^{\mathrm{i}(\omega_{kg}-\omega)t}}{(\omega_{mg}-\omega)\omega_{ng}(\omega_{kg}-\omega)} + \frac{G_{kn}F_{nm}F_{mg}\,\mathrm{e}^{\mathrm{i}(\omega_{kg}-\omega)t}}{(\omega_{mg}-\omega)(\omega_{ng}-2\omega)(\omega_{kg}-\omega)} \\ & + \frac{F_{kn}G_{nm}G_{mg}\,\mathrm{e}^{\mathrm{i}(\omega_{kg}+\omega)t}}{(\omega_{mg}+\omega)(\omega_{ng}+2\omega)(\omega_{kg}+\omega)} + \frac{G_{kn}F_{nm}G_{mg}\,\mathrm{e}^{\mathrm{i}(\omega_{kg}+\omega)t}}{(\omega_{mg}+\omega)\omega_{ng}(\omega_{kg}+\omega)} \\ & \left. + \frac{G_{kn}G_{nm}F_{mg}\,\mathrm{e}^{\mathrm{i}(\omega_{kg}+\omega)t}}{(\omega_{mg}-\omega)\omega_{ng}(\omega_{kg}+\omega)} + \frac{G_{kn}G_{nm}G_{mg}\,\mathrm{e}^{\mathrm{i}(\omega_{kg}+3\omega)t}}{(\omega_{mg}+\omega)(\omega_{ng}+2\omega)(\omega_{kg}+3\omega)}\right] \end{aligned} \tag{9.12}$$

Now we can calculate the expectation value of the current density operator $\hat{\mathbf{j}}$:

$$\mathbf{j}(t) = \int \mathrm{d}\mathbf{r}\psi^*(\mathbf{r}, t)\hat{\mathbf{j}}\psi(\mathbf{r}, t) \tag{9.13}$$

Substituting equation (9.4) into (9.13) gives a formula for $\mathbf{j}(t)$ in terms of $a_n(t)$ and the matrix elements $\mathbf{j}_{nk}$ of the current density operator:

$$\mathbf{j}(t) = \sum_{kn} a_n^*(t)a_k(t)\mathbf{j}_{nk}\,\mathrm{e}^{\mathrm{i}\omega_{nk}t} \tag{9.14}$$

Using formulae (9.10) to (9.12) we can obtain the current density as a power series expansion in the external perturbation $|H_{nk}^{\mathrm{int}}|/(\hbar\omega)$:

$$\mathbf{j}(t) = \mathbf{j}^{(0)}(t) + \mathbf{j}^{(1)}(t) + \mathbf{j}^{(2)}(t) + \cdots \tag{9.15}$$

where

$$\mathbf{j}^{(0)}(t) = \sum_k (a_k^{(0)*}\mathbf{j}_{kg}\,\mathrm{e}^{\mathrm{i}\omega_{kg}t} + a_k^{(0)}\mathbf{j}_{gk}\,\mathrm{e}^{-\mathrm{i}\omega_{kg}t}) \tag{9.16}$$

$$\mathbf{j}^{(1)}(t) = \sum_k (a_k^{(1)*} \mathbf{j}_{kg} \, \mathrm{e}^{\mathrm{i}\omega_{kg} t} + a_k^{(1)} \mathbf{j}_{gk} \, \mathrm{e}^{-\mathrm{i}\omega_{kg} t}) \tag{9.17}$$

$$\mathbf{j}^{(2)}(t) = \sum_k (a_k^{(2)*} \mathbf{j}_{kg} \, \mathrm{e}^{\mathrm{i}\omega_{kg} t} + a_k^{(2)} \mathbf{j}_{gk} \, \mathrm{e}^{-\mathrm{i}\omega_{kg} t}) + \sum_{kn} a_n^{(1)*} a_k^{(1)} \mathbf{j}_{nk} \, \mathrm{e}^{\mathrm{i}\omega_{nk} t} \tag{9.18}$$

$$\mathbf{j}^{(3)}(t) = \sum_k (a_k^{(3)*} \mathbf{j}_{kg} \, \mathrm{e}^{\mathrm{i}\omega_{kg} t} + a_k^{(3)} \mathbf{j}_{gk} \, \mathrm{e}^{-\mathrm{i}\omega_{kg} t}) + \sum_{kn} (a_n^{(1)*} a_k^{(2)} + a_n^{(2)*} a_k^{(1)}) \mathbf{j}_{nk} \, \mathrm{e}^{\mathrm{i}\omega_{nk} t} \tag{9.19}$$

These formulae will be used below to calculate the optical susceptibilities. However, we will first need to specify the interaction Hamiltonian.

9.2 THE INTERACTION HAMILTONIAN AND CURRENT DENSITY OPERATOR

We will derive the interaction Hamiltonian in the nonrelativistic approximation, ignoring the spin–orbit interaction and other effects of similar magnitude, such as the relativistic mass correction. In the first-order spatial dispersion approximation, the total nonrelativistic Hamiltonian of the molecular electrons in the presence of the light field has the following form [2, 3] (see Appendix 1):

$$\hat{H}(\mathbf{r}_0) = \sum_{s=1}^{N} \left\{ \frac{[\hat{\mathbf{p}}_s + (e/\mathrm{c})(\mathbf{r}_s \cdot \nabla)\mathcal{A}]^2}{2m_\mathrm{e}} - \frac{e}{\mathrm{c}} \left(\mathbf{r}_s \cdot \frac{\partial \mathcal{A}}{\partial t} \right) - \hat{\boldsymbol{\mu}}^s \cdot \mathbf{B} + U(\mathbf{r}_s) \right\} \tag{9.20}$$

In this equation $\mathbf{r}_0$ is the coordinate of the molecular centre and subscript s enumerates the molecular electrons, $-e$ and m_e are the electrons electric charge and mass, $\mathbf{r}_s$ and $\hat{\mathbf{p}}_s$ are the coordinate vector with respect to the molecular centre and the canonic momentum of the sth electron, $\mathcal{A}$ and $\mathbf{B}$ are the vector potential and the magnetic field associated with the light wave at the centre of the molecule, $U(\mathbf{r}_s)$ is the potential energy of an electron in the field created by the rest of the molecule and $\hat{\boldsymbol{\mu}}^s$ is the operator of the intrinsic magnetic moment of the electrons.

Hamiltonian (9.20) is presented in the Coulomb gauge of the electromagnetic fields where $\operatorname{div} \mathcal{A} = 0$ [4, 5]. This gauge imposes the following conditions on **E**, **B** and $\mathcal{A}$:

$$\mathbf{B} = \operatorname{curl} \mathcal{A}, \qquad \mathbf{E} = -\frac{1}{c}\frac{\partial \mathcal{A}}{\partial t} \tag{9.21}$$

By subtracting the energy of the molecule itself,

$$\hat{H}_0 = \sum_{s=1}^{N} \frac{\hat{\mathbf{p}}_s^2}{2m} + U(\mathbf{r}_s) \tag{9.22}$$

from equation (9.20), the following light–matter interaction Hamiltonian $\hat{H}^{\text{int}}$ can be obtained:

$$\begin{aligned}\hat{H}^{\text{int}} &= \hat{H}(\mathbf{r}_0) - \hat{H}_0 \\ &= \sum_{s=1}^{N}\left\{\frac{e}{2m_{\text{e}}c}[\hat{\mathbf{p}}_s\cdot(\mathbf{r}_s\cdot\nabla)\mathcal{A} + (\mathbf{r}_s\cdot\nabla)\mathcal{A}\cdot\hat{\mathbf{p}}_s]\right. \\ &\left. - \frac{e}{c}\left(\mathbf{r}_s\cdot\frac{\partial\mathcal{A}}{\partial t}\right) - \hat{\boldsymbol{\mu}}^s\cdot\mathbf{B} + \frac{e^2[(r_s\cdot\nabla)\mathcal{A}]^2}{2m_{\text{e}}c^2}\right]\end{aligned} \tag{9.23}$$

Using the relationship between the electrons momentum in the electromagnetic field and its velocity $m_{\text{e}}v_s = \mathbf{p}_s + (e/c)(\mathbf{r}_s\cdot\nabla)\mathcal{A}$ (see Appendix 1), we can rewrite the interaction Hamiltonian (9.23) in terms of the multipole moments of the molecule:

$$\hat{H}^{\text{int}} = \frac{1}{c}\hat{\mathbf{d}}\cdot\frac{\partial\mathcal{A}}{\partial t} - \frac{1}{c}\left(\frac{\text{d}\hat{q}^{il}}{\text{d}t} - \text{e}_{ilk}\hat{\mathbf{m}}^k\right)\nabla_l\mathcal{A}_i \tag{9.24}$$

where indices i, l and k label the Cartesian coordinates. In formula (9.24),

$$\hat{\mathbf{d}} = (-e)\sum_{s=1}^{N}\mathbf{r}_s \tag{9.25}$$

$$\hat{q}^{il} = (-e)\sum_{s=1}^{N}\frac{r_{si}r_{sl}}{2} \tag{9.26}$$

and

$$\hat{\mathbf{m}} = \sum_{s=1}^{N}\left\{-\frac{e}{2m_{\text{e}}c}[\mathbf{r}_s \times \hat{\mathbf{p}}_s] + \hat{\boldsymbol{\mu}}^s\right\} \tag{9.27}$$

are the operators of the electric dipole, electric quadrupole and magnetic dipole moments of a molecule.

To obtain the interaction Hamiltonian for a plane light wave of amplitude $\mathbf{A}$, frequency ω and wave vector $\mathbf{k}$:

$$\mathbf{E}(\mathbf{r},\, t) = \mathbf{A}\,\text{e}^{-\text{i}\omega t + i\mathbf{k}\cdot\mathbf{r}} + \text{c.c.} \tag{9.28}$$

we will use the relationship (9.21), between the vector potential and the electric field, and rearrange the interaction Hamiltonian (9.24) as follows:

$$\hat{H}^{\text{int}} = -\left(\hat{d}^j + \frac{k_l}{\omega}\hat{R}^{jl}\right)A_j\,\text{e}^{-\text{i}\omega t + i\mathbf{k}\cdot\mathbf{r}_0} + \text{h.c.} \tag{9.29}$$

In this, and any subsequent expressions, the abbreviation h.c. denotes the Hermitian conjugate. In equation (9.29) we have introduced the operator

$$\hat{R}^{il} = \frac{d\hat{q}^{il}}{dt} - ce_{ilk}\hat{m}^k \tag{9.30}$$

which accounts for both the electroquadrupole and magnetodipole interactions. It plays a very important role in the theory of nonlocal optical interactions [2] and will be referred to below as the spatial dispersion operator.

One can see that the interaction Hamiltonian (9.29) has the form of equation (9.6), $\hat{H}^{\text{int}}(t) = \hat{F}\,e^{-i\omega t} + \hat{G}\,e^{i\omega t}$, and is therefore suitable for use with time-dependent perturbation theory. Indeed, formula (9.6) will be returned by taking

$$\begin{aligned} \hat{F} &= \hat{V}^j A_j\, e^{i\mathbf{k}\cdot\mathbf{r}_0} \\ \hat{G} &= \hat{V}^j A_j^*\, e^{-i\mathbf{k}\cdot\mathbf{r}_0} \end{aligned} \tag{9.31}$$

where

$$\hat{V}^j = -\left(\hat{d}^j + \frac{k_l}{\omega}\hat{R}^{jl}\right) \tag{9.32}$$

The current density operator of moving charges in the molecule can also be presented in terms of the dipole and spatial dispersion operators [6]:

$$\begin{aligned} \hat{\mathbf{j}}(\mathbf{r}, \mathbf{r}_0, t) = \sum_{s=1}^{N}\Bigg\{&-\frac{e}{2}\left[\frac{d\mathbf{r}_s}{dt}\delta(\mathbf{r}-\mathbf{r}_0-\mathbf{r}_s) + \delta(\mathbf{r}-\mathbf{r}_0-\mathbf{r}_s)\frac{d\mathbf{r}_s}{dt}\right] \\ &+\frac{c}{2}\,\text{curl}\,[\hat{\boldsymbol{\mu}}^s\delta(\mathbf{r}-\mathbf{r}_0-\mathbf{r}_s) + \delta(\mathbf{r}-\mathbf{r}_0-\mathbf{r}_s)\,\hat{\boldsymbol{\mu}}^s]\Bigg\} \end{aligned} \tag{9.33}$$

(compare with classical formula (8.58)). Here the first two terms give the electric current due to the kinetic motion of the electrons and the other two terms give the spin component of the current. The spatial Fourier spectrum of the current density,

$$\hat{\mathbf{j}}(\mathbf{K}, \mathbf{r}_0, t) = \int_{-\infty}^{+\infty} \hat{\mathbf{j}}(\mathbf{r}, \mathbf{r}_0, t) e^{-i\mathbf{K}\cdot\mathbf{r}}\, d\mathbf{r} \tag{9.34}$$

is

$$\hat{\mathbf{j}}(\mathbf{K}, \mathbf{r}_0, t) = \sum_{s=1}^{N}\left\{-\frac{e}{2}\left(\frac{d\mathbf{r}_s}{dt}e^{-i\mathbf{K}\cdot\mathbf{r}_s} + e^{-i\mathbf{K}\cdot\mathbf{r}_s}\frac{d\mathbf{r}_s}{dt}\right) - \frac{ic}{2}[\mathbf{K}\times\hat{\boldsymbol{\mu}}^s]\,e^{-i\mathbf{K}\cdot\mathbf{r}_s}\right\} e^{-i\mathbf{K}\cdot\mathbf{r}_0} \tag{9.35}$$

Since the currents are confined within a molecule of typical size $a = \max|\mathbf{r}_s|$, which is much smaller than the wavelength λ, formula (9.35) can be expanded over a/λ. Retaining only the linear terms in a/λ constitutes the first-order spatial dispersion approximation. In this approximation $\exp(-i\mathbf{K}\cdot\mathbf{r}_s) = 1 - i\mathbf{K}\cdot\mathbf{r}_s$ and

$$\hat{j}_i(\mathbf{K}, \mathbf{r}_0, t) = \sum_{s=1}^{N}\left\{-\mathrm{e}\frac{\mathrm{d}r_{si}}{\mathrm{d}t} + iK_l\frac{e}{2}\left(\frac{\mathrm{d}r_{si}}{\mathrm{d}t}r_{sl} + r_{sl}\frac{\mathrm{d}r_{si}}{\mathrm{d}t}\right) - ic[\mathbf{K}\times\hat{\boldsymbol{\mu}}^s]_i\right\}\mathrm{e}^{-i\mathbf{K}\cdot\mathbf{r}_0} \tag{9.36}$$

Using the algebraic identity

$$\frac{\mathrm{d}r_{si}}{\mathrm{d}t}r_{sj} + r_{sj}\frac{\mathrm{d}r_{si}}{\mathrm{d}t} = \frac{\mathrm{d}(r_{si}r_{sj})}{\mathrm{d}t} - \mathrm{e}_{ijk}\left[\mathbf{r}_s\times\frac{\mathrm{d}\mathbf{r}_s}{\mathrm{d}t}\right]_k \tag{9.37}$$

formula (9.36) for the spectral component of the current density operator can be rewritten as

$$\hat{\mathbf{j}}(\mathbf{K}, \mathbf{r}_0, t) = \hat{\mathbf{f}}\,\mathrm{e}^{-\mathrm{i}\mathbf{K}\cdot\mathbf{r}_0} \tag{9.38}$$

where

$$\hat{f}^{\,i} = \frac{\mathrm{d}}{\mathrm{d}t}\hat{d}^i - iK_l\hat{R}^{il} \tag{9.39}$$

Note that both the interaction Hamiltonian (9.29) and the current density are presented in terms of the same operators $\hat{d}^i$ and $\hat{R}^{il}$.

In the next section we will calculate the expectation values of the Fourier components of the current density as series in the light waves electric field. Integration over molecular positions and orientations will then give us the constitutive equation.

9.3 LINEAR SUSCEPTIBILITIES

In this section we will calculate linear optical susceptibilities using the quantum perturbation theory outlined in Section 9.1. Using the light–matter interaction Hamiltonian (9.29), we will obtain the expectation values of the current density operator (9.38) as series (9.16) to (9.19), but we will only retain those terms that are linear in the strength of the light field, i.e. those that are first-order terms in spatial dispersion. This will enable us to calculate the components of the dielectric and nonlocality tensors and to write down the linear constitutive equation.

In the expectation value of the current density operator, linear terms in the field of the light wave will appear in both the zero- and first-order terms of expansion (9.15). It is interesting to note that the field linear terms emerge even when the current density is calculated for unperturbed wavefunctions. This is because the current density operator itself depends on the vector potential.

The expectation value of the current density operator (9.38) for unperturbed wavefunctions is given by equation (9.16) with $a_k^{(0)} = \delta_{kg}$:

$$\hat{j}_i^{(0)}(\mathbf{K}, \mathbf{r}_0, t) = \left[\left(\frac{\mathrm{d}}{\mathrm{d}t} \hat{d}^i \right)_{gg} - iK_l R_{gg}^{il} \right] \mathrm{e}^{-i\mathbf{K}\cdot\mathbf{r}_0} \tag{9.40}$$

We are concerned with the light field dependent part of this expectation value. The dipole term, first on the right-hand side, can be rearranged using the relationship $m_\mathrm{e}\mathbf{v}_s = \mathbf{p}_s + (e/c)(\mathbf{r}_s \cdot \nabla)\mathcal{A}$:

$$\left(\frac{\mathrm{d}}{\mathrm{d}t} \hat{\mathbf{d}} \right)_{gg} = -\frac{\mathrm{e}}{m_\mathrm{e}} \sum_s \left[\hat{\mathbf{p}}_s + \frac{e}{c} (\mathbf{r}_s \cdot \nabla)\mathcal{A} \right]_{gg} \tag{9.41}$$

The first term on the right-hand side of equation (9.41) does not depend on $\mathscr{R}$ and therefore does not contribute to the light-induced current density. The second term here can be rearranged, using the definition of the dipole momentum operator (9.25), to give

$$\frac{e^2}{mc} \sum_s [(\mathbf{r}_s \cdot \nabla)\mathcal{A}]_{gg} = -\frac{e}{mc} (\mathbf{d}_{gg} \cdot \nabla)\mathcal{A} \tag{9.42}$$

This is a first-order spatial dispersion term of order a/λ. Indeed, in equation (9.42) $\nabla \propto ik \propto 1/\lambda$ and $d \propto a$, where a is the molecule characteristic size. However, the nonlocal, second term in equation (9.40), which contains $\hat{R}^{il}$ is a second-order term in a/λ. Indeed, the matrix element of the quadrupole part of $\hat{R}^{il} = \mathrm{d}\hat{q}^{il}/\mathrm{d}t - ce_{ilk}\hat{m}^k$ can be presented as follows:

$$\left(\frac{\mathrm{d}\hat{q}^{il}}{\mathrm{d}t} \right)_{gg} = -\frac{e}{2m_\mathrm{e}} \sum_s (\hat{p}_s^i r_s^l + r_s^i \hat{p}_s^l)_{gg} - \frac{e^2}{2m_\mathrm{e}c} \sum_s \{ r_s^l (\mathbf{r}_s \cdot \nabla)\mathcal{A}_i + [r_s^i (\mathbf{r}_s \cdot \nabla)\mathcal{A}_l]_{gg} \tag{9.43}$$

In formula (9.43) the first sum on the right-hand side does not depend on electromagnetic filed and does not contribute to the optically induced current. It is of no importance to us. The second sum depends on the gradient of the vector potential $\mathcal{A}$ and therefore is an (a/λ)-order term. Indeed, $\nabla \propto 1/\lambda$ and $|\mathbf{r}_s| \propto a$. Returning to formula (9.40), since $|\mathbf{K}| \propto 1/\lambda$, $K_l(\mathrm{d}\hat{q}^{il}/\mathrm{d}t)_{gg}$ is an $(a/\lambda)^2$ term and since we are deriving the current density in the first-order spatial dispersion approximation, it will be dropped. The magnetic dipole term in the spatial dispersion operator $\hat{R}^{il} = \mathrm{d}\hat{q}^{il}/\mathrm{d}t - ce_{ilk}\hat{m}^k$ does not depend on the vector potential and will not contribute to the induced current density in the zero-order approximation. Thus in the following calculations, the second term in equation (9.40) will be ignored.

Combining equations (9.41) and (9.43) we arrive at the current density in the zero-order approximation:

$$\mathbf{j}^{(0)}(\mathbf{K}, \mathbf{r}_0, t) = -\frac{e}{m_e\omega}(\mathbf{k}\cdot\mathbf{d}_{gg})(\mathbf{A}\,\mathrm{e}^{-\mathrm{i}\omega t+i\mathbf{k}\cdot\mathbf{r}_0} + \mathbf{A}^*\,\mathrm{e}^{\mathrm{i}\omega t-i\mathbf{k}\cdot\mathbf{r}_0})\,\mathrm{e}^{-i\mathbf{K}\cdot\mathbf{r}_0} \tag{9.44}$$

where $\mathbf{d}_{gg}$ is the electric dipole moment in the ground state.

By substituting form (9.31) into (9.10) and resorting (9.17) we can now calculate the first-order term of the current density:

$$\begin{aligned} j_i^{(1)}(\mathbf{K}, \mathbf{r}_0, t) = &-\sum_a\left[\frac{V_{ga}^j f_{ag}^i}{\hbar(\omega_{ag}+\omega)} + \frac{f_{ga}^i V_{ag}^j}{\hbar(\omega_{ag}-\omega)}\right]A_j\,\mathrm{e}^{-\mathrm{i}\omega t+i(\mathbf{k}-\mathbf{K})\cdot\mathbf{r}_0} \\ &-\sum_a\left[\frac{(V_{ga}^j)^* f_{ga}^i}{\hbar(\omega_{ag}+\omega)} + \frac{f_{ag}^i (V_{ag}^j)^*}{\hbar(\omega_{ag}-\omega)}\right]A_j^*\,\mathrm{e}^{\mathrm{i}\omega t-i(\mathbf{k}+\mathbf{K})\cdot\mathbf{r}_0} \end{aligned} \tag{9.45}$$

where subscript a enumerates all the molecular energy levels apart from the ground state g.

Both $\mathbf{j}^{(0)}$ and $\mathbf{j}^{(1)}$ contribute to the expectation value of the linear in $\mathbf{A}$ light-induced current density in the medium. As in the oscillator model, the macroscopic current density is an integral over the positions of the molecules in a unit volume V:

$$\mathbf{J}^{(1)}(\mathbf{K}, t) = N_0\int_V \mathrm{d}\mathbf{r}_0(\mathbf{j}^{(0)}(\mathbf{K}, \mathbf{r}_0, t) + \mathbf{j}^{(1)}(\mathbf{K}, \mathbf{r}_0, t)) \tag{9.46}$$

From now on all molecules are assumed to be identically oriented. The inverse Fourier transform of equation (9.46) gives the macroscopic current density in the coordinate representation:

$$\mathbf{J}^{(1)}(\omega, \mathbf{r}) = \frac{1}{8\pi^3}\int \mathrm{d}t\int \mathrm{d}\mathbf{K}\;\mathbf{J}^{(1)}(\mathbf{K}, t)\,\mathrm{e}^{-i\mathbf{K}\cdot\mathbf{r}+\mathrm{i}\omega t} \tag{9.47}$$

By resorting $\mathbf{P}^{(1)}(\omega, \mathbf{r}) = -\mathbf{J}^{(1)}(\omega, \mathbf{r})/(\mathrm{i}\omega)$, we obtain, from formula (9.47), the linear constitutive equation (compare with the phenomenologically derived constitutive equation (2.14)):

$$P_i^{(1)}(\omega, \mathbf{r}) = \frac{1}{4\pi}[(\epsilon_{ij}(\omega) - \delta_{ij})E_j(\omega, \mathbf{r}) + \Gamma_{ijn}^{(1)}(\omega)\nabla_n E_j(\omega, \mathbf{r})] \tag{9.48}$$

where the dielectric tensor is given by

$$\epsilon_{ij}(\omega) = \delta_{ij} + \frac{4\pi\mathrm{i}N_0}{\hbar\omega}\sum_a\left[\frac{d_{ga}^j(\partial d^i/\partial t)_{ag}}{\omega_{ag}+\omega} + \frac{(\partial d^i/\partial t)_{ga}d_{ag}^j}{\omega_{ag}-\omega}\right] \tag{9.49}$$

and the tensor of nonlocal optical response is

$$\Gamma^{(1)}_{ijn}(\omega) = -\frac{4\pi e^2 N_0}{m_e \omega^2} d^n_{gg}\delta_{ij} + \frac{4\pi N_0}{\hbar\omega^2}\sum_a \left[\frac{R^{jn}_{ga}(\partial d^i/\partial t)_{ag} - i\omega d^j_{ga}R^{in}_{ag}}{\omega_{ag}+\omega} + \frac{(\partial d^i/\partial t)_{ga}R^{jl}_{ag} - i\omega R^{il}_{ga}d^j_{ag}}{\omega_{ag}-\omega}\right] \tag{9.50}$$

Here N_0 is number density. The time derivatives in equations (9.49) and (9.50) can be excluded by applying the relationship $(\partial \mathbf{d}/\partial t)_{ab} = i\omega_{ab}\mathbf{d}_{ab}$ [1]. For the dielectric tensor this gives

$$\epsilon_{ij}(\omega) = \delta_{ij} - \frac{4\pi N_0}{\hbar\omega}\sum_a (d^j_{ga}d^i_{ag} - d^i_{ga}d^j_{ag}) + \frac{4\pi N_0}{\hbar}\sum_a \left[\frac{\omega_{ag}(d^j_{ga}d^i_{ag} + d^i_{ga}d^j_{ag})}{\omega^2_{ag}-\omega^2}\right] + \frac{4\pi\omega N_0}{\hbar}\sum_a \left(\frac{d^i_{ga}d^j_{ag} - d^j_{ga}d^i_{ag}}{\omega^2_{ag}-\omega^2}\right) \tag{9.51}$$

Importantly, the first sum in the right-hand side of formula (9.51) is zero:

$$\sum_a (d^j_{ga}d^i_{ag} - d^i_{ga}d^j_{ag}) = (\hat{d}^j\hat{d}^i - \hat{d}^i\hat{d}^j)_{gg} = e^2\left(\sum_s r^j_s \sum_s r^i_s - \sum_s r^i_s \sum_s r^j_s\right)_{gg} = 0 \tag{9.52}$$

Consequently, the dielectric tensor can be presented as $\epsilon_{ij}(\omega) = \epsilon^s_{ij}(\omega) + \epsilon^a_{ij}(\omega)$, i.e. as a sum of symmetric and antisymmetric parts with respect to the permutation of Cartesian indices i and j. The symmetric part is

$$\epsilon^s_{ij}(\omega) = \epsilon^s_{ji}(\omega) = \delta_{ij} + \frac{4\pi N_0}{\hbar}\sum_a \left[\frac{\omega_{ag}(d^j_{ga}d^i_{ag} + d^i_{ga}d^j_{ag})}{\omega^2_{ag}-\omega^2}\right] \tag{9.53}$$

and the part that is antisymmetric with respect to the permutation of i and j is

$$\epsilon^a_{ij}(\omega) = -\epsilon^a_{ji}(\omega) = \frac{4\pi\omega N_0}{\hbar}\sum_a \left(\frac{d^j_{ga}d^i_{ag} - d^i_{ga}d^j_{ag}}{\omega^2_{ag}-\omega^2}\right) \tag{9.54}$$

In a similar way, we can simplify formula (9.50) for the nonlocal susceptibility tensor:

$$\Gamma^{(1)}_{ijn}(\omega) = \frac{4\pi \mathrm{i} N_0}{\hbar\omega^2}\left[\frac{e^2}{m_\mathrm{e}} \mathrm{i}\hbar d^n_{gg}\delta_{ij} + \sum_a (R^{jn}_{ga} d^i_{ag} - d^i_{ga} R^{jn}_{ag})\right]$$
$$- \frac{4\pi \mathrm{i} N_0}{\hbar\omega}\sum_a \left[\frac{\omega_{ag}(R^{jn}_{ga} d^i_{ag} + d^i_{ga} R^{jn}_{ag} + R^{in}_{ga} d^j_{ag} + d^j_{ga} R^{in}_{ag})}{\omega^2_{ag} - \omega^2}\right]$$
$$+ \frac{4\pi \mathrm{i} N_0}{\hbar}\sum_a \left(\frac{R^{jn}_{ga} d^i_{aa} - d^i_{ga} R^{jn}_{ag} + R^{in}_{ga} d^j_{ag} - d^j_{ga} R^{in}_{ag}}{\omega^2_{ag} - \omega^2}\right) \tag{9.55}$$

The quantum mechanical operators of the components of coordinate r_i and momentum $\hat{p}_j$ only commute if $i \neq j$, that is $[r_i, \hat{p}_j] = -i\hbar\delta_{ij}$. From here it can be shown that $[\hat{R}^{in}, \hat{d}^j]_{gg} = -(i\hbar e^2/m_\mathrm{e})d^n_{gg}\delta_{ij}$ and so the first two terms of equation (9.55) cancel. Hence, $\Gamma^{(1)}_{ijn}$ can be presented as $\Gamma^{(1)}_{ijn} = \Gamma^{(1)s}_{ijn} + \Gamma^{(1)a}_{ijn}$, where the third term of equation (9.55) is symmetrical with respect to permutation of the first two indices:

$$\Gamma^{(1)s}_{ijn}(\omega) = \Gamma^{(1)s}_{jin}(\omega)$$
$$= -\frac{4\pi \mathrm{i} N_0}{\hbar\omega}\sum_a \left[\frac{\omega_{ag}(R^{jn}_{ga} d^i_{ag} + d^i_{ga} R^{jn}_{ag} + R^{in}_{ga} d^j_{ag} + d^j_{ga} R^{in}_{ag})}{\omega^2_{ag} - \omega^2}\right] \tag{9.56}$$

and the last term of equation (9.55) is antisymmetric with respect to the following permutation:

$$\Gamma^{(1)a}_{ijn}(\omega) = -\Gamma^{(1)a}_{jin}(\omega)$$
$$= \frac{4\pi \mathrm{i} N_0}{\hbar}\sum_a \left(\frac{R^{jn}_{ga} d^i_{ag} - d^i_{ga} R^{jn}_{ag} + R^{in}_{ga} d^j_{ag} - d^j_{ga} R^{in}_{ag}}{\omega^2_{ag} - \omega^2}\right) \tag{9.57}$$

Thus, using quantum mechanics we have calculated the dielectric and nonlocality tensors which were phenomenologically introduced in Chapter 2 and used for the description of linear optical effects in Chapter 3.

If all of the molecular wavefunctions are real, the matrix elements of the dipole moment operator $\hat{\mathbf{d}}$ are real while those of the spatial dispersion operator $\hat{R}$ are purely imaginary. With purely real $\hat{\mathbf{d}}$ and imaginary $\hat{R}$ the antisymmetric part of the dielectric tensor $\epsilon_{ij}(\omega)$ and the symmetric part of the nonlocality tensor $\Gamma^{(1)}_{ijn}(\omega)$ are zero. This rule holds for the majority of optical media. However, there is a the reason for the complexity of the wavefunctions, formulae (9.54) and (9.56) can be used to calculate $\epsilon^a_{ij}(\omega)$ and $\Gamma^{(1)s}_{ijn}(\omega)$. The same result, that is $\epsilon^a_{ij}(\omega) = 0$ and $\Gamma^{(1)s}_{ijn}(\omega) = 0$, can be obtained using thermodynamic arguments from the principle of time-reversibility of the light–matter interaction [7, 8]. However, the time-reversibility of this interaction is not always assured.

For example, $\epsilon^a_{ij}(\omega)$ and $\Gamma^{(1)s}_{ijn}(\omega)$ are permitted in magnetized materials [9] and antiferromagnetic materials [10]. Time-reversibility is broken and $\epsilon^a_{ij}(\omega)$ allowed in media with fractional carrier statistics. Such fractional, semi Boson–semi Fermion statistics have been predicted for high-T_c superconductors (see Chapter 1). $\Gamma^{(1)s}_{ijn}(\omega)$ is permitted in crystals with nonlocal response that lack an inversion centre. In this case the light–matter interaction lacks time-reversibility and the more rigid symmetry of simultaneous space and time inversion holds instead [11] (see Appendix 4).

If the molecules of the medium are not identically oriented, the calculation of the optical susceptibilities must include the taking of an average over the molecular orientation distribution function. This may impose symmetrical limitations on the existence of some components of the susceptibility tensors. In particular, the nonlocality tensor may vanish in some crystal classes [12]. In isotropic media, only the antisymmetric part of the nonlocality tensor and the symmetric part of the dielectric tensor survive averaging and may only be nonzero if the molecules are chiral. In isotropic medium as a result of orientational averaging, the contribution of quadrupole transitions to the nonlocal susceptibility vanishes and only the magnetic dipole term in the spatial dispersion operator is important [13, 14]. In most crystals, however, both quadrupole and magnetic dipole transitions are equally significant.

9.4 DISSIPATION OF ENERGY AND FREQUENCY DISPERSION OF LINEAR OPTICAL SUSCEPTIBILITIES

So far our model has not included dissipation. When the frequency of the incident light wave is close to a natural transition frequency ω_{ag}, resonant absorption becomes important. We will now incorporate absorbtion into our model in the way described in detail in Ref. [14]. Until now the electronic energy levels have been regarded as strictly discrete. According to the uncertainty principle this implies that their lifetime is infinite. To account for resonance phenomena we need to incorporate a finite energy width of the energy levels, thereby allowing for a finite lifetime. If the total probability for transition from, say, the exited state a to all lower states is small, then this state is called quasi-discrete and the probability $\psi_a\psi^*_a$ decays exponentially with time: $\psi_a(t)\psi^*_a(t) = \psi_a(0)\psi^*_a(0)\exp\{-\gamma_a t\}$. Here $1/\gamma_a$ is called the lifetime of the excited state a. Thus, the wavefunction of state a is

$$\psi_a(t) = u_a e^{-i\omega_{ag}t-\gamma_a t/2} \tag{9.58}$$

The lifetimes of states can be incorporated into the formalism of perturbation theory by changing to complex optical transition frequencies. Since the lifetime of the ground state is infinite ($\gamma_g = 0$), we can make the replacement

$\omega_{ag} \Rightarrow \omega_{ag} - \mathrm{i}\gamma_a/2$ in the denominators of formulae (9.53) and (9.54) and (9.56) and (9.57):

$$\frac{1}{\omega_{ag}^2 - \omega^2} \Rightarrow \frac{1}{(\omega_{ag}^2 - \omega^2) - \mathrm{i}\omega_{ag}\gamma_a - \gamma_a^2/4} \tag{9.59}$$

Because $\gamma_a \ll \omega_{ag}$, in the proximity of a resonance $\omega_{ag} \simeq \omega$ we approximate

$$\frac{1}{(\omega_{ag}^2 - \omega^2) - \mathrm{i}\omega_{ag}\gamma_a - \gamma_a^2/4} \approx f_a(\omega) + \mathrm{i}g_a(\omega) \tag{9.60}$$

which introduces the dispersion and absorption line shape functions $f_a(\omega)$ and $g_a(\omega)$:

$$\begin{aligned} f_a(\omega) &= \frac{\omega_{ag}^2 - \omega^2}{(\omega_{ag}^2 - \omega^2)^2 + \omega^2\gamma_a^2} \\ g_a(\omega) &= \frac{\omega\gamma_a}{(\omega_{ag}^2 - \omega^2)^2 + \omega^2\gamma_a^2} \end{aligned} \tag{9.61}$$

(Compare with the frequency dispersion functions (8.34) used in the classical oscillator theory.)

Thus, near an absorption resonance $\omega \approx \omega_{ag}$ the formulae (9.53) and (9.54) for the symmetric and antisymmetric parts of the dielectric tensor reduce to

$$\begin{pmatrix} \mathrm{Re} \\ \mathrm{Im} \end{pmatrix} \{\epsilon_{ij}^{\mathrm{s}}(\omega) - \delta_{ij}\}|_{\omega\approx\omega_{ag}} = \frac{8\pi\omega N_0}{\hbar} \begin{pmatrix} f_a(\omega) \\ g_a(\omega) \end{pmatrix} \mathrm{Re}\,\{d_{ga}^j d_{ag}^i\} \tag{9.62}$$

$$\begin{pmatrix} \mathrm{Re} \\ \mathrm{Im} \end{pmatrix} \{\epsilon_{ij}^{\mathrm{a}}(\omega)\}|_{\omega\approx\omega_{ag}} = \frac{8\pi\omega N_0}{\hbar} \begin{pmatrix} -g_a(\omega) \\ f_a(\omega) \end{pmatrix} \mathrm{Im}\,\{d_{ga}^j d_{ag}^i\} \tag{9.63}$$

Similarly, resonance formulae for the nonlocality tensor components will be

$$\begin{pmatrix} \mathrm{Re} \\ \mathrm{Im} \end{pmatrix} \{\Gamma_{ijn}^{(1)\mathrm{s}}(\omega)\}|_{\omega\approx\omega_{ga}} = \frac{8\pi N_0}{\hbar} \begin{pmatrix} -g_a(\omega) \\ f_a(\omega) \end{pmatrix} \mathrm{Re}\,\{R_{ga}^{jn} d_{ag}^i + d_{ag}^j R_{ga}^{in}\} \tag{9.64}$$

and

$$\begin{pmatrix} \mathrm{Re} \\ \mathrm{Im} \end{pmatrix} \{\Gamma_{ijn}^{(1)\mathrm{a}}(\omega)\}|_{\omega\approx\omega_{ga}} = \frac{8\pi N_0}{\hbar} \begin{pmatrix} f_a(\omega) \\ g_a(\omega) \end{pmatrix} \mathrm{Im}\,\{R_{ga}^{jn} d_{ag}^i + d_{ag}^j R_{ga}^{in}\} \tag{9.65}$$

Here we have used the fact that the dipole moment operator $\hat{\mathbf{d}}$ and the spatial dispersion operator $\hat{R}^{il}$ are Hermitian, $d_{ag}^i = (d_{ga}^i)^*$; $R_{ag}^{il} = (R_{ga}^{il})^*$.

We will now illustrate the results of quantum mechanical calculations of the dielectric and nonlocality tensors using an elementary model. We consider a two-level system with a ground state 0 and an upper state 1. In order to contribute to the frequency dispersion of the dielectric and nonlocality tensors, the matrix elements of the dipole moment and spatial dispersion operators, $\mathbf{d}_{10}$

and R_{10}^{il}, must be nonzero. Therefore, the wavefunction of the upper state, for example, should have mixed parity [14]. Let the dipole moment of the molecule be oriented along the Y direction of the laboratory Cartesian system: $\mathbf{d}_{10} = \{0; d_{10}^{y}; 0\}$ (see Fig. 9.1). For the wave propagating along Z-direction ($k_z \neq 0$) the nonlocal contribution of the interaction Hamiltonian (9.31) is proportional to the convolution $k_z R_{10}^{jz}$. This convolution is a vector which we present on Fig. 9.1 alongisde with the vector of dipole moment. For simplicity we assume that only one its component, namely $k_z R_{10}^{xz}$ has a non-zero value. We must now use formulae (9.62) to (9.65) to find the frequency dispersion of $\epsilon_{ij}(\omega)$ and $\Gamma_{ijl}^{(1)}(\omega)$ near the resonance $\omega \simeq \omega_{10}$. This gives

$$\begin{pmatrix} \mathrm{Re} \\ \mathrm{Im} \end{pmatrix} \{\epsilon_{yy}^{s}(\omega) - 1\} = \frac{8\pi\omega N_0}{\hbar} \begin{pmatrix} f_1(\omega) \\ g_1(\omega) \end{pmatrix} \mathrm{Re}\,\{d_{01}^{y} d_{10}^{y}\} \tag{9.66}$$

where $\epsilon_{xx}(\omega) = \epsilon_{zz}(\omega) = 1$ and all other components of the dielectric tensor are zero. In this model the antisymmetric part of the dielectric tensor vanishes because $\mathrm{Im}\,\{d_{01}^{y} d_{10}^{y}\} = \mathrm{Im}\,\{d_{01}^{y}(d_{01}^{y})^{*}\} = 0$.

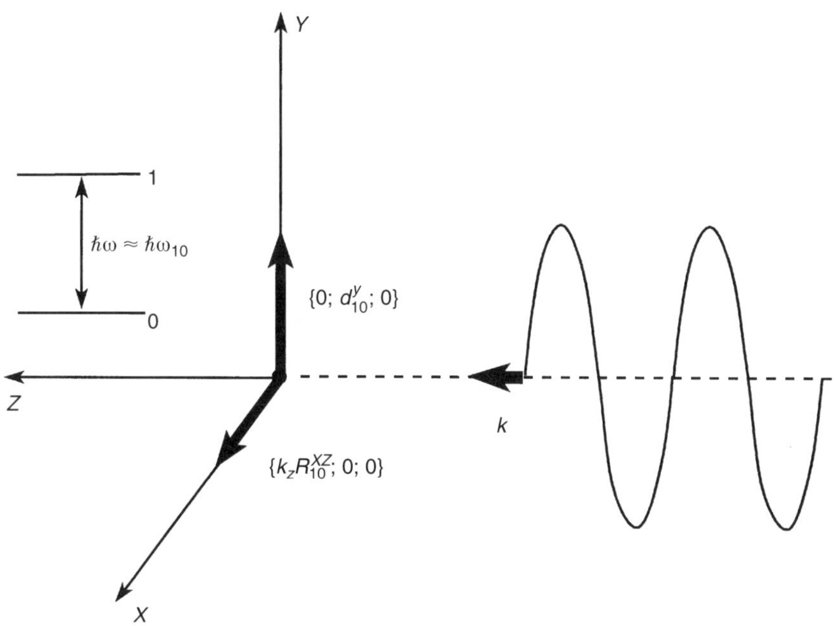

Fig. 9.1 An elementary example of a two-level system showing local and nonlocal optical responses. An electromagnetic wave at frequency ω propagating along the Z direction causes transitions between the ground state 0 and upper state 1. The dipole moment of the transition between states 1 and 0, $\{0, d_{10}^{y}, 0\}$ is oriented along the Y direction and only one component, R_{10}^{xz}, of the spatial dispersion operator's matrix element is nonzero

For this elementary example the antisymmetric part of the nonlocality tensor $\Gamma^{(1)a}_{ijn}(\omega) = -\Gamma^{(1)a}_{jin}(\omega)$ has the following nonzero components:

$$\begin{pmatrix} \text{Re} \\ \text{Im} \end{pmatrix} \{\Gamma^{(1)a}_{yxz}(\omega)\} = -\frac{8\pi N_0}{\hbar} \begin{pmatrix} f_1(\omega) \\ g_1(\omega) \end{pmatrix} \text{Im}\,\{R^{xz}_{01} d^{y}_{10}\} \tag{9.67}$$

All other components apart from $\Gamma^{(1)a}_{xyz} = -\Gamma^{(1)a}_{yxz}$ are zero. As we saw in Chapter 3, a medium with these non-zero nonlocality tersor components will show optical activity.

Similarly, the model gives the following nonzero components for the symmetric part of the nonlocality tensor $\Gamma^{(1)s}_{ijn}(\omega) = \Gamma^{(1)s}_{jin}(\omega)$:

$$\begin{pmatrix} \text{Re} \\ \text{Im} \end{pmatrix} \{\Gamma^{(1)s}_{yxz}(\omega)\} = \frac{8\pi N_0}{\hbar} \begin{pmatrix} -g_1(\omega) \\ f_1(\omega) \end{pmatrix} \text{Re}\,\{R^{xz}_{01} d^{y}_{10}\} \tag{9.68}$$

A medium with these nonzero nonlocality tensor components will show nonlocal birefringence (see Appendix 4).

From equations (9.66) with (9.67) we find the following relationship between the components of the dielectric and nonlocality tensors (compare with relationship (8.40) and (8.50) derived within the classical oscillator approach):

$$\frac{|\Gamma^{(1)a}_{xyz}(\omega)|}{|\epsilon_{yy}(\omega) - 1|} = \frac{1}{\omega} \frac{\text{Im}\,\{R^{xz}_{01}\}}{d^{y}_{01}} \tag{9.69}$$

A simple approximation follows from definitions (9.30), (9.25), (9.26) and (9.27) relating the matrix elements of the dipole and spatial dispersion operators with the typical size a of the molecule: $|R^{xz}_{01}| \approx a\omega|\mathbf{d}_{01}|$. Accordingly, from equation (9.69) we obtain

$$|\Gamma(\omega)| \approx a|\epsilon(\omega) - 1| \tag{9.70}$$

This simple two-level model gives a frequency dispersion for the nonlocality tensor which is similar to that given by the classical oscillator model for $\xi \gg \gamma\omega_0$ (compare with formula (8.40)). The condition $\xi \gg \gamma\omega_0$ corresponds to a strong split of the natural frequencies in the Born molecular model. The opposite situation of $\xi \leqslant \gamma\omega_0$ can be modelled in quantum mechanics by a three-level system with two close upper states [14].

The origin of optical susceptibilities can be illustrated using Feynman diagrams [15, 16]. In Feynman diagrams the molecular state is represented by a line (see Fig. 9.2). An arrow directed towards the state line indicates a photon absorbed by the molecule; an arrow directed away from the state line indicates a photon emitted by the molecule. A solid line arrow indicates that the transition happens via local interactions described by the electrodipole operator $\hat{\mathbf{d}}$; a dashed line indicates transitions that occur via nonlocal interaction described by the spatial dispersion operator $\hat{R}^{il}$. Orthogonal light polarizations are represented by arrows on opposite sides of the state line. As a rule, especially in nonlinear interactions, there are several Feynman diagrams behind each susceptibility: they describe different terms in the sums like (9.53) to (9.57) and therefore describe different sequences of transitions which lead to the same optical response. As examples, Fig. 9.2 shows Feynman diagrams for the linear susceptibilities discussed in this chapter.

9.5 NONLINEAR SUSCEPTIBILITIES

Nonlinear optical susceptibilities emerge from the higher orders of perturbation theory. Considerable effort has been concentrated on the development of the quantum mechanical theory of local and nonlocal optical susceptibilities [12, 17–24]. We will give formulae, below, for the second- and third-order local

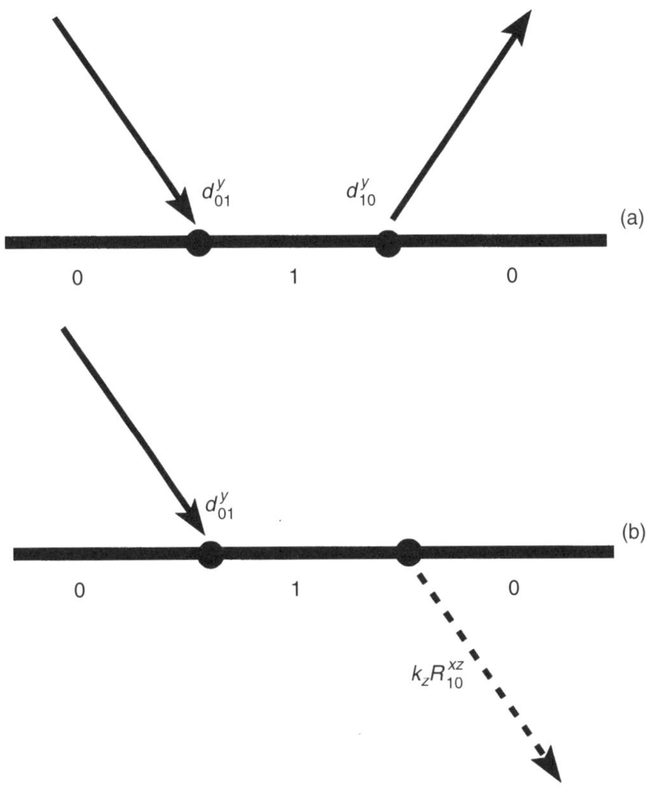

Fig. 9.2 Feynman diagrams illustrating processes in the elementary molecular model of Fig. 9.1: (a) linear local optical response (see formula (9.66)); (b) linear nonlocal optical response (see formula (9.67))

and nonlocal nonlinear susceptibilities. When calculating nonlinear susceptibilities, we will assume, for the sake of simplicity, that all of the molecular wavefunctions are real. This implies that the matrix elements of the dipole moment operator are real, $\mathbf{d}_{ag} = (\mathbf{d}_{ag})^* = \mathbf{d}_{ga}$, while those of the spatial dispersion operator are purely imaginary, $R^{il}_{ag} = -(R^{il}_{ag})^* = -R^{il}_{ga}$. Quantum theory outlined in Sections 9.1 and 9.2 gives the following second-order susceptibilities for constitutive equation (2.51):

$$\chi^{(2)}_{ijk}(2\omega, \omega, \omega) = \frac{N_0}{\hbar^2} \sum_{a,b} \left[\frac{(\omega_{ag}\omega_{bg} - \omega^2) d^k_{ga} d^i_{ab} d^j_{bg}}{(\omega^2_{ag} - \omega^2)(\omega^2_{bg} - \omega^2)} + \frac{(\omega_{ag}\omega_{bg} + 2\omega^2)(d^k_{ga} d^j_{ab} + d^j_{ga} d^k_{ab}) d^i_{bg}}{(\omega^2_{bg} - \omega^2)(\omega^2_{ag} - 4\omega^2)} \right] \tag{9.71}$$

and

$$\Gamma^{s}_{ijkn}(2\omega, \omega, \omega) = \frac{N_0}{\hbar^2} \sum_{a,b} \left\{ \frac{\omega_{ab}(R^{kn}_{ga} d^i_{ab} d^j_{bg} + \frac{1}{2} d^j_{ga} R^{in}_{ab} d^k_{bg})}{(\omega^2_{ag} - \omega^2)(\omega^2_{bg} - \omega^2)} + \frac{(\omega_{ag} + 2\omega_{bg})[(R^{jn}_{gb} d^k_{ba} + d^k_{gb} R^{jn}_{ba}) d^i_{ag} + \frac{1}{2}(d^j_{gb} d^k_{ba} + d^k_{gb} d^j_{ba}) R^{in}_{ag}]}{(\omega^2_{bg} - \omega^2)(\omega^2_{ag} - 4\omega^2)} \right\} \tag{9.72}$$

(see Ref. [19] for more details). Similarly, the perturbation theory gives the following third-order susceptibilities for constitutive equation (2.56):

$$\begin{aligned} \chi_{ijkl} = \frac{N_0}{\hbar^3} \sum_{a,b,c} \Bigg\{ & (\omega^2_{bg} - \omega^2)^{-1} (\omega^2_{cg} - \omega^2)^{-1} \\ & \times \Bigg[\frac{\omega_{bg}\omega_{cg} + \omega^2}{\omega_{ag}} d^i_{gc} d^l_{ab} (d^j_{ca} d^k_{bg} + d^k_{ca} d^j_{bg}) \\ & + \frac{\omega_{cg}\omega_{bg} - \omega^2}{\omega_{ag}} d^i_{gc} d^j_{ab} (d^k_{ca} d^l_{bg} + d^l_{ca} d^k_{bg}) \\ & + \frac{\omega_{ag}(\omega_{bg}\omega_{cg} + \omega^2) + 2\omega^2(\omega_{bg} + \omega_{cg})}{\omega^2_{ag} - 4\omega^2} d^k_{gb} d^j_{ba} (d^i_{ac} d^l_{cg} + d^l_{gc} d^i_{ca}) \Bigg] \Bigg\} \\ & + [j \leftrightarrow k] \end{aligned} \tag{9.73}$$

and

$$\Gamma_{ijlmn} = -\frac{N_0}{\hbar^3}\sum_{a,b,c}\Bigg\{(\omega_{bg}^2-\omega^2)^{-1}(\omega_{cg}^2-\omega^2)^{-1}$$
$$\times\left[\frac{\omega_{cg}+\omega_{bg}}{\omega_{ag}}d_{gc}^i d_{ca}^j(R_{ab}^{ln}d_{bg}^k + d_{ab}^l R_{bg}^{kn})\right.$$
$$\left.+\frac{\omega_{cb}}{\omega_{ag}}d_{gc}^i d_{ca}^k(R_{ab}^{ln}d_{bg}^j + d_{ab}^l R_{bg}^{jn})\right]$$
$$+\frac{\omega_{ag}(\omega_{cg}+\omega_{bg})+2(\omega^2+\omega_{cg}\omega_{bg})}{\omega_{ag}^2-4\omega^2}$$
$$\times\left[d_{ca}^l d_{ab}^k(R_{gc}^{in}d_{bg}^j + d_{gc}^i R_{bg}^{jn}) + d_{gc}^i d_{bg}^j(R_{ca}^{ln}d_{ab}^k + d_{ca}^l R_{ab}^{kn})\right]\Bigg\}$$
$$-\{i\leftrightarrow j,\ k\leftrightarrow l\}+\{j\leftrightarrow k\}-\{i\rightarrow j,\ j\rightarrow l,\ l\rightarrow k,\ k\rightarrow i\} \tag{9.74}$$

(also see Ref. [19]). Thus, equations (9.71) to (9.73) give quantum-mechanical formulae for the local and nonlocal nonlinear optical susceptibilities which were phenomenologically introduced in Chapter 2 and used for the description of nonlinear optical effects in Chapters 4 to 6.

We will now illustrate the results of quantum-mechanical calculations of optical nonlinearities using an elementary molecular model similar to that presented in Fig. 9.1. In order to see two-photon resonances in nonlinearity, a new level 2, with opposite parity to the ground state, will be added to the model (see Fig. 9.3). For simplicity we assume that the dipole moments $\mathbf{d}_{10}$ and $\mathbf{d}_{12}$ are oriented along the Y direction: $\mathbf{d}_{10} = \{0, d_{10}^y, 0\}$; $\mathbf{d}_{12} = \{0, d_{12}^y, 0\}$. Here again, like in the linear case, we will assume for simplicity that only matrix elements R_{10}^{jz} and R_{12}^{jz} are non-zero. Relevant vectors $k_z R_{10}^{jz}$ and $k_z R_{12}^{jz}$ are present on Fig. 9.3 alongisde with the vectors of dipole moments of the transitions involved.

As formulae (9.74) and (9.75) show, the cubic local and nonlocal susceptibilities rise near one- and two-photon resonances. The later appears due to the terms containing denominator $\omega_{ag}^2 - 4\omega^2$. We will examine the frequency dispersion of the nonlinear optical susceptibilities in the proximity of one- and two-photon resonances separately. Resonance absorption will be accounted for similarly as it was in the case of linear response (see formulae (9.59) and (60)).

Within the simple model presented in Fig. 9.3, in the proximity of one-photon resonance $\omega \approx \omega_{10}$, formula (9.73) returns only one nonzero component of the local cubic nonlinearity tensor:

$$\begin{pmatrix}\mathrm{Re}\\ \mathrm{Im}\end{pmatrix}\chi_{yyyy} = \frac{16\pi N_0}{\hbar^3}\begin{pmatrix} f_1^2(\omega) - g_1^2(\omega) \\ 2f_1(\omega)g_1(\omega)\end{pmatrix}\omega_{10}|d_{10}^y|^4 \tag{9.75}$$

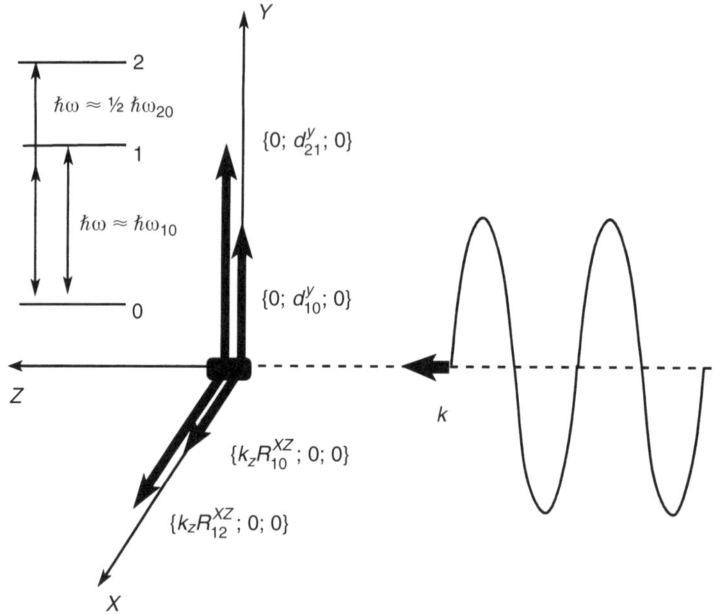

Fig. 9.3 An example of a three-level system showing local and nonlocal nonlinear optical responses. An electromagnetic wave propagates along the Z direction. The dipole moments of the transitions between states 0 and 1 and between states 1 and 2 are oriented along the Y direction. Only one component, R^{xz}_{10}, of the matrix element tensor of the spatial dispersion operator between states 1 and 0 and only one component, R^{xz}_{12}, of the matrix element tensor of the spatial dispersion operator between states 1 and 2 are nonzero

while for the nonlocal cubic nonlinearity tensor formula (9.74) gives

$$\begin{pmatrix}\mathrm{Re}\\ \mathrm{Im}\end{pmatrix}\Gamma_{xyyyz} = \frac{16\pi N_0}{\hbar^3}\begin{pmatrix} f_1^2(\omega) - g_1^2(\omega) \\ 2f_1(\omega)g_1(\omega)\end{pmatrix}$$
$$\times\left[\mathrm{Im}\,\{(d^y_{01})^3 R^{xz}_{10}\} - \frac{2\omega_{10}d^y_{01}d^y_{12}}{(\omega_{20}-2\omega_{10})\omega_{20}} \times \mathrm{Im}\,\{\omega_{21}d^y_{21}R^{xz}_{10} + \omega_{10}d^y_{01}R^{xz}_{12}\}\right] \tag{9.76}$$

The dependence on frequency of the nonlinear susceptibilities in the proximity of a one-photon resonance are shown in Fig. 9.4. Note that here the frequency dispersions of χ_{xxxx} and Γ_{xyyyz} are similar.

In the proximity of a two-photon resonance, i.e. for frequencies $\omega \approx \omega_{20}/2$, formula (9.73) gives

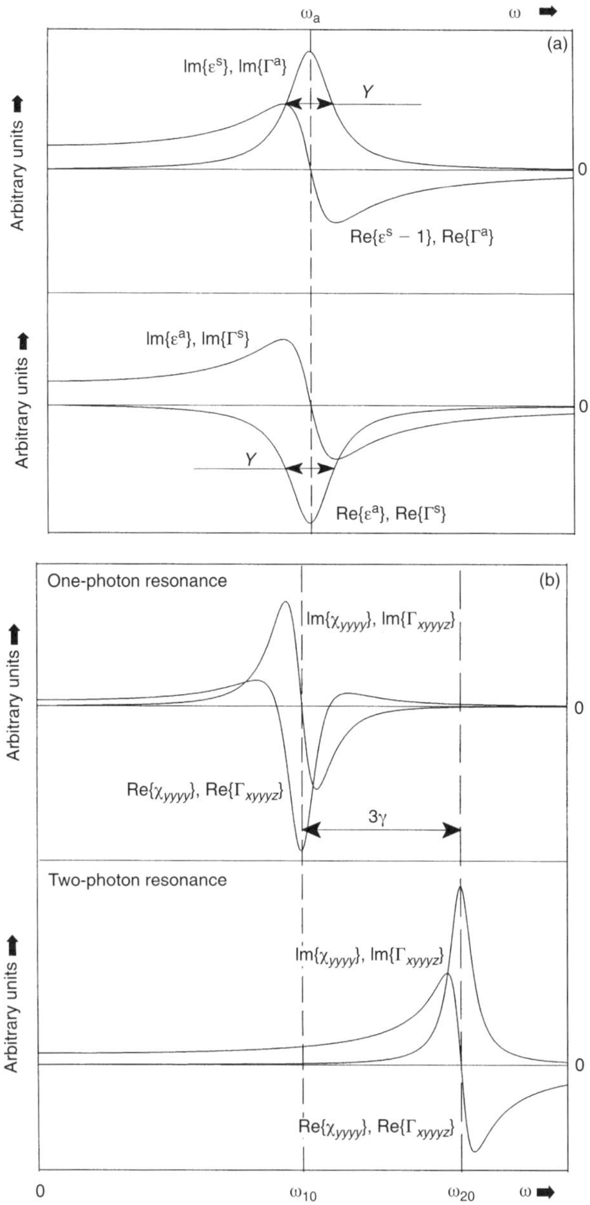

Fig. 9.4 (a) Sketches of the frequency dispersion of the real and imaginary parts of the dielectric tensor and nonlocality tensor as given by components of the formulae (9.62)–(9.65); (b) sketches of the frequency dispersion of the real and imaginary parts of the local and nonlocal cubic susceptibilities in the proximity of one- and two-photon resonance at $\omega_{20} - \omega_{10} = 3\gamma$ (see formulae (9.75)–(9.78))

$$\begin{pmatrix} \mathrm{Re} \\ \mathrm{Im} \end{pmatrix} \chi_{yyyy} = \frac{32\pi N_0}{\hbar^3} \begin{pmatrix} f_2(2\omega) \\ g_2(2\omega) \end{pmatrix} \frac{(d_{10}^y)^2 (d_{21}^y)^2}{\omega_{20} - 2\omega_{10}} \tag{9.77}$$

while for the nonlocal cubic nonlinearity, formula (9.74) gives

$$\begin{pmatrix} \mathrm{Re} \\ \mathrm{Im} \end{pmatrix} \Gamma_{xyyyz} = -\frac{16\pi N_0}{\hbar^3} \begin{pmatrix} f_2(2\omega) \\ g_2(2\omega) \end{pmatrix} \frac{\mathrm{Im}\,\{d_{01}^y d_{12}^y (d_{21}^y R_{10}^{xz} + d_{01}^y R_{12}^{xz})\}}{(\omega_{10} - \omega_{20}/2)^2} \tag{9.78}$$

The dependence on frequency of the nonlinear susceptibilities in the proximity of a two-photon resonance are shown in Fig. 9.4. At two-photon resonance, the frequency dispersions of χ_{xxxx} and Γ_{xyyyz} are again similar.

Assuming that dipole matrix elements are all of the same order of magnitude $|\mathbf{d}_{10}| \simeq |\mathbf{d}_{21}| \simeq d$, and likewise the matrix elements of the spatial dispersion

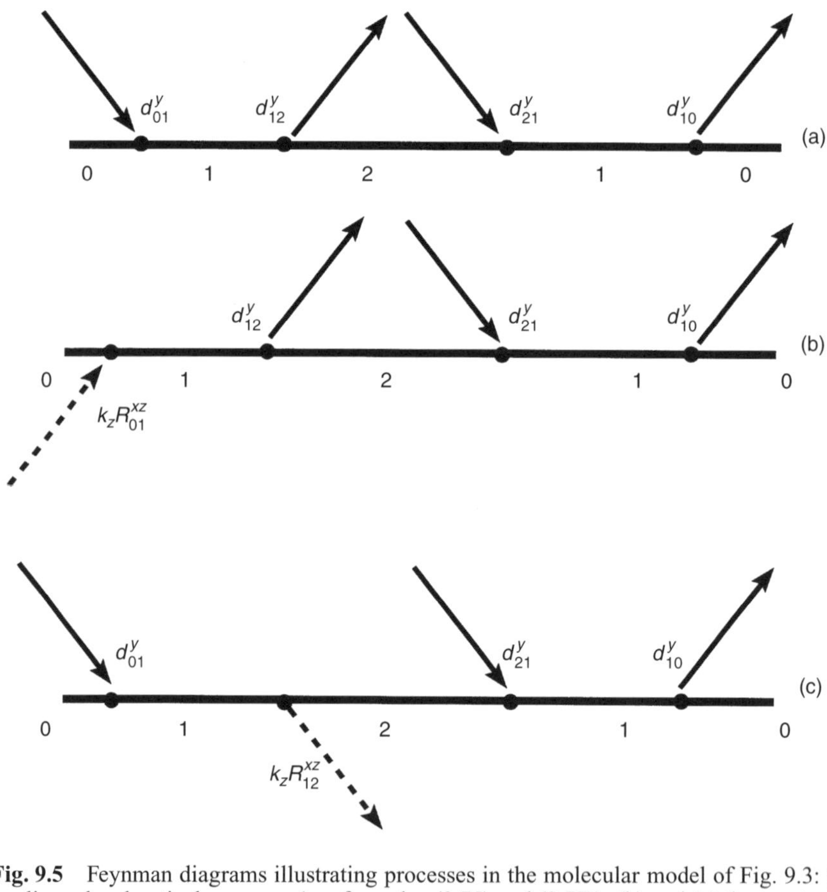

Fig. 9.5 Feynman diagrams illustrating processes in the molecular model of Fig. 9.3: (a) nonlinear local optical response (see formulae (9.75) and (9.77)); (b) and (c) the two terms of equations (9.76) and (9.78) leading to nonlinear nonlocal response

operator, $|R_{10}^{xz}| \simeq |R_{21}^{xz}| \simeq R$, we can readily arrive at the following relationship between local and nonlocal cubic susceptibilities:

$$\frac{|\Gamma_{xyyyz}|}{|\chi_{yyyy}|} \approx \frac{1}{\omega}\frac{R}{d} \tag{9.79}$$

or, using $R = awd$, we obtain

$$|\Gamma| \approx a|\chi| \tag{9.80}$$

(Compare with equation (9.70) and the similar formula (8.55) derived using the classical oscillator model.)

Feynman diagrams relating to the nonlinearities given by formulae (9.73) to (9.79) are presented in Fig. 9.5.

REFERENCES

1. L.D. Landau and E.M. Lifshitz, *Quantum Mechanics*, Pergamon Press, Oxford, 1977.
2. Yu.P. Svirko, *Soviet Phys. Lebedev Institute Rep.*, **11**, 32 (1988).
3. Yu.P. Svirko, *J. Quant. Semiclas. Opt.*, **10**, 303 (1998).
4. E.A. Power and S. Zienau, *Phil. Trans. R. Soc.*, **251**, 427 (1959).
5. R.G. Woolley, *Adv. Phys.*, **25**, 27 (1975).
6. I.A.Frenkel, *Wave Mechanics; Advanced General Theory*, Dover Publication, New York, 1950.
7. L.D. Landau, E.M. Lifshitz and L.P. Pitaevski, *Electrodynamics of Continuous Media*, Pergamon Press, Oxford, 1984.
8. A.R. Bungay, Y.P. Svirko and N.I. Zheludev, *Phys. Rev. B*, **47**, 16141 (1993).
9. R.M. Hornreich and S. Shtrikman, *Phys. Rev.*, **171**, 1065 (1968).
10. I. Dzyaloshinskii and E.V. Papamichail, *Phys. Rev. Lett.*, **75**, 3004 (1995).
11. Yu.P. Svirko and N. I. Zheludev, *Opt. Lett.*, **20**, 1809 (1995).
12. S.V. Popov, N.I. Zheludev and Yu.P. Svirko, *Tensors for Nonlinear Optical Susceptibilities*, IOP Publishing, Bristol and Phyladelphia, 1995.
13. M.V. Vol'kenshtein, *Molecular Optics*, Gostekhizdat, Moscow, 1951.
14. L.D. Barron, *Molecular Light Scattering and Optical Activity*, Cambridge University Press, Cambridge, 1982.
15. A. Yariv, *Quantum Electronics*, John Wiley & Sons, New York, 1975.
16. P.M. Atkins and L.D. Barron, *Proc. Roy. Soc.*, **304A**, 303 (1968).
17. E. Adler. *Phys. Rev.*, **134**, A728 (1964).
18. G. Wagniere, *Chem. Phys.*, **54**, 441 (1981).
19. G.A. Lyakhov and Yu.P. Svirko, in F.V. Bunkin (ed.), *Nonlinear Optics and Nonlinear Acoustics of Fluids*, Nova Publishers, New York, 1990, p.31.
20. D. Mukherjee and M. Chowdhury, *Physica*, **58**, 109 (1972).
21. I. Tinoko, *J. Chem. Phys.*, **62**, 1006 (1975).
22. R.A. Harris, *J. Chem. Phys.*, **65**, 5294 (1976).
23. L.D. Barron and A.D. Buckingham, *J. Phys. B: Atom. Molec. Phys.*, **6**, 1295 (1983).
24. A.A. Golubkov and V.A. Makarov, *Vestnik Moskovskogo Universiteta Seriya 3 Fizika Astronomiya*, **31**, 49 (1990).

FURTHER READING

N.B. Baranova and B.Ya. Zel'dovich, *Soviet Phys. Uspekhy*, **23**, 143 (1979).
R.W. Boyd, *Nonlinear Optics*, Academic Press, San Diego, 1992.
A.D. Buckingham, *Adv. Chem. Phys.*, **12**, 107 (1967).
P.N. Butcher and D. Cotter, *The Elements of Nonlinear Optics*, Cambridge University Press, Cambridge, 1990.
G.A. Lyakhov and Yu.P. Svirko, *JETP*, **53**, 669 (1981).
J.J. Sakurai, *Modern Quantum Mechanics*, Addison-Wesley, Reading, Massachusets, 1994.

Appendix 1 The Tensor of Local Cubic Optical Nonlinearity

This appendix gives tables for the cubic optical nonlinearity tensor $\chi_{ijkl} = \chi_{ikjl}$, which is symmetric with respect to permutation of the second and third indices. This nonlinearity is used throughout the book, in particular in constitutive equation (2.56). For compactness of presentation, the letter χ is dropped from the tensor component symbols and only the tensor indices are retained. All independent components of the tensor are numbered. Nonzero dependent components are presented as linear combinations of the independent components if they are not given by the basic rule $\chi_{ijkl} = \chi_{ikjl}$.

Tables for other optical nonlinearity tensors up to the seventh rank may be found in Ref. [1].

Classes: 1, $\bar{1}$

1. *xxxx*; **2.** *xxxy*; **3.** *xxxz*; **4.** *xxyx*; **5.** *xxyy*; **6.** *xxyz*; **7.** *xxzx*; **8.** *xxzy*; **9.** *xxzz*; **10.** *xyyx*; **11.** *xyyy*; **12.** *xyyz*; **13.** *xyzx*; **14.** *xyzy*; **15.** *xyzz*; **16.** *xzzx*; **17.** *xzzy*; **18.** *xzzz*; **19.** *yxxx*; **20.** *yxxy*; **21.** *yxxz*; **22.** *yxyx*; **23.** *yxyy*; **24.** *yxyz*; **25.** *yxzx*; **26.** *yxzy*; **27.** *yxzz*; **28.** *yyyx*; **29.** *yyyy*; **30.** *yyyz*; **31.** *yyzx*; **32.** *yyzy*; **33.** *yyzz*; **34.** *yzzx*; **35.** *yzzy*; **36.** *yzzz*; **37.** *zxxx*; **38.** *zxxy*; **39.** *zxxz*; **40.** *zxyx*; **41.** *zxyy*; **42.** *zxyz*; **43.** *zxzx*; **44.** *zxzy*; **45.** *zxzz*; **46.** *zyyx*; **47.** *zyyy*; **48.** *zyyz*; **49.** *zyzx*; **50.** *zyzy*; **51.** *zyzz*; **52.** *zzzx*; **53.** *zzzy*; **54.** *zzzz*

Classes: 2 (2$\|$*Y*), *m*(*m*$\perp$*Y*), 2/*m* (*m*$\perp$*Y*)

1. *xxxx*; **2.** *xxxz*; **3.** *xxyy*; **4.** *xxzx*; **5.** *xxzz*; **6.** *xyyx*; **7.** *xyyz*; **8.** *xyzy*; **9.** *xzzx*; **10.** *xzzz*; **11.** *yxxy*; **12.** *yxyx*; **13.** *yxyz*; **14.** *yxzy*; **15.** *yyyy*; **16.** *yyzx*; **17.** *yyzz*; **18.** *yzzy*; **19.** *zxxx*; **20.** *zxxz*; **21.** *zxyy*; **22.** *zxzx*; **23.** *zxzz*; **24.** *zyyx*; **25.** *zyyz*; **26.** *zyzy*; **27.** *zzzx*; **28.** *zzzz*

Classes: 2 (2$\|$*Z*), *m*(*m*$\perp$*Z*), 2/*m* (*m*$\perp$*Z*)

1. *xxxx*; **2.** *xxxy*; **3.** *xxyx*; **4.** *xxyy*; **5.** *xxzz*; **6.** *xyyx*; **7.** *xyyy*; **8.** *xyzz*; **9.** *xzzx*; **10.** *xzzy*; **11.** *yxxx*; **12.** *yxxy*; **13.** *yxyx*; **14.** *yxyy*; **15.** *yxzz*; **16.** *yyyx*; **17.** *yyyy*; **18.** *yyzz*; **19.** *yzzx*; **20.** *yzzy*; **21.** *zxxz*; **22.** *zxyz*; **23.** *zxzx*; **24.** *zxzy*; **25.** *zyyz*; **26.** *zyzx*; **27.** *zyzy*; **28.** *zzzz*

Classes: 222, *mm*2, *mmm*

1. *xxxx*; **2.** *xxyy*; **3.** *xxzz*; **4.** *xyyx*; **5.** *xzzx*; **6.** *yxxy*; **7.** *yxyx*; **8.** *yyyy*; **9.** *yyzz*; **10.** *yzzy*; **11.** *zxxz*; **12.** *zxzx*; **13.** *zyyz*; **14.** *zyzy*; **15.** *zzzz*

Classes: 3, $\bar{3}$

1. *yxyx*; **2.** *yyyy*; *xxxx* = *yyyy*; *xxyy* = *yxyx*; *xyyx* = -2*yxyx* + *yyyy*; *yxxy* =

$-2yxyx + yyyy$; **3.** $yxyy$; **4.** $yyyx$; $xxxy = -yyyx$; $xxyx = -yxyy$; $xyyy = -2yxyy - yyyx$; $yxxx = 2yxyy + yyyx$; **5.** $yxyz$; $xxxz = -yxyz$; $xyyz = yxyz$; **6.** $yyyz$; $xxyz = -yyyz$; $yxxz = -yyyz$; **7.** $yyzx$; $xxzx = -yyzx$; $xyzy = yyzx$; $yxzy = yyzx$; **8.** $yyzy$; $xxzy = -yyzy$; $xyzx = -yyzy$; $yxzx = -yyzy$; **9.** $zyyx$; $zxxx = -zyyx$; $zxyy = zyyx$; **10.** $zyyy$; $zxxy = -zyyy$; $zxyx = -zyyy$; **11.** $xxzz$; $xxzz = yyzz$; **12.** $xyzz$; $xyzz = -yxzz$; **13.** $xzzx$; $xzzx = yzzy$; **14.** $xzzy$; $xzzy = -yzzx$; **15.** $zxxz$; $zxxz = zyyz$; **16.** $zxzx$; $zxzx = zyzy$; **17.** $zxzy$; $zxzy = -zyzx$; **18.** $zzzz$

Classes: 32 ($2\|X$), $m(m\perp X)$, $\bar{3}/m$ ($m\perp X$)

1. $yxyx$; **2.** $yyyy$; $xxxx = yyyy$; $xxyy = yxyx$; $xyyx = -2yxyx + yyyy$; $yxxy = -2yxyx + yyyy$ **3.** $yyyz$; $xxyz = -yyyz$; $yxxz = -yyyz$ **4.** $yyzy$; $xxzy = -yyzy$; $xyzx = -yyzy$; $yxzx = -yyzy$; **5.** $zyyy$; $zxxy = -zyyy$; $zxyx = -zyyy$; **6.** $xxzz$; $xxzz = yyzz$; **7.** $xzzx$; $xzzx = yzzy$ **8.** $zxxz$; $zxxz = zyyz$; **9.** $zxzx = zyzy$; **10.** $zzzz$

Classes: 32 ($2\|Y$), $m(m\perp Y)$, $\bar{3}/m$ ($m\perp Y$)

1. $yxyx$; **2.** $yyyy$; $xxxx = yyyy$; $xxyy = yxyx$; $xyyx = -2yxyx + yyyy$; $yxxy = -2yxyx + yyyy$; **3.** $yxyz$; $xxxz = -yxyz$; $xyyz = yxyz$; **4.** $yyzx$; $xxzx = -yyzx$; $xyzy = yyzx$; $yxzy = yyzx$; **5.** $zyyx$; $zxxx = -zyyx$; $zxyy = zyyx$; **6.** $xxzz$; $xxzz = yyzz$; **7.** $xzzx$; $xzzx = yzzy$; **8.** $zxxz$; $zxxz = zyyz$; **9.** $zxzx$; $zxzx = zyzy$; **10.** $zzzz$

Classes: 4 $\bar{4}$, $4/m$

1. $yxxx = -xyyy$; **2.** $yxxy = xyyx$; **3.** $yxyx = xxyy$; **4.** $yxyy = -xxyx$; **5.** $yxzz = -xyzz$; **6.** $yyyx = -xxxy$; **7.** $yyyy = xxxx$; **8.** $yyzz = xxzz$; **9.** $yzzx = -xzzy$; **10.** $yzzy = xzzx$; **11.** $zyyz = zxxz$; **12.** $zyzx = -zxzy$; **13.** $zyzy = zxzx$; **14.** $zzzz$

Classes: 442 $4mm$, $\bar{4}2m$, $4/mmm$

1. $yxxy = xyyx$; **2.** $yxyx = xxyy$; **3.** $yyyy = xxxx$; **4.** $yyzz = xxzz$; **5.** $yzzy = xzzx$; **6.** $zyyz= zxxz$; **7.** $zyzy = zxzx$; **8.** $zzzz$

Classes: 6, $\bar{6}$, $6/m$, ∞, ∞/m

1. $yxyx$; **2.** $yyyy$; $xxxx = yyyy$; $xxyy = yxyx$; $xyyx = -2yxyx + yyyy$; $yxxy = -2yxyx + yyyy$; **3.** $yxyy$; **4.** $yyyx$; $xxxy = -yyyx$; $xxyx = -yxyy$; $xyyy = -2yxyy - yyyx$; $yxxx = 2yxyy + yyyx$; **5.** $xxzz$; $xxzz = yyzz$; **6.** $xyzz$; $xyzz = -yxzz$; **7.** $xzzx$; $xzzx = yzzy$; **8.** $xzzy$; $xzzy = -yzzx$; **9.** $zxxz$; $zxxz = zyyz$; **10.** $zxzx$; $zxzx = zyzy$; **11.** $zxzy$; $zxzy = -zyzx$; **12.** $zzzz$

Classes: 622, $6mm$, $\bar{6}2m$, $6/mmm$, $\infty 2$, ∞m, ∞/mm

1. $yxyx$; **2.** $yyyy$; $xxxx = yyyy$; $xxyy = yxyx$; $xyyx = -2yxyx + yyyy$; $yxxy = -2yxyx + yyyy$; **3.** $xxzz$; $xxzz = yyzz$; **4.** $xzzx$; $xzzx = yzzy$; **5.** $zxxz$; $zxxz = zyyz$; **6.** $zxzx$; $zxzx = zyzy$; **7.** $zzzz$

Classes: 23, $m3$

1. $zxxz = xyyx = yzzy$; **2.** $zxzx = xxyy = yyzz$; **3.** $zyyz = xzzx = yxxy$; **4.** $zyzy = xxzz = yxyx$; **5.** $zzzz = xxxx = yyyy$

Classes: 432, $\bar{4}3m$, $m3m$
1. *zyyz* = *xzzx* = *zxxz* = *yxxy* = *yzzy* = *xyyx*; **2.** *zyzy* = *xxzz* = *zxzx* = *yxyx* = *yyzz* = *xxyy*; **3.** *zzzz* = *xxxx* = *yyyy*

Classes: $\infty\infty$, $\infty\infty m$
1. *xxyy*; **2.** *xyyx*; $\chi_{ijkl} = \chi_{xxyy}(\delta_{ij}\delta_{kl} + \delta_{ik}\delta_{jl}) + \chi_{xyyx}\delta_{il}\delta_{jk}$

REFERENCE

1. S.V. Popov, Yu.P. Svirko and N.I. Zheludev, *Susceptibility Tensors for Nonlinear Optics*, IOP Publishing, Bristol and Philadelphia, 1995.

Appendix 2 The Tensor of Nonlocal Cubic Optical Nonlinearity

This appendix gives tables for the nonlocal cubic optical nonlinearity tensor $\Gamma_{ijklm} = \Gamma_{ikjlm}$, which is symmetric with respect to permutation of the second and third indices. This nonlinearity is used throughout the book, in particular in constitutive equation (2.56). For compactness of presentation, the letter Γ is dropped from the tensor component symbols and only the tensor indices are retained. All independent components of the tensor are numbered. Nonzero dependent components are presented as linear combinations of the independent components if they are not given by the basic rule $\Gamma_{ijklm} = \Gamma_{ikjlm}$.

Tables for other optical nonlinearity tensors up to the seventh rank may be found in Ref. [1].

The tensor vanishes in the following classes: $\bar{1}$, $2/m$, mmm, $\bar{3}$, $\bar{3}m$, $6/m$, $6/mmm$, $4/m$, $4/mmm$, $m3$, $m3m$, ∞/m, ∞/m, $\infty\infty m$

Class: 2 ($2\|Y$)

1. *xxxxy*; **2.** *xxxyx*; **3.** *xxxyz*; **4.** *xxxzy*; **5.** *xxyxx*; **6.** *xxyxz*; **7.** *xxyyy*; **8.** *xxyzx*; **9.** *xxyzz*; **10.** *xxzxy*; **11.** *xxzyx*; **12.** *xxzyz*; **13.** *xxzzy*; **14.** *xyyxy*; **15.** *xyyyx*; **16.** *xyyyz*; **17.** *xyyzy*; **18.** *xyzxx*; **19.** *xyzxz*; **20.** *xyzyy*; **21.** *xyzzx*; **22.** *xyzzz*; **23.** *xzzxy*; **24.** *xzzyx*; **25.** *xzzyz*; **26.** *xzzzy*; **27.** *yxxxx*; **28.** *yxxxz*; **29.** *yxxyy*; **30.** *yxxzx*; **31.** *yxxzz*; **32.** *yxyxy*; **33.** *yxyyx*; **34.** *yxyyz*; **35.** *yxyzy*; **36.** *yxzxx*; **37.** *yxzxz*; **38.** *yxzyy*; **39.** *yxzzx*; **40.** *yxzzz*; **41.** *yyyxx*; **42.** *yyyxz*; **43.** *yyyyy*; **44.** *yyyzx*; **45.** *yyyzz*; **46.** *yyzxy*; **47.** *yyzyx*; **48.** *yyzyz*; **49.** *yyzzy*; **50.** *yzzxx*; **51.** *yzzxz*; **52.** *yzzyy*; **53.** *yzzzx*; **54.** *yzzzz*; **55.** *zxxxy*; **56.** *zxxyx*; **57.** *zxxyz*; **58.** *zxxzy*; **59.** *zxyxx*; **60.** *zxyxz*; **61.** *zxyyy*; **62.** *zxyzx*; **63.** *zxyzz*; **64.** *zxzxy*; **65.** *zxzyx*; **66.** *zxzyz*; **67.** *zxzzy*; **68.** *zyyxy*; **69.** *zyyyx*; **70.** *zyyyz*; **71.** *zyyzy*; **72.** *zyzxx*; **73.** *zyzxz*; **74.** *zyzyy*; **75.** *zyzzx*; **76.** *zyzzz*; **77.** *zzzxy*; **78.** *zzzyx*; **79.** *zzzyz*; **80.** *zzzzy*

Class: 2 ($2\|Z$)

1. *xxxxz*; **2.** *xxxyz*; **3.** *xxxzx*; **4.** *xxxzy*; **5.** *xxyxz*; **6.** *xxyyz*; **7.** *xxyzx*; **8.** *xxyzy*; **9.** *xxzxx*; **10.** *xxzxy*; **11.** *xxzyx*; **12.** *xxzyy*; **13.** *xxzzz*; **14.** *xyyxz*; **15.** *xyyyz*; **16.** *xyyzx*; **17.** *xyyzy*; **18.** *xyzxx*; **19.** *xyzxy*; **20.** *xyzyx*; **21.** *xyzyy*; **22.** *xyzzz*; **23.** *xzzxz*; **24.** *xzzyz*; **25.** *xzzzx*; **26.** *xzzzy*; **27.** *yxxxz*; **28.** *yxxyz*; **29.** *yxxzx*; **30.** *yxxzy*; **31.** *yxyxz*; **32.** *yxyy*; **33.** *yxyzx*; **34.** *yxyzy*; **35.** *yxzxx*; **36.** *yxzxy*; **37.** *yxzyx*; **38.** *yxzyy*; **39.** *yxzzz*; **40.** *yyyxz*; **41.** *yyyyz*; **42.** *yyyzx*; **43.** *yyyzy*; **44.** *yyzx*; **45.** *yyzxy*; **46.** *yyzyx*; **47.** *yyzyy*; **48.** *yyzzz*; **49.** *yzzxz*; **50.** *yzzyz*;

51. *yzzzx*; **52.** *yzzzy*; **53.** *zxxxx*; **54.** *zxxxy*; **55.** *zxxyx*; **56.** *zxxyy*; **57.** *zxxzz*; **58.** *zxyxx*; **59.** *zxyxy*; **60.** *zxyyx*; **61.** *zxyyy*; **62.** *zxyzz*; **63.** *zxzxz*; **64.** *zxzyz*; **65.** *zxzzx*; **66.** *zxzzy*; **67.** *zyyxx*; **68.** *zyyxy*; **69.** *zyyyx*; **70.** *zyyyy*; **71.** *zyyzz*; **72.** *zyzxz*; **73.** *zyzyz*; **74.** *zyzzx*; **75.** *zyzzy*; **76.** *zzzxx*; **77.** *zzzxy*; **78.** *zzzyx*; **79.** *zzzyy*; **80.** *zzzzz*

Class: *m* ($m \perp Y$)

1. *xxxxx*; **2.** *xxxxz*; **3.** *xxxyy*; **4.** *xxxzx*; **5.** *xxxzz*; **6.** *xxyxy*; **7.** *xxyyx*; **8.** *xxyyz*; **9.** *xxyzy*; **10.** *xxzxx*; **11.** *xxzxz*; **12.** *xxzyy*; **13.** *xxzzx*; **14.** *xxzzz*; **15.** *xyyxx*; **16.** *xyyxz*; **17.** *xyyyy*; **18.** *xyyzx*; **19.** *xyyzz*; **20.** *xyzxy*; **21.** *xyzyx*; **22.** *xyzyz*; **23.** *xyzzy*; **24.** *xzzxx*; **25.** *xzzxz*; **26.** *xzzyy*; **27.** *xzzzx*; **28.** *xzzzz*; **29.** *yxxxy*; **30.** *yxxyx*; **31.** *yxxyz*; **32.** *yxxzy*; **33.** *yxyxx*; **34.** *yxyxz*; **35.** *yxyyy*; **36.** *yxyzx*; **37.** *yxyzz*; **38.** *yxzxy*; **39.** *yxzyx*; **40.** *yxzyz*; **41.** *yxzzy*; **42.** *yyyxy*; **43.** *yyyyx*; **44.** *yyyyz*; **45.** *yyyzy*; **46.** *yyzxx*; **47.** *yyzxz*; **48.** *yyzyy*; **49.** *yyzzx*; **50.** *yyzzz*; **51.** *yzzxy*; **52.** *yzzyx*; **53.** *yzzyz*; **54.** *yzzzy*; **55.** *zxxxx*; **56.** *zxxxz*; **57.** *zxxyy*; **58.** *zxxzx*; **59.** *zxxzz*; **60.** *zxyxy*; **61.** *zxyyx*; **62.** *zxyyz*; **63.** *zxyzy*; **64.** *zxzxx*; **65.** *zxzxz*; **66.** *zxzyy*; **67.** *zxzzx*; **68.** *zxzzz*; **69.** *zyyxx*; **70.** *zyyxz*; **71.** *zyyyy*; **72.** *zyyzx*; **73.** *zyyzz*; **74.** *zyzxy*; **75.** *zyzyx*; **76.** *zyzyz*; **77.** *zyzzy*; **78.** *zzzxx*; **79.** *zzzxz*; **80.** *zzzyy*; **81.** *zzzzx*; **82.** *zzzzz*

Class: *m* ($m \perp Z$)

1. *xxxxx*; **2.** *xxxxy*; **3.** *xxxyx*; **4.** *xxxyy*; **5.** *xxxzz*; **6.** *xxyxx*; **7.** *xxyxy*; **8.** *xxyyx*; **9.** *xxyyy*; **10.** *xxyzz*; **11.** *xxzxz*; **12.** *xxzyz*; **13.** *xxzzx*; **14.** *xxzzy*; **15.** *xyyxx*; **16.** *xyyxy*; **17.** *xyyyx*; **18.** *xyyyy*; **19.** *xyyzz*; **20.** *xyzxz*; **21.** *xyzyz*; **22.** *xyzzx*; **23.** *xyzzy*; **24.** *xzzxx*; **25.** *xzzxy*; **26.** *xzzyx*; **27.** *xzzyy*; **28.** *xzzzz*; **29.** *yxxxx*; **30.** *yxxxy*; **31.** *yxxyx*; **32.** *yxxyy*; **33.** *yxxzz*; **34.** *yxyxx*; **35.** *yxyxy*; **36.** *yxyyx*; **37.** *yxyyy*; **38.** *yxyzz*; **39.** *yxzxz*; **40.** *yxzyz*; **41.** *yxzzx*; **42.** *yxzzy*; **43.** *yyyxx*; **44.** *yyyxy*; **45.** *yyyyx*; **46.** *yyyyy*; **47.** *yyyzz*; **48.** *yyzxz*; **49.** *yyzyz*; **50.** *yyzzx*; **51.** *yyzzy*; **52.** *yzzxx*; **53.** *yzzxy*; **54.** *yzzyx*; **55.** *yzzyy*; **56.** *yzzzz*; **57.** *zxxxz*; **58.** *zxxyz*; **59.** *zxxzx*; **60.** *zxxzy*; **61.** *zxyxz*; **62.** *zxyyz*; **63.** *zxyzx*; **64.** *zxyzy*; **65.** *zxzxx*; **66.** *zxzxy*; **67.** *zxzyx*; **68.** *zxzyy*; **69.** *zxzzz*; **70.** *zyyxz*; **71.** *zyyyz*; **72.** *zyyzx*; **73.** *zyyzy*; **74.** *zyzxx*; **75.** *zyzxy*; **76.** *zyzyx*; **77.** *zyzyy*; **78.** *zyzzz*; **79.** *zzzxz*; **80.** *zzzyz*; **81.** *zzzzx*; **82.** *zzzzy*

Class: 222

1. *xxxyz*; **2.** *xxxzy*; **3.** *xxyxz*; **4.** *xxyzx*; **5.** *xxzxy*; **6.** *xxzyx*; **7.** *xyyyz*; **8.** *xyyzy*; **9.** *xyzxx*; **10.** *xyzyy*; **11.** *xyzzz*; **12.** *xzzyz*; **13.** *xzzzy*; **14.** *yxxxz*; **15.** *yxxzx*; **16.** *yxyyz*; **17.** *yxyzy*; **18.** *yxzxx*; **19.** *yxzyy*; **20.** *yxzzz*; **21.** *yyyxz*; **22.** *yyyzx*; **23.** *yyzxy*; **24.** *yyzyx*; **25.** *yzzxz*; **26.** *yzzzx*; **27.** *zxxxy*; **28.** *zxxyx*; **29.** *zxyxx*; **30.** *zxyyy*; **31.** *zxyzz*; **32.** *zxzyz*; **33.** *zxzzy*; **34.** *zyyxy*; **35.** *zyyyx*; **36.** *zyzxz*; **37.** *zyzzx*; **38.** *zzzxy*; **39.** *zzzyx*

Class: *mm*2

1. *xxxxz*; **2.** *xxxzx*; **3.** *xxyyz*; **4.** *xxyzy*; **5.** *xxzxx*; **6.** *xxzyy*; **7.** *xxzzz*; **8.** *xyyxz*; **9.** *xyyzx*; **10.** *xyzxy*; **11.** *xyzyx*; **12.** *xzzxz*; **13.** *xzzzx*; **14.** *yxxyz*; **15.** *yxxzy*;

16. *yxyxz*; **17.** *yxyzx*; **18.** *yxzxy*; **19.** *yxzyx*; **20.** *yyyyz*; **21.** *yyyzy*; **22.** *yyzxx*; **23.** *yyzyy*; **24.** *yyzzz*; **25.** *yzzyz*; **26.** *yzzzy*; **27.** *zxxxx*; **28.** *zxxyy*; **29.** *zxxzz*; **30.** *zxyxy*; **31.** *zxyyx*; **32.** *zxzxz*; **33.** *zxzzx*; **34.** *zyyxx*; **35.** *zyyyy*; **36.** *zyyzz*; **37.** *zyzyz*; **38.** *zyzzy*; **39.** *zzzxx*; **40.** *zzzyy*; **41.** *zzzzz*

Class: 3

1. *yxyxx*; **2.** *yxyyy*; **3.** *yyyxy*; **4.** *yyyyx*; $xxxxx = -\frac{1}{2}yxyxx - \frac{1}{2}yxyyy - \frac{1}{2}yyyxy - \frac{1}{2}yyyyx$; $xxxyy = -\frac{1}{2}yxyxx - \frac{1}{2}yxyyy + \frac{1}{2}yyyxy + \frac{1}{2}yyyyx$; $xxyxy = -\frac{1}{2}yxyxx + \frac{1}{2}yxyyy - \frac{1}{2}yyyxy + \frac{1}{2}yyyyx$; $xxyyx = -\frac{1}{2}yxyxx + \frac{1}{2}yxyyy + \frac{1}{2}yyyxy - \frac{1}{2}yyyyx$; $xyyxx = -\frac{1}{2}yxyxx + yxyyy - \frac{1}{2}yyyxy - \frac{1}{2}yyyyx$; $xyyyy = +yxyxx - \frac{1}{2}yxyyy + \frac{1}{2}yyyxy + \frac{1}{2}yyyyx$; $yxxxy = +yxyxx - yxyyy + yyyyx$; $yxxyx = +yxyxx - yxyyy + yyyxy$; **5.** *yxyxy*; **6.** *yxyyx*; **7.** *yyyxx*; **8.** *yyyyy*; $xxxxy = -\frac{1}{2}yxyxy + \frac{1}{2}yxyyx + \frac{1}{2}yyyxx - \frac{1}{2}yyyyy$; $xxxyx = +\frac{1}{2}yxyxy - \frac{1}{2}yxyyx + \frac{1}{2}yyyxx - \frac{1}{2}yyyyy$; $xxyxx = +\frac{1}{2}yxyxy + \frac{1}{2}yxyyx - \frac{1}{2}yyyxx - \frac{1}{2}yyyyy$; $xxyyy = -\frac{1}{2}yxyxy - \frac{1}{2}yxyyx - \frac{1}{2}yyyxx - \frac{1}{2}yyyyy$; $xyyxy = -\frac{1}{2}yxyxy - yxyyx + \frac{1}{2}yyyxx - \frac{1}{2}yyyyy$; $xyyyx = -yxyxy - \frac{1}{2}yxyyx + \frac{1}{2}yyyxx - \frac{1}{2}yyyyy$; $yxxxx = -yxyxy - yxyyx - yyyyy$; $yxxyy = +yxyxy + yxyyx - yyyxx$; **9.** *yxyxz*; **10.** *yyyyz*; $xxxxz = +yyyyz$; $xxyyz = +yxyxz$; $xyyxz = -2yxyxz + yyyyz$; $yxxyz = -2yxyxz + yyyyz$; **11.** *yxyyz*; **12.** *yyyxz*; $xxxyz = -yyyxz$; $xxyxz = -yxyyz$; $xyyyz = -2yxyyz - yyyxz$; $yxxxz = +2yxyyz + yyyxz$; **13.** *yxyzx*; **14.** *yyyzy*; $xxxzx = +yyyzy$; $xxyzy = +yxyzx$; $xyyzx = -2yxyzx + yyyzy$; $yxxzy = -2yxyzx + yyyzy$; **15.** *yxyzy*; **16.** *yyyzx*; $xxxzy = -yyyzx$; $xxyzx = -yxyzy$; $xyyzy = -2yxyzy - yyyzx$; $yxxzx = +2yxyzy + yyyzx$; **17.** *yxyzz*; $xxxzz = -yxyzz$; $xyyzz = +yxyzz$; **18.** *yyyzz*; $xxyzz = -yyyzz$; $yxxzz = -yyyzz$; **19.** *yxzyx*; **20.** *yyzxx*; **21.** *yyzyy*; $xxzxx = +yyzyy$; $xxzyy = +yyzxx$; $xyzxy = +yxzyx$; $xyzyx = -yxzyx - yyzxx + yyzyy$; $yxzxy = -yxzyx - yyzxx + yyzyy$; **22.** *yxzyy*; **23.** *yyzxy*; **24.** *yyzyx*; $xxzxy = -yyzyx$; $xxzyx = -yyzxy$; $xyzxx = -yxzyy$; $xyzyy = -yxzyy - yyzxy - yyzyx$; $yxzxx = +yxzyy + yyzxy + yyzyx$; **25.** *yyzxz*; $xxzxz = -yyzxz$; $xyzyz = +yyzxz$; $yxzyz = +yyzxz$; **26.** *yyzyz*; $xxzyz = -yyzyz$; $xyzxz = -yyzyz$; $yxzxz = -yyzyz$; **27.** *yyzzx*; $xxzzx = -yyzzx$; $xyzzy = +yyzzx$; $yxzzy = +yyzzx$; **28.** *yyzzy*; $xxzzy = -yyzzy$; $xyzzx = -yyzzy$; $yxzzx = -yyzzy$; **29.** *yzzyx*; $xzzxx = -yzzyx$; $xzzyy = +yzzyx$; $yzzxy = +yzzyx$; **30.** *yzzyy*; $xzzxy = -yzzyy$; $xzzyx = -yzzyy$; $yzzxx = -yzzyy$; **31.** *zyyxx*; **32.** *zyyyy*; $zxxxx = +zyyyy$; $zxxyy = +zyyxx$; $zxyxy = -\frac{1}{2}zyyxx + \frac{1}{2}zyyyy$; $zxyyx = -\frac{1}{2}zyyxx + \frac{1}{2}zyyyy$; **33.** *zyyxy*; **34.** *zyyyx*; $zxxxy = -zyyyx$; $zxxyx = -zyyxy$; $zxyxx = +\frac{1}{2}zyyxy + \frac{1}{2}zyyyx$; $zxyyy = -\frac{1}{2}zyyxy - \frac{1}{2}zyyyx$; **35.** *zyyxz*; $zxxxz = -zyyxz$; $zxyyz = +zyyxz$; **36.** *zyyyz*; $zxxyz = -zyyyz$; $zxyxz = -zyyyz$; **37.** *zyyzx*; $zxxzx = -zyyzx$; $zxyzy = +zyyzx$; **38.** *zyyzy*; $zxxzy = -zyyzy$; $zxyzx = -zyyzy$; **39.** *zyzyx*; $zxzxx = -zyzyx$; $zxzyy = +zyzyx$; $zyzxy = +zyzyx$; **40.** *zyzyy*; $zxzxy = -zyzyy$; $zxzyx = -zyzyy$; $zyzxx = -zyzyy$; **41.** *xxzzz*; $xxzzz = +yyzzz$; **42.** *xyzzz*; $xyzzz = -yxzzz$; **43.** *xzzxz*; $xzzxz = +yzzyz$; **44.** *xzzyz*; $xzzyz = -yzzxz$; **45.** *xzzzx*; $xzzzx = +yzzzy$; **46.** *xzzzy*; $xzzzy = -yzzzx$; **47.** *zxxzz*; $zxxzz = +zyyzz$; **48.** *zxzxz*; $zxzxz = +zyzyz$; **49.** *zxzyz*; $zxzyz = -zyzxz$; **50.** *zxzzx*; $zxzzx = +zyzzy$; **51.** *zxzzy*; $zxzzy = -zyzzx$; **52.** *zzzxx*; $zzzxx = +zzzyy$; **53.** *zzzxy*; $zzzxy = -zzzyx$; **54.** *zzzzz*

Class: 32 (2$\|X$)

1. $yxyxx$; **2.** $yxyyy$; **3.** $yyyxy$; **4.** $yyyyx$; $xxxxx = -\frac{1}{2}yxyxx - \frac{1}{2}yxyyy - \frac{1}{2}yyyxy - \frac{1}{2}yyyyx$; $xxxyy = -\frac{1}{2}yxyxx - \frac{1}{2}yxyyy + \frac{1}{2}yyyxy + \frac{1}{2}yyyyx$; $xxyxy = -\frac{1}{2}yxyxx + \frac{1}{2}yxyyy - \frac{1}{2}yyyxy + \frac{1}{2}yyyyx$; $xxyyx = -\frac{1}{2}yxyxx + \frac{1}{2}yxyyy + \frac{1}{2}yyyxy - \frac{1}{2}yyyyx$; $xyyxx = -\frac{1}{2}yxyxx + yxyyy - \frac{1}{2}\mathrm{yyyxy} - \frac{1}{2}yyyyx$; $xyyyy = +yxyxx - \frac{1}{2}yxyyy + \frac{1}{2}yyyxy + \frac{1}{2}yyyyx$; $yxxxy = +yxyxx - yxyyy + yyyyx$; $yxxyx = +yxyxx - yxyyy + yyyxy$; **5.** $yxyyz$; **6.** $yyyxz$; $xxxyz = -yyyxz$; $xxyxz = -yxyyz$; $xyyyz = -2yxyyz - yyyxz$; $yxxxz = +2yxyyz + yyyxz$; **7.** $yxyzy$; **8.** $yyyzx$; $xxxzy = -yyyzx$; $xxyzx = -yxyzy$; $xyyzy = -2yxyzy - yyyzx$; $yxxzx = +2yxyzy + yyyzx$; **9.** $yxyzz$; $xxxzz = -yxyzz$; $xyyzz = +yxyzz$; **10.** $yxzyy$; **11.** $yyzxy$; **12.** $yyzyx$; $xxzxy = -yyzyx$; $xxzyx = -yyzxy$; $xyzxx = -yxzyy$; $xyzyy = -yxzyy - yyzxy - yyzyx$; $yxzxx = +yxzyy + yyzxy + yyzyx$; **13.** $yyzxz$; $xxzxz = -yyzxz$; $xyzyz = +yyzxz$; $yxzyz = +yyzxz$; **14.** $yyzzx$; $xxzzx = -yyzzx$; $xyzzy = +yyzzx$; $yxzzy = +yyzzx$; **15.** $yzzyx$; $xzzxx = -yzzyx$; $xzzyy = +yzzyx$; $yzzxy = +yzzyx$; **16.** $zyyxy$; **17.** $zyyyx$; $zxxxy = -zyyyx$; $zxxyx = -zyyxy$; $zxyxx = +\frac{1}{2}zyyxy + \frac{1}{2}zyyyx$; $zxyyy = -\frac{1}{2}zyyxy - \frac{1}{2}zyyyx$; **18.** $zyyxz$; $zxxxz = -zyyxz$; $zxyyz = +zyyxz$; **19.** $zyyzx$; $zxxzx = -zyyzx$; $zxyzy = +zyyzx$; **20.** $zyzyx$; $zxzxx = -zyzyx$; $zxzyy = +zyzyx$; $zyzxy = +zyzyx$; **21.** $xyzzz$; $xyzzz = -yxzzz$; **22.** $xzzyz$; $xzzyz = -yzzxz$; **23.** $xzzzy$; $xzzzy = -yzzzx$; **24.** $zxzyz$; $zxzyz = -zyzxz$; **25.** $zxzzy$; $zxzzy = -zyzzx$; **26.** $zzzxy$; $zzzxy = -zzzyx$

Class: 32 (2$\|Y$)

1. $yxyxy$; **2.** $yxyyx$; **3.** $yyyxx$; **4.** $yyyyy$; $xxxxy = -\frac{1}{2}yxyxy + \frac{1}{2}yxyyx + \frac{1}{2}yyyxx - \frac{1}{2}yyyyy$; $xxxyx = +\frac{1}{2}yxyxy - \frac{1}{2}yxyyx + \frac{1}{2}yyyxx - \frac{1}{2}yyyyy$; $xxyxx = +\frac{1}{2}yxyxy + \frac{1}{2}yxyyx - \frac{1}{2}yyyxx - \frac{1}{2}yyyyy$; $xxyyy = -\frac{1}{2}yxyxy - \frac{1}{2}yxyyx - \frac{1}{2}yyyxx - \frac{1}{2}yyyyy$; $xyyxy = -\frac{1}{2}yxyxy - yxyyx + \frac{1}{2}yyyxx - \frac{1}{2}yyyyy$; $xyyyx = -yxyxy - \frac{1}{2}yxyyx + \frac{1}{2}yyyxx - \frac{1}{2}yyyyy$; $yxxxx = -yxyxy - yxyyx - yyyyy$; $yxxyy = +yxyxy + yxyyx - yyyxx$; **5.** $yxyyz$; **6.** $yyyxz$; $xxxyz = -yyyxz$; $xxyxz = -yxyyz$; $xyyyz = -2yxyyz - yyyxz$; $yxxxz = +2yxyyz + yyyxz$; **7.** $yxyzy$; **8.** $yyyzx$; $xxxzy = -yyyzx$; $xxyzx = -yxyzy$; $xyyzy = -2yxyzy - yyyzx$; $yxxzx = +2yxyzy + yyyzx$; **9.** $yyyzz$; $xxyzz = -yyyzz$; $yxxzz = -yyyzz$; **10.** $yxzyy$; **11.** $yyzxy$; **12.** $yyzyx$; $xxzxy = -yyzyx$; $xxzyx = -yyzxy$; $xyzxx = -yxzyy$; $xyzyy = -yxzyy - yyzxy - yyzyx$; $yxzxx = +yxzyy + yyzxy + yyzyx$; **13.** $yyzyz$; $xxzyz = -yyzyz$; $xyzxz = -yyzyz$; $yxzxz = -yyzyz$; **14.** $yyzzy$; $xxzzy = -yyzzy$; $xyzzx = -yyzzy$; $yxzzx = -yyzzy$; **15.** $yzzyy$; $xzzxy = -yzzyy$; $xzzyx = -yzzyy$; $yzzxx = -yzzyy$; **16.** $zyyxy$; **17.** $zyyyx$; $zxxxy = -zyyyx$; $zxxyx = -zyyxy$; $zxyxx = +\frac{1}{2}zyyxy + \frac{1}{2}zyyyx$; $zxyyy = -\frac{1}{2}zyyxy - \frac{1}{2}zyyyx$; **18.** $zyyyz$; $zxxyz = -zyyyz$; $zxyxz = -zyyyz$; **19.** $zyyzy$; $zxxzy = -zyyzy$; $zxyzx = -zyyzy$; **20.** $zyzyy$; $zxzxy = -zyzyy$; $zxzyx = -zyzyy$; $zyzxx = -zyzyy$; **21.** $xyzzz$; $xyzzz = -yxzzz$; **22.** $xzzyz$; $xzzyz = -yzzxz$; **23.** $xzzzy$; $xzzzy = -yzzzx$; **24.** $zxzyz$; $zxzyz = -zyzxz$; **25.** $zxzzy$; $zxzzy = -zyzzx$; **26.** $zzzxy$; $zzzxy = -zzzyx$

Class: 3m ($m \perp X$)

1. $yxyxy$; **2.** $yxyyx$; **3.** $yyyxx$; **4.** $yyyyy$; $xxxxy = -\frac{1}{2}yxyxy + \frac{1}{2}yxyyx + \frac{1}{2}yyyxx -$

$\frac{1}{2}yyyyy$; $xxxyx = +\frac{1}{2}yxyxy - \frac{1}{2}yxyyx + \frac{1}{2}yyyxx - \frac{1}{2}yyyyy$; $xxyxx = +\frac{1}{2}yxyxy + \frac{1}{2}yxyyx - \frac{1}{2}yyyxx - \frac{1}{2}yyyyy$; $xxyyy = -\frac{1}{2}yxyxy - \frac{1}{2}yxyyx - \frac{1}{2}yyyxx - \frac{1}{2}yyyyy$; $xyyxy = -\frac{1}{2}yxyxy - yxyyx + \frac{1}{2}yyyxx - \frac{1}{2}yyyyy$; $xyyyx = -yxyxy - \frac{1}{2}yxyyx + \frac{1}{2}yyyxx - \frac{1}{2}yyyyy$; $yxxxx = -yxyxy - yxyyx - yyyyy$; $yxxyy = +yxyxy + yxyyx - yyyxx$; **5.** $yxyxz$; **6.** $yyyyz$; $xxxxz = +yyyyz$; $xxyyz = +yxyxz$; $xyyxz = -2yxyxz + yyyyz$; $yxxyz = -2yxyxz + yyyyz$; 7. $yxyzx$; **8.** $yyyzy$; $xxxzx = +yyyzy$; $xxyzy = +yxyzx$; $xyyzx = -2yxyzx + yyyzy$; $yxxzy = -2yxyzx + yyyzy$; **9.** $yyyzz$; $xxyzz = -yyyzz$; $yxxzz = -yyyzz$; **10.** $yxzyx$; **11.** $yyzxx$; **12.** $yyzyy$; $xxzxx = +yyzyy$; $xxzyy = +yyzxx$; $xyzxy = +yxzyx$; $xyzyx = -yxzyx - yyzxx + yyzyy$; $yxzxy = -yxzyx - yyzxx + yyzyy$; **13.** $yyzyz$; $xxzyz = -yyzyz$; $xyzxz = -yyzyz$; $yxzxz = -yyzyz$; **14.** $yyzzy$; $xxzzy = -yyzzy$; $xyzzx = -yyzzy$; $yxzzx = -yyzzy$; **15.** $yzzyy$; $xzzxy = -yzzyy$; $xzzyx = -yzzyy$; $yzzxx = -yzzyy$; **16.** $zyyxx$; **17.** $zyyyy$; $zxxxx = +zyyyy$; $zxxyy = +zyyxx$; $zxyxy = -\frac{1}{2}zyyxx + \frac{1}{2}zyyyy$; $zxyyx = -\frac{1}{2}zyyxx + \frac{1}{2}zyyyy$; **18.** $zyyyz$; $zxxyz = -zyyyz$; $zxyxz = -zyyyz$; **19.** $zyyzy$; $zxxzy = -zyyzy$; $zxyzx = -zyyzy$; **20.** $zyzyy$; $zxzxy = -zyzyy$; $zxzyx = -zyzyy$; $zyzxx = -zyzyy$; **21.** $xxzzz$; $xxzzz = +yyzzz$; **22.** $xzzxz$; $xzzxz = +yzzyz$; **23.** $xzzzx$; $xzzzx = +yzzzy$; **24.** $zxxzz$; $zxxzz = +zyyzz$; **25.** $zxzxz$; $zxzxz = +zyzyz$; **26.** $zxzzx$; $zxzzx = +zyzzy$; **27.** $zzzxx$; $zzzxx = +zzzyy$; **28.** $zzzzz$

Class: 3*m* (*m*⊥*Y*)

1. $yxyxx$; **2.** $yxyyy$; **3.** $yyyxy$; **4.** $yyyyx$; $xxxxx = -\frac{1}{2}yxyxx - \frac{1}{2}yxyyy - \frac{1}{2}yyyxy - \frac{1}{2}yyyyx$; $xxxyy = -\frac{1}{2}yxyxx - \frac{1}{2}yxyyy + \frac{1}{2}yyyxy + \frac{1}{2}yyyyx$; $xxyxy = -\frac{1}{2}yxyxx + \frac{1}{2}yxyyy - \frac{1}{2}yyyxy + \frac{1}{2}yyyyx$; $xxyyx = -\frac{1}{2}yxyxx + \frac{1}{2}yxyyy + \frac{1}{2}yyyxy - \frac{1}{2}yyyyx$; $xyyxx = -\frac{1}{2}yxyxx + yxyyy - \frac{1}{2}yyyxy - \frac{1}{2}yyyyx$; $xyyyy = +yxyxx - \frac{1}{2}yxyyy + \frac{1}{2}yyyxy + \frac{1}{2}yyyyx$; $yxxxy = +yxyxx - yxyyy + yyyyx$; $yxxyx = +yxyxx - yxyyy + yyyxy$; **5.** $yxyxz$; **6.** $yyyyz$; $xxxxz = +yyyyz$; $xxyyz = +yxyxz$; $xyyxz = -2yxyxz + yyyyz$; $yxxyz = -2yxyxz + yyyyz$; 7. $yxyzx$; **8.** $yyyzy$; $xxxzx = +yyyzy$; $xxyzy = +yxyzx$; $xyyzx = -2yxyzx + yyyzy$; $yxxzy = -2yxyzx + yyyzy$; **9.** $yxyzz$; $xxxzz = -yxyzz$; $xyyzz = +yxyzz$; **10.** $yxzyx$; **11.** $yyzxx$; **12.** $yyzyy$; $xxzxx = +yyzyy$; $xxzyy = +yyzxx$; $xyzxy = +yxzyx$; $xyzyx = -yxzyx - yyzxx + yyzyy$; $yxzxy = -yxzyx - yyzxx + yyzyy$; **13.** $yyzxz$; $xxzxz = -yyzxz$; $xyzyz = +yyzxz$; $yxzyz = +yyzxz$; **14.** $yyzzx$; $xxzzx = -yyzzx$; $xyzzy = +yyzzx$; $yxzzy = +yyzzx$; **15.** $yzzyx$; $xzzxx = -yzzyx$; $xzzyy = +yzzyx$; $yzzxy = +yzzyx$; **16.** $zyyxx$; **17.** $zyyyy$; $zxxxx = +zyyyy$; $zxxyy = +zyyxx$; $zxyxy = -\frac{1}{2}zyyxx + \frac{1}{2}zyyyy$; $zxyyx = -\frac{1}{2}zyyxx + \frac{1}{2}zyyyy$; **18.** $zyyxz$; $zxxxz = -zyyxz$; $zxyyz = +zyyxz$; **19.** $zyyzx$; $zxxzx = -zyyzx$; $zxyzy = +zyyzx$; **20.** $zyzyx$; $zxzxx = -zyzyx$; $zxzyy = +zyzyx$; $zyzxy = +zyzyx$; **21.** $xxzzz$; $xxzzz = +yyzzz$; **22.** $xzzxz$; $xzzxz = +yzzyz$; **23.** $xzzzx$; $xzzzx = +yzzzy$; **24.** $zxxzz$; $zxxzz = +zyyzz$; **25.** $zxzxz$; $zxzxz = +zyzyz$; **26.** $zxzzx$; $zxzzx = +zyzzy$; **27.** $zzzxx$; $zzzxx = +zzzyy$; **28.** $zzzzz$

Class: 4

1. $yxxxz = -xyyyz$; **2.** $yxxyz = xyyxz$; **3.** $yxxzx = -xyyzy$; **4.** $yxxzy = xyyzx$; **5.** $yxyxz = xxyyz$; **6.** $yxyyz = -xxyxz$; **7.** $yxyzx = xxyzy$; **8.** $yxyzy = -xxyzx$; **9.** $yxzxx = -xyzyy$; **10.** $yxzxy = xyzyx$; **11.** $yxzyx = xyzxy$; **12.** $yxzyy = -xyzxx$;

13. *yxzzz* = −*xyzzz*; **14.** *yyyxz* = −*xxxyz*; **15.** *yyyyz* = *xxxxz*; **16.** *yyyzx* = −*xxxzy*; **17.** *yyyzy* = *xxxzx*; **18.** *yyzxx* = *xxzyy*; **19.** *yyzxy* = −*xxzyx*; **20.** *yyzyx* = −*xxzxy*; **21.** *yyzyy* = *xxzxx*; **22.** *yyzzz* = *xxzzz*; **23.** *yzzxz* = −*xzzyz*; **24.** *yzzyz* = *xzzxz*; **25.** *yzzzx* = −*xzzzy*; **26.** *yzzzy* = *xzzzx*; **27.** *zxyyx* = *zxyxy*; **28.** *zxyyy* = −*zxyxx*; **29.** *zyyxx* = *zxxyy*; **30.** *zyyxy* = −*zxxyx*; **31.** *zyyyx* = −*zxxxy*; **32.** *zyyyy* = *zxxxx*; **33.** *zyyzz* = *zxxzz*; **34.** *zyzxz* = −*zxzyz*; **35.** *zyzyz* = *zxzxz*; **36.** *zyzzx* = −*zxzzy*; **37.** *zyzzy* = *zxzzx*; **38.** *zzzyx* = −*zzzxy*; **39.** *zzzyy* = *zzzxx*; **40.** *zzzzz*

Class: $\bar{4}$

1. *yxxxz* = *xyyyz*; **2.** *yxxyz* = −*xyyxz*; **3.** *yxxzx* = *xyyzy*; **4.** *yxxzy* = −*xyyzx*; **5.** *yxyxz* = −*xxyyz*; **6.** *yxyyz* = *xxyxz*; **7.** *yxyzx* = −*xxyzy*; **8.** *yxyzy* = *xxyzx*; **9.** *yxzxx* = *xyzyy*; **10.** *yxzxy* = −*xyzyx*; **11.** *yxzyx* = −*xyzxy*; **12.** *yxzyy* = *xyzxx*; **13.** *yxzzz* = *xyzzz*; **14.** *yyyxz* = *xxxyz*; **15.** *yyyyz* = −*xxxxz*; **16.** *yyyzx* = *xxxzy*; **17.** *yyyzy* = −*xxxzx*; **18.** *yyzxx* = −*xxzyy*; **19.** *yyzxy* = *xxzyx*; **20.** *yyzyx* = *xxzxy*; **21.** *yyzyy* = −*xxzxx*; **22.** *yyzzz* = −*xxzzz*; **23.** *yzzxz* = *xzzyz*; **24.** *yzzyz* = −*xzzxz*; **25.** *yzzzx* = *xzzzy*; **26.** *yzzzy* = −*xzzzx*; **27.** *zxyyx* = −*zxyxy*; **28.** *zxyyy* = *zxyxx*; **29.** *zxyzz*; **30.** *zyyxx* = −*zxxyy*; **31.** *zyyxy* = *zxxyx*; **32.** *zyyyx* = *zxxxy*; **33.** *zyyyy* = −*zxxxx*; **34.** *zyyzz* = −*zxxzz*; **35.** *zyzxz* = *zxzyz*; **36.** *zyzyz* = −*zxzxz*; **37.** *zyzzx* = *zxzzy*; **38.** *zyzzy* = −*zxzzx*; **39.** *zzzyx* = *zzzxy*; **40.** *zzzyy* = −*zzzxx*

Class: 422

1. *yxxxz* = −*xyyyz*; **2.** *yxxzx* = −*xyyzy*; **3.** *yxyyz* = −*xxyxz*; **4.** *yxyzy* = −*xxyzx*; **5.** *yxzxx* = −*xyzyy*; **6.** *yxzyy* = −*xyzxx*; **7.** *yxzzz* = −*xyzzz*; **8.** *yyyxz* = −*xxxyz*; **9.** *yyyzx* = −*xxxzy*; **10.** *yyzxy* = −*xxzyx*; **11.** *yyzyx* = −*xxzxy*; **12.** *yzzxz* = −*xzzyz*; **13.** *yzzzx* = −*xzzzy*; **14.** *zxyyy* = −*zxyxx*; **15.** *zyyxy* = −*zxxyx*; **16.** *zyyyx* = −*zxxxy*; **17.** *zyzxz* = −*zxzyz*; **18.** *zyzzx* = −*zxzzy*; **19.** *zzzyx* = −*zzzxy*

Class: 4*mm*

1. *yxxyz* = *xyyxz*; **2.** *yxxzy* = *xyyzx*; **3.** *yxyxz* = *xxyyz*; **4.** *yxyzx* = *xxyzy*; **5.** *yxzxy* = *xyzyx*; **6.** *yxzyx* = *xyzxy*; **7.** *yyyyz* = *xxxxz*; **8.** *yyyzy* = *xxxzx*; **9.** *yyzxx* = *xxzyy*; **10.** *yyzyy* = *xxzxx*; **11.** *yyzzz* = *xxzzz*; **12.** *yzzyz* = *xzzxz*; **13.** *yzzzy* = *xzzzx*; **14.** *zxyyx* = *zxyxy*; **15.** *zyyxx* = *zxxyy*; **16.** *zyyyy* = *zxxxx*; **17.** *zyyzz* = *zxxzz*; **18.** *zyzyz* = *zxzxz*; **19.** *zyzzy* = *zxzzx*; **20.** *zzzyy* = *zzzxx*; **21.** *zzzzz*

Class: $\bar{4}2m$ $(2\|X)$

1. *yxxxz* = *xyyyz*; **2.** *yxxzx* = *xyyzy*; **3.** *yxyyz* = *xxyxz*; **4.** *yxyzy* = *xxyzx*; **5.** *yxzxx* = *xyzyy*; **6.** *yxzyy* = *xyzxx*; **7.** *yxzzz* = *xyzzz*; **8.** *yyyxz* = *xxxyz*; **9.** *yyyzx* = *xxxzy*; **10.** *yyzxy* = *xxzyx*; **11.** *yyzyx* = *xxzxy*; **12.** *yzzxz* = *xzzyz*; **13.** *yzzzx* = *xzzzy*; **14.** *zxyyy* = *zxyxx*; **15.** *zxyzz*; **16.** *zyyxy* = *zxxyx*; **17.** *zyyyx* = *zxxxy*; **18.** *zyzxz* = *zxzyz*; **19.** *zyzzx* = *zxzzy*; **20.** *zzzyx* = *zzzxy*

Class: $\bar{4}2m$ $(m\perp X)$

1. *yxxyz* = −*xyyxz*; **2.** *yxxzy* = −*xyyzx*; **3.** *yxyxz* = −*xxyyz*; **4.** *yxyzx* = −*xxyzy*; **5.** *yxzxy* = −*xyzyx*; **6.** *yxzyx* = −*xyzxy*; **7.** *yyyyz* = −*xxxxz*; **8.** *yyyzy* = −*xxxzx*; **9.** *yyzxx* = −*xxzyy*; **10.** *yyzyy* = −*xxzxx*; **11.** *yyzzz* = −*xxzzz*; **12.** *yzzyz* =

−xzzxz; **13.** yzzzy = −xzzzx; **14.** zxyyx = −zxyxy; **15.** zyyxx = −zxxyy; **16.** zyyyy = −zxxxx; **17.** zyyzz = −zxxzz; **18.** zyzyz = −zxzxz; **19.** zyzzy = −zxzz; **20.** zzzyy = −zzzxx

Class: 6 and ∞

1. yxyxz; **2.** yyyyz; xxxxz = +yyyyz; xxyyz = +yxyxz; xyyxz = −2yxyxz + yyyyz; yxxyz = −2yxyxz + yyyyz; **3.** yxyyz; **4.** yyyxz; xxxyz = − yyyxz; xxyxz = −yxyyz; xyyyz = −2yxyyz − yyyxz; yxxxz = +2yxyyz + yyyxz; **5.** yxyzx; **6.** yyyzy; xxxzx = +yyyzy; xxyzy = +yxyzx; xyyzx = −2yxyzx + yyyzy; yxxzy = −2yxyzx + yyyzy; **7.** yxyzy; **8.** yyyzx; xxxzy = −yyyzx; xxyzx = −yxyzy; xyyzy = −2yxyzy − yyyzx; yxxzx = +2yxyzy + yyyzx; **9.** yxzyx; **10.** yyzxx; **11.** yyzyy; xxzxx = +yyzyy; xxzyy = +yyzxx; xyzxy = +yxzyx; xyzyx = −yxzyx − yyzxx + yyzyy; yxzxy = −yxzyx − yyzxx + yyzyy; **12.** yxzyy; **13.** yyzxy; **14.** yyzyx; xxzxy = −yyzyx; xxzyx = −yyzxy; xyzxx = −yxzyy; xyzyy = −yxzyy − yyzxy − yyzyx; yxzxx = +yxzyy + yyzxy + yyzyx; **15.** zyyxx; **16.** zyyyy; zxxxx = +zyyyy; zxxyy = +zyyxx; zxyxy = $-\frac{1}{2}$zyyxx + $\frac{1}{2}$zyyyy; zxyyx = $-\frac{1}{2}$zyyxx + $\frac{1}{2}$zyyyy; **17.** zyyxy; **18.** zyyyx; zxxxy = −zyyyx; zxxyx = −zyyxy; zxyxx = $+\frac{1}{2}$zyyxy + $\frac{1}{2}$zyyyx; zxyyy = $-\frac{1}{2}$zyyxy − $\frac{1}{2}$zyyyx; **19.** xxzzz; xxzzz = +yyzzz; **20.** xyzzz; xyzzz = −yxzzz; **21.** xzzxz; xzzxz = +yzzyz; **22.** xzzyz; xzzyz = −yzzxz; **23.** xzzzx; xzzzx = +yzzzy; **24.** xzzzy; xzzzy = −yzzzx; **25.** zxxzz; zxxzz = +zyyzz; **26.** zxzxz; zxzxz = +zyzyz; **27.** zxzyz; zxzyz = −zyzxz; **28.** zxzzx; zxzzx = +zyzzy; **29.** zxzzy; zxzzy = −zyzzx; **30.** zzzxx; zzzxx = +zzzyy; **31.** zzzxy; zzzxy = −zzzyx; **32.** zzzzz

Class: $\bar{6}$

1. yxyxx; **2.** yxyyy; **3.** yyyxy; **4.** yyyyx; xxxxx = $-\frac{1}{2}$yxyxx − $\frac{1}{2}$yxyyy − $\frac{1}{2}$yyyxy − $\frac{1}{2}$yyyyx; xxxyy = $-\frac{1}{2}$yxyxx − $\frac{1}{2}$yxyyy + $\frac{1}{2}$yyyxy + $\frac{1}{2}$yyyyx; xxyxy = $-\frac{1}{2}$yxyxx + $\frac{1}{2}$yxyyy − $\frac{1}{2}$yyyxy + $\frac{1}{2}$yyyyx; xxyyx = $-\frac{1}{2}$yxyxx + $\frac{1}{2}$yxyyy + $\frac{1}{2}$yyyxy − $\frac{1}{2}$yyyyx; xyyxx = $-\frac{1}{2}$yxyxx + yxyyy − $\frac{1}{2}$yyyxy − $\frac{1}{2}$yyyyx; xyyyy = +yxyxx − $\frac{1}{2}$yxyyy + $\frac{1}{2}$yyyxy + $\frac{1}{2}$yyyyx; yxxxy = +yxyxx − yxyyy + yyyyx; yxxyx = +yxyxx − yxyyy + yyyxy; **5.** yxyxy; **6.** yxyyx; **7.** yyyxx; **8.** yyyyy; xxxxy = $-\frac{1}{2}$yxyxy + $\frac{1}{2}$yxyyx + $\frac{1}{2}$yyyxx − $\frac{1}{2}$yyyyy; xxxyx = $+\frac{1}{2}$yxyxy − $\frac{1}{2}$yxyyx + $\frac{1}{2}$yyyxx − $\frac{1}{2}$yyyyy; xxyxx = $+\frac{1}{2}$yxyxy + $\frac{1}{2}$yxyyx − $\frac{1}{2}$yyyxx − $\frac{1}{2}$yyyyy; xxyyy = $-\frac{1}{2}$yxyxy − $\frac{1}{2}$yxyyx − $\frac{1}{2}$yyyxx − $\frac{1}{2}$yyyyy; xyyxy = $-\frac{1}{2}$yxyxy − yxyyx + $\frac{1}{2}$yyyxx − $\frac{1}{2}$yyyyy; xyyyx = −yxyxy − $\frac{1}{2}$yxyyx + $\frac{1}{2}$yyyxx − $\frac{1}{2}$yyyyy; yxxxx = −yxyxy − yxyyx − yyyyy; yxxyy = +yxyxy + yxyyx − yyyxx; **9.** yxyzz; xxxzz = −yxyzz; xyyzz = +yxyzz; **10.** yyyzz; xxyzz = −yyyzz; yxxzz = −yyyzz; **11.** yyzxz; xxzxz = −yyzxz; xyzyz = +yyzxz; yxzyz = +yyzxz; **12.** yyzyz; xxzyz = −yyzyz; xyzxz = −yyzyz; yxzxz = −yyzyz; **13.** yyzzx; xxzzx = −yyzzx; xyzzy = +yyzzx; yxzzy = +yyzzx; **14.** yyzzy; xxzzy = −yyzzy; xyzzx = −yyzzy; yxzzx = −yyzzy; **15.** yzzyx; xzzxx = −yzzyx; xzzyy = +yzzyx; yzzxy = +yzzyx; **16.** yzzyy; xzzxy = −yzzyy; xzzyx = −yzzyy; yzzxx = −yzzyy; **17.** zyyxz; zxxxz = −zyyxz; zxyyz = +zyyxz; **18.** zyyyz; zxxyz = −zyyyz; zxyxz = −zyyyz; **19.** zyyzx; zxxzx = −zyyzx; zxyzy = +zyyzx; **20.** zyyzy; zxxzy = −zyyzy; zxyzx = −zyyzy; **21.** zyzyx; zxzxx = −zyzyx; zxzyy = +zyzyx; zyzxy = +zyzyx; **22.** zyzyy; zxzxy = −zyzyy; zxzyx = −zyzyy; zyzxx = −zyzyy

Class: 622 and $\infty 2$

1. *yxyyz*; **2.** *yyyxz*; *xxxyz* = −*yyyxz*; *xxyxz* = −*yxyyz*; *xyyyz* = −2*yxyyz* − *yyyxz*; *yxxxz* = +2*yxyyz* + *yyyxz*; **3.** *yxyzy*; **4.** *yyyzx*; *xxxzy* = −*yyyzx*; *xxyzx* = −*yxyzy*; *xyyzy* = −2*yxyzy* − *yyyzx*; *yxxzx* = +2*yxyzy* + *yyyzx*; **5.** *yxzyy*; **6.** *yyzxy*; **7.** *yyzyx*; *xxzxy* = −*yyzyx*; *xxzyx* = −*yyzxy*; *xyzxx* = −*yxzyy*; *xyzyy* = −*yxzyy* − *yyzxy* − *yyzyx*; *yxzxx* = +*yxzyy* + *yyzxy* + *yyzyx*; **8.** *zyyxy*; **9.** *zyyyx*; *zxxxy* = −*zyyyx*; *zxxyx* = −*zyyxy*; *zxyxx* = $+\frac{1}{2}$*zyyxy* + $\frac{1}{2}$*zyyyx*; *zxyyy* = $-\frac{1}{2}$*zyyxy* − $\frac{1}{2}$*zyyyx*; **10.** *xyzzz*; *xyzzz* = −*yxzzz*; **11.** *xzzyz*; *xzzyz* = −*yzzxz*; **12.** *xzzzy*; *xzzzy* = −*yzzzx*; **13.** *zxzyz*; *zxzyz* = −*zyzxz*; **14.** *zxzzy*; *zxzzy* = −*zyzzx*; **15.** *zzzxy*; *zzzxy* = −*zzzyx*

Class: 6*mm* and ∞/mm

1. *yxyxz*; **2.** *yyyyz*; *xxxxz* = +*yyyyz*; *xxyyz* = +*yxyxz*; *xyyxz* = −2*yxyxz* + *yyyyz*; *yxxyz* = −2*yxyxz* + *yyyyz*; **3.** *yxyzx*; **4.** *yyyzy*; *xxxzx* = +*yyyzy*; *xxyzy* = +*yxyzx*; *xyyzx* = −2*yxyzx* + *yyyzy*; *yxxzy* = −2*yxyzx* + *yyyzy*; **5.** *yxzyx*; **6.** *yyzxx*; **7.** *yyzyy*; *xxzxx* = +*yyzyy*; *xxzyy* = +*yyzxx*; *xyzxy* = +*yxzyx*; *xyzyx* = −*yxzyx* − *yyzxx* + *yyzyy*; *yxzxy* = −*yxzyx* − *yyzxx* + *yyzyy*; **8.** *zyyxx*; **9.** *zyyyy*; *zxxxx* = +*zyyyy*; *zxxyy* = +*zyyxx*; *zxyxy* = $-\frac{1}{2}$*zyyxx* + $\frac{1}{2}$*zyyyy*; *zxyyx* = $-\frac{1}{2}$*zyyxx* + $\frac{1}{2}$*zyyyy*; **10.** *xxzzz*; *xxzzz* = +*yyzzz*; **11.** *xzzxz*; *xzzxz* = +*yzzyz*; **12.** *xzzzx*; *xzzzx* = +*yzzzy*; **13.** *zxxzz*; *zxxzz* = +*zyyzz*; **14.** *zxzxz*; *zxzxz* = +*zyzyz*; **15.** *zxzzx*; *zxzzx* = +*zyzzy*; **16.** *zzzxx*; *zzzxx* = +*zzzyy*; **17.** *zzzzz*

Class: $\bar{6}2m$ ($m \perp X$)

1. *yxyxy*; **2.** *yxyyx*; **3.** *yyyxx*; **4.** *yyyyy*; *xxxxy* = $-\frac{1}{2}$*yxyxy* + $\frac{1}{2}$*yxyyx* + $\frac{1}{2}$*yyyxx* − $\frac{1}{2}$*yyyyy*; *xxxyx* = $+\frac{1}{2}$*yxyxy* − $\frac{1}{2}$*yxyyx* + $\frac{1}{2}$*yyyxx* − $\frac{1}{2}$*yyyyy*; *xxyxx* = $+\frac{1}{2}$*yxyxy* + $\frac{1}{2}$*yxyyx* − $\frac{1}{2}$*yyyxx* − $\frac{1}{2}$*yyyyy*; *xxyyy* = $-\frac{1}{2}$*yxyxy* − $\frac{1}{2}$*yxyyx* − $\frac{1}{2}$*yyyxx* − $\frac{1}{2}$*yyyyy*; *xyyxy* = $-\frac{1}{2}$*yxyxy* − *yxyyx* + $\frac{1}{2}$*yyyxx* − $\frac{1}{2}$*yyyyy*; *xyyyx* = −*yxyxy* − $\frac{1}{2}$*yxyyx* + $\frac{1}{2}$*yyyxx* − $\frac{1}{2}$*yyyyy*; *yxxxx* = −*yxyxy* − *yxyyx* − *yyyyy*; *yxxyy* = +*yxyxy* + *yxyyx* − *yyyxx*; **5.** *yyyzz*; *xxyzz* = −*yyyzz*; *yxxzz* = −*yyyzz*; **6.** *yyzyz*; *xxzyz* = −*yyzyz*; *xyzxz* = − *yyzyz*; *yxzxz* = −*yyzyz*; **7.** *yyzzy*; *xxzzy* = −*yyzzy*; *xyzzx* = −*yyzzy*; *yxzzx* = −*yyzzy*; **8.** *yzzyy*; *xzzxy* = −*yzzyy*; *xzzyx* = −*yzzyy*; *yzzxx* = −*yzzyy*; **9.** *zyyyz*; *zxxyz* = −*zyyyz*; *zxyxz* = −*zyyyz*; **10.** *zyyzy*; *zxxzy* = −*zyyzy*; *zxyzx* = −*zyyzy*; **11.** *zyzyy*; *zxzxy* = −*zyzyy*; *zxzyx* = −*zyzyy*; *zyzxx* = −*zyzyy*

Class: $\bar{6}2m$ ($2 \parallel X$)

1. *yxyxx*; **2.** *yxyyy*; **3.** *yyyxy*; **4.** *yyyyx*; *xxxxx* = $-\frac{1}{2}$*yxyxx* − $\frac{1}{2}$*yxyyy* − $\frac{1}{2}$*yyyxy* − $\frac{1}{2}$*yyyyx*; *xxxyy* = $-\frac{1}{2}$*yxyxx* − $\frac{1}{2}$*yxyyy* + $\frac{1}{2}$*yyyxy* + $\frac{1}{2}$*yyyyx*; *xxyxy* = $-\frac{1}{2}$*yxyxx* + $\frac{1}{2}$*yxyyy* − $\frac{1}{2}$*yyyxy* + $\frac{1}{2}$*yyyyx*; *xxyyx* = $-\frac{1}{2}$*yxyxx* + $\frac{1}{2}$*yxyyy* + $\frac{1}{2}$*yyyxy* − $\frac{1}{2}$*yyyyx*; *xyyxx* = $-\frac{1}{2}$*yxyxx* + *yxyyy* − $\frac{1}{2}$*yyyxy* − $\frac{1}{2}$*yyyyx*; *xyyyy* = +*yxyxx* − $\frac{1}{2}$*yxyyy* + $\frac{1}{2}$*yyyxy* + $\frac{1}{2}$*yyyyx*; *yxxxy* = +*yxyxx* − *yxyyy* + *yyyyx*; *yxxyx* = +*yxyxx* − *yxyyy* + *yyyxy*; **5.** *yxyzz*; *xxxzz* = −*yxyzz*; *xyyzz* = +*yxyzz*; **6.** *yyzxz*; *xxzxz* = −*yyzxz*; *xyzyz* = +*yyzxz*; *yxzyz* = +*yyzxz*; **7.** *yyzzx*; *xxzzx* = −*yyzzx*; *xyzzy* = +*yyzzx*; *yxzzy* = +*yyzzx*; **8.** *yzzyx*; *xzzxx* = −*yzzyx*; *xzzyy* = +*yzzyx*; *yzzxy* = +*yzzyx*; **9.** *zyyxz*; *zxxxz* = −*zyyxz*; *zxyyz* = +*zyyxz*; **10.** *zyyzx*; *zxxzx* = −*zyyzx*; *zxyzy* = +*zyyzx*; **11.** *zyzyx*; *zxzxx* = −*zyzyx*; *zxzyy* = +*zyzyx*; *zyzxy* = +*zyzyx*

Class: 23

1. *zxxxy* = *xyyyz* = *yzzzx*; **2.** *zxxyx* = *xyyzy* = *yzzxz*; **3.** *zxyxx* = *xyzyy* = *yxzzz*; **4.** *zxyyy* = *xyzzz* = *yxzxx*; **5.** *zxyzz* = *xyzxx* = *yxzyy*; **6.** *zxzyz* = *xxyzx* = *yyzxy*; **7.** *zxzzy* = *xxyxz* = *yyzyx*; **8.** *zyyxy* = *xzzyz* = *yxxzx*; **9.** *zyyyx* = *xzzzy* = *yxxxz*; **10.** *zyzxz* = *xxzyx* = *yxyzy*; **11.** *zyzzx* = *xxzxy* = *yxyyz*; **12.** *zzzxy* = *xxxyz* = *yyyzx*; **13.** *zzzyx* = *xxxzy* = *yyyxz*

Class: 432

1. *zxyyy* = *xyzzz* = −*zxyxx* = *yxzxx* = −*yxzzz* = −*xyzyy*; **2.** *zyyxy* = *xzzyz* = −*zxxyx* = *yxxzx* = −*yzzxz* = −*xyyzy*; **3.** *zyyyx* = *xzzzy* = − *zxxxy* = *yxxxz* = −*yzzzx* = −*xyyyz*; **4.** *zyzxz* = *xxzyx* = −*zxzyz* = *yxyzy* = −*yyzxy* = −*xxyzx*; **5.** *zyzzx* = *xxzxy* = −*zxzzy* = *yxyyz* = −*yyzyx* = −*xxyxz*; **6.** *zzzyx* = *xxxzy* = −*zzzxy* = *yyyxz* = −*yyyzx* = −*xxxyz*

Class: $\bar{4}3m$

1. *zxyyy* = *xyzzz* = *yxzzz* = *zxyxx* = *yxzxx* = *xyzyy*; **2.** *zxyzz* = *xyzxx* = *yxzyy*; **3.** *zyyxy* = *xzzyz* = *yzzxz* = *zxxyx* = *yxxzx* = *xyyzy*; **4.** *zyyyx* = *xzzzy* = *yzzzx* = *zxxxy* = *yxxxz* = *xyyyz*; **5.** *zyzxz* = *xxzyx* = *yyzxy* = *zxzyz* = *yxyzy* = *xxyzx*; **6.** *zyzzx* = *xxzxy* = *yyzyx* = *zxzzy* = *yxyyz* = *xxyxz*; **7.** *zzzyx* = *xxxzy* = *yyyzx* = *zzzxy* = *yyyxz* = *xxxyz*

Class: ∞∞

1. *xxxyz*; **2.** *xxyzx*; **3.** *xxyxz*; $\Gamma_{ijklm} = -\frac{1}{2}\Gamma_{xxyxz}(\delta_{kl}e_{jim} + \delta_{jl}e_{kim} + \delta_{ik}e_{jlm} + \delta_{ji}e_{klm}) + (\Gamma_{xxxyz} + \Gamma_{xxyxz})\delta_{kj}e_{ilm} - \frac{1}{2}(\Gamma_{xxyxz} + 2\Gamma_{xxyzx})(\delta_{jm}e_{ilk} + \delta_{km}e_{ilj})$

REFERENCE

1. S.V. Popov, Yu.P. Svirko and N.I. Zheludev, *Susceptibility Tensors for Nonlinear Optics*, IOP Publishing, Bristol and Philadelphia, 1995.

Appendix 3 The Hamiltonian of Molecular Electrons

In this appendix we will derive the Hamiltonian (9.20) of molecular electrons in the presence of an electromagnetic wave. This Hamiltonian is used in Chapter 9 for calculating, the optical nonlinearities of molecular electrons using quantum mechanics.

The equation of motion of the sth optical electron in the presence of an external electromagnetic field can be written as follows:

$$m_e \frac{d\mathbf{v}_s}{dt} = -\frac{\partial U}{\partial \mathbf{r}_s} + \frac{e}{c}\left\{\frac{\partial \mathcal{A}(\mathbf{r}_0 + \mathbf{r}_s)}{\partial t} - [\mathbf{v}_s \times \operatorname{curl} \mathcal{A}(\mathbf{r}_0 + \mathbf{r}_s)]\right\} \tag{A3.1}$$

where $\mathcal{A}(\mathbf{r}_0 + \mathbf{r}_s)$ is the vector potential at the location of the sth electron and the Coulomb gauge (formula (9.21)) is presumed. If in equation (A3.1) the Lorentz force $-e\mathbf{E} - (e/c)[\mathbf{v}_s \times \mathbf{B}]$ is expanded over $\max |\mathbf{r}_s|/\lambda$ and only the first two terms of the expansion are kept, we arrive at an equation suitable for the analysis of the first-order spatial dispersion effects:

$$m_e \frac{d\mathbf{v}_s}{dt} = -\frac{\partial U}{\partial \mathbf{r}_s} + \frac{e}{c}\left[\frac{\partial \mathcal{A}}{\partial t} + (\mathbf{r}_s \cdot \nabla)\frac{\partial \mathcal{A}}{\partial t}\right] - \frac{e}{c}[\mathbf{v}_s \times \operatorname{curl} \mathcal{A}] \tag{A3.2}$$

where $\mathcal{A}$ is the vector potential at the centre of the molecule at $\mathbf{r}_0$. This equation of motion corresponds to the following Lagrange function of the molecular electrons:

$$L = \sum_{s=1}^{N}\left\{\frac{m_e \mathbf{v}_s^2}{2} + \frac{e}{c}\left(\mathbf{r}_s \cdot \frac{\partial \mathcal{A}}{\partial t}\right) - \frac{e}{c}[\mathbf{v}_s \cdot (\mathbf{r}_s \nabla)\mathcal{A}] - U(\mathbf{r}_s)\right\} \tag{A3.3}$$

This Lagrange function will return equation (A3.2) when substituted into the Lagrange equation [1]:

$$\frac{\partial L}{\partial \mathbf{r}_s} = \frac{d}{dt}\frac{\partial L}{\partial \mathbf{v}_s} \tag{A3.4}$$

Since $\mathbf{p}_s = \partial L/\partial \mathbf{v}_s$, it follows from equation (A3.3) that

$$\mathbf{p}_s = m_e \mathbf{v}_s - \frac{e}{c}(\mathbf{r}_s \cdot \nabla)\mathcal{A} \tag{A3.5}$$

Now, using the following relationship between the Hamiltonian and Lagrange functions [1]:

$$H = \sum_{n=1}^{N} (\mathbf{p}_s \cdot \mathbf{v}_s) - L \tag{A3.6}$$

we derive the Hamiltonian of the molecular electrons in the presence of a light field:

$$\hat{H}(\mathbf{r}_0) = \sum_{s=1}^{N} \left\{ \frac{[\hat{\mathbf{p}}_s + (e/c)(\mathbf{r}_s \cdot \nabla)\mathcal{A}]^2}{2m_\mathrm{e}} - \frac{e}{c}\left(\mathbf{r}_s \cdot \frac{\partial \mathcal{A}}{\partial t}\right) + U(\mathbf{r}_s) \right\} \tag{A3.7}$$

Finally, by adding to equation (A3.7) the energy of the interaction between the electon spins and the magnetic field of the light wave $\hat{H}^{\mathrm{spin}} = -\sum(\hat{\boldsymbol{\mu}}^s \cdot \mathbf{B})$ [2], we arrive at the Hamiltonian of molecular electrons in the presence of the light wave (equation (9.20)).

REFERENCES

1. L.D. Landau and E.M. Lifshitz, *Mechanics*, Pergamon Press, Oxford, 1960.
2. L.D. Landau and E.M. Lifshitz, *Quantum Mechanics. Non-relativistic Theory*, Pergamon Press, Oxford, 1977.

Appendix 4 Nonlocal Birefringence

In this section we will consider the wave picture of the nonlocal birefringence effect, also known as gyrotropic linear dichroism [1–3]. The manifestations of this effect are similar to those of conventional birefringence, as described in Section 3.3. However, in nonlocal birefringence the origin of optical anisotropy is different—it is due to the nonlocality of optical response. We will examine the main features of the effect using cubic crystals of the $\bar{4}3m$ and 23 point groups as examples.

A4.1 NONLOCAL BIREFRINGENCE IN $\bar{4}3m$ CRYSTALS

In cubic crystals of the $\bar{4}3m$ point group the dielectric tensor is totally isotropic, $\epsilon_{ij} = \epsilon\delta_{ij}$. In this case, as in isotropic liquids, we cannot expect any polarization effects due to local optical response. However, despite the absence of any polarization effects due to dielectric tensor anisotropy, some crystals with cubic symmetry can show polarization effects due to nonlocality. Nonlocality is introduced into constitutive equation (2.16) by the term containing $\Gamma^{(1)}_{ijk}$. A trivial manifestation of nonlocality is the effect of optical activity which we considered in Section 3.3. Optical activity is due to the *antisymmetric* part of the nonlocality tensor $\Gamma^{(1,a)}_{ijk} = -\Gamma^{(1,a)}_{jik}$. In contrast, the effect of nonlocal birefringence is due to the *symmetric* part of the nonlocality tensor $\Gamma^{(1,s)}_{ijk} = \Gamma^{(1,s)}_{jik}$.

According to a fundamental thermodynamical argument known as the Onzager principle, the symmetric part of the nonlocality tensor, even when permitted by the crystal point group, can only emerge if the microscopic time-reversibility of the light–matter interaction is broken [4]. A discussion of the circumstances that allow broken time reversibility can be found at the end of Section 9.3. In particular, time reversal is broken in magnetized crystals [8]. It also has been suggested that reversibility can be broken in crystals lacking an inversion centre. In accordance with arguments of paper [5] light–matter interaction Hamiltonians are only symmetrical with respect to simultaneous inversion on time and space rather than just in time. This is seen as a consequence of the fact that the propagating electromagnetic wave does not have a definite parity with respect to space inversion or a definite symmetry with respect to time reversal: reversal of time t in the argument of an oscillatory propagating function $f(\omega t - \mathbf{k}\cdot\mathbf{r})$ yields not only reversal of the propagation direction but also a phase lag, which is important in the nonlocal light–matter interactions of noncentrosymmetric molecules of finite size.

From now on it will be presumed, for whatever reason, that the time-reversibility of the light–matter interaction process is broken and the symmetric part of the nonlocality tensor is not banned by the Onzager principle. The polarization effect on nonlocal birefringence relevant to such symmetric part of the nonlocal tensor will be examined below.

In noncentrosymmetric cubic crystals of the $\bar{4}3m$ point group, symmetry allows only symmetric nonlocality tensors ($\Gamma^{(\mathrm{l,s})}_{ijk} = \Gamma^{(\mathrm{l,s})}_{jik}$). Therefore, these crystals do not show conventional optical activity. The anisotropic effect due to the symmetric part of the nonlocality tensor manifests itself as birefringence and dichroism for linearly polarized light, rather than birefringence/dichroism for circularly polarized light as in conventional optical activity. This is why it is called gyrotropic linear dichroism. 'Gyrotropic' refers to its nonlocal nature.

Let us consider a wave picture of the polarization effect associated with gyrotropic linear dichroism for light propagating along the Z axis of the natural frame in a crystal of the $\bar{4}3m$ point group symmetry. By symmetry, the following components of the nonlocality tensor are nonzero in this point group: $\Gamma^{(\mathrm{l})}_{xyz} = \Gamma^{(\mathrm{l})}_{yxz} = \Gamma^{(\mathrm{l})}_{yzx} = \Gamma^{(\mathrm{l})}_{xzy} = \Gamma^{(\mathrm{l})}_{zxy} = \Gamma^{(\mathrm{l})}_{zyx} = \Gamma^{\mathrm{s}}$. From formula (3.19) we get

$$\begin{aligned} u_{xy} &= u_{yx} = \frac{\mathrm{i}\omega}{c}\Gamma^{\mathrm{s}} \\ u_{xx} &= u_{yy} = 0 \end{aligned} \tag{A4.1}$$

This gives us $\mathbf{\Omega}^{\mathrm{L}}_0 = 0$ and

$$\mathbf{\Omega}^{\mathrm{L}} = \left\{0, \frac{2\mathrm{i}\omega\Gamma^{\mathrm{s}}}{c}, 0\right\} \tag{A4.2}$$

The wave propagation process can easily be visualized if Γ^{s} is purely imaginary and, therefore, $\mathbf{\Omega}^{\mathrm{L}}$ is real. The evolution equations (3.30) will describe the precession of the Stokes vector around vector $\mathbf{\Omega}^{\mathrm{L}}$. This situation is similar to that presented in Fig. 3.4, but with the vector $\mathbf{\Omega}^{\mathrm{L}}$ directed along the S_2 axis. Thus, from any initial position, the end of the Stokes vector of the propagating wave will trace out a vertical circle on the Poincaré sphere and both the ellipticity and the polarization azimuth will oscillate.

In the general case, where $\mathbf{\Omega}^{\mathrm{L}}$ is given by formula (A4.2), the set of equations (3.30) can be rewritten as follows:

$$\begin{aligned} \frac{\mathrm{d}S_1}{\mathrm{d}z} &= -2\,\mathrm{Im}\,\{k\}S_1 + \frac{\omega^2}{c^2}\,\mathrm{Im}\,\{\Gamma^{\mathrm{s}}\}S_3 \\ \frac{\mathrm{d}S_2}{\mathrm{d}z} &= -2\,\mathrm{Im}\,\{k\}S_2 - \frac{\omega^2}{c^2}\,\mathrm{Re}\,\{\Gamma^{\mathrm{s}}\}S_0 \\ \frac{\mathrm{d}S_3}{\mathrm{d}z} &= -2\,\mathrm{Im}\,\{k\}S_3 - \frac{\omega^2}{c^2}\,\mathrm{Im}\,\{\Gamma^{\mathrm{s}}\}S_1 \\ \frac{\mathrm{d}S_0}{\mathrm{d}z} &= -2\,\mathrm{Im}\,\{k\}S_0 - \frac{\omega^2}{c^2}\,\mathrm{Re}\,\{\Gamma^{\mathrm{s}}\}S_2 \end{aligned} \tag{A4.3}$$

As a rule, gyrotropic linear dichroism is a small effect and so it makes sense to consider only small changes in polarization state, assuming $(\omega/c)^2|\Gamma^s|L \ll 1$. In this case, for an initially linearly polarized wave ($S_{30} = 0$) with initial polarization azimuth $\Theta(0)$, the polarization state of the wave changes as follows:

$$\begin{pmatrix} \Theta(L) - \Theta(0) \\ \eta(L) \end{pmatrix} = -\frac{\omega^2 L}{2c^2} \begin{pmatrix} \mathrm{Re} \\ \mathrm{Im} \end{pmatrix} \{\Gamma^s\} \cos[2\Theta(0)] \tag{A4.4}$$

It can easily be seen from equation (A4.4) that, if the initial linear polarization is along the bisector between the X and Y axes, $\Theta(0) = \pi/4$, or perpendicular to it, $\Theta(0) = -\pi/4$, the polarization state will not change with propagation. Therefore such polarizations are linear eigenpolarizations (see Fig. A4.1).

Polarization rotation of approximately 2×10^{-3} rad, as described by formula

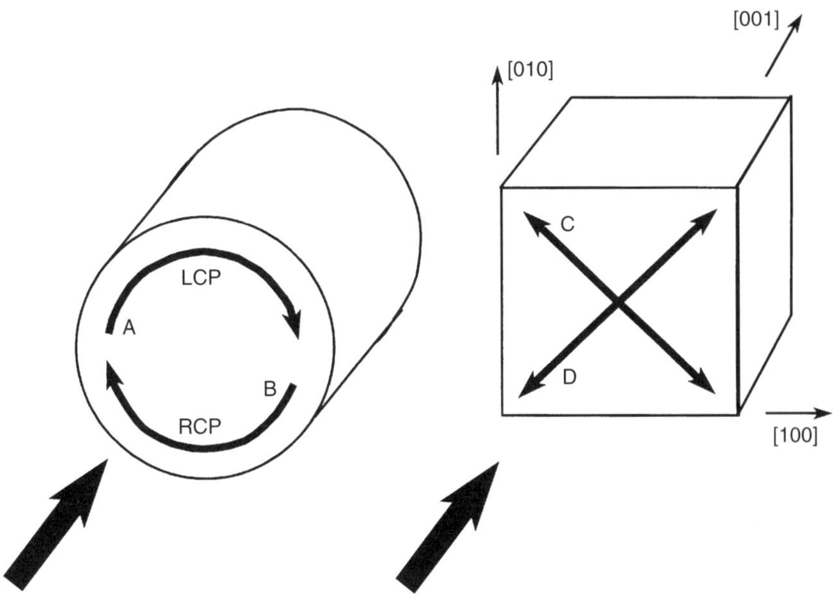

Fig. A4.1 (a) In isotropic gyrotropic media the nonlocality tensor is antisymmetric, $\Gamma^{(l)}_{ijn} = -\Gamma^{(l)}_{jin}$. The eigenpolarizations A and B are right and left circular polarizations. On the Poincaré sphere they are represented by points at the 'north' and 'south' poles. (b) If in a cubic crystal of the $\bar{4}3m$ point group the nonlocality tensor is permitted, it is symmetric, $\Gamma^{(l)}_{ijn} = \Gamma^{(l)}_{jin}$. The eigenpolarizations C and D are two orthogonal linear polarizations along the [110] and [$1\bar{1}0$] directions in the natural coordinate frame of the crystal. On the Poincare sphere they are represented by the two antipodal points where the S_2 axis crosses the equator

(3.53), is seen at $\lambda = 880$ nm in a 20 μm thick crystal of GaAs which belongs to $\bar{4}3$m point symmetry group. This gives $\mathrm{Re}\{\Gamma^{\mathrm{s}}\} \simeq 4 \times 10^{-10}$ cm. Absorptive anisotropy due to nonlocal birefringence could be one of the factors forcing predominant polarization orientation of the output light in the GaAs based vertical cavity surface emitting semiconductor lasers (VCSELs) to be along the [110] direction of the crystal, see Refs. [12–14].

A4.2 NONRECIPROCITY OF NONLOCAL BIREFRINGENCE

In this section we will describe the nonreciprocal properties of nonlocal birefringence using a crystal which belongs to the 23 symmetry point group, as an example. This crystal symmetry group does not support birefringence, $\epsilon_{ij}(\omega) = \epsilon(\omega)\delta_{ij}$, but does support conventional optical activity, described by the antisymmetric part of the nonlocality tensor $\Gamma^{\mathrm{a}}_{ijn} = -\Gamma^{\mathrm{a}}_{jin}$. With this diagonal dielectric tensor and antisymmetric nonlocality tensor, the light polarization azimuth rotation is reciprocal, i.e. is compensated for when the light beam is reflected back through the sample. This reciprocity is a well-known characteristic feature of natural optical activity.

The 23 symmetry group also supports the symmetric part of the nonlocality tensor, $\Gamma^{(1,\mathrm{s})}_{ijn} = \Gamma^{(1,\mathrm{s})}_{jin}$. We will illustrate that this causes the plane of polarization of a wave propagating along the [001] direction to rotate nonreciprocally. Such nonreciprocity is a characteristic feature of nonlocal birefringence [1].

The spatial evolution of the wave's **E** field is governed by the following set of equations, see [7]:

$$\frac{\mathrm{d}^2 E_x}{\mathrm{d}z^2} + \frac{\omega^2}{c^2}\left(\epsilon E_x + \Gamma^{(1)}_{xyz}\frac{\mathrm{d}E_y}{\mathrm{d}z}\right) = 0$$
$$\frac{\mathrm{d}^2 E_y}{\mathrm{d}z^2} + \frac{\omega^2}{c^2}\left(\epsilon E_y + \Gamma^{(1)}_{yxz}\frac{\mathrm{d}E_x}{\mathrm{d}z}\right) = 0 \tag{A4.5}$$

We solve equation (A4.5) by substituting $\mathbf{E} = \mathbf{A}\,\mathrm{e}^{\mathrm{i}(\omega/c)nz}$, where $n = \epsilon^{1/2}$ and **A** is a slowly varying amplitude, so $|\mathrm{d}\mathbf{A}/\mathrm{d}z| \ll |\omega\mathbf{A}/c|$. If the incident wave is linearly polarized at an angle Θ_0 to the [100] direction, that is $A_x(0) = A_0\cos\Theta_0$ and $A_y(0) = A_0\sin\Theta_0$, the Cartesian components of **A** for the transmitted light are:

$$\begin{pmatrix} A_x(z) \\ A_y(z) \end{pmatrix} = A_0\begin{pmatrix} \cos\Theta_0 \\ \sin\Theta_0 \end{pmatrix}\cosh(\mu z) - A_0\frac{v_{\mathrm{s}} \pm v_{\mathrm{a}}}{\mu}\begin{pmatrix} \sin\Theta_0 \\ \cos\Theta_0 \end{pmatrix}\sinh(\mu z) \tag{A4.6}$$

Here, $v_{\mathrm{s,a}} = [\omega^2/(4c^2)](\Gamma^{(1)}_{xyz} \pm \Gamma^{(1)}_{yxz})$ and $\mu^2 = v_{\mathrm{s}}^2 - v_{\mathrm{a}}^2$. We make the justifiable assumption that the symmetric part of the nonlocality tensor is a small fraction of the antisymmetric part, that is $|v_s| \ll |v_a|$. Therefore, $\mu \approx iv_a$ and $\mu/(v_a + v_s) \approx i(1 - v_s/v_a)$, and for the polarization azimuths $\Theta_{\rightarrow}$ of the forward propagating wave and $\Theta_{\leftarrow}$ of the backward wave, we have

$$\Theta_{\rightarrow} = (\Theta_0 + \Theta_{OA}) - \mathrm{Re}\left\{\frac{v_s}{v_a}\sin(v_a L)\cos(2\Theta_0 + v_a L)\right\}$$
$$\Theta_{\leftarrow} = (\Theta_{\rightarrow} - \Theta_{OA}) - \mathrm{Re}\left\{\frac{v_s}{v_a}\sin(v_a L)\cos(2\Theta_{\rightarrow} - v_a L)\right\} \qquad \text{(A4.7)}$$

where $\Theta_{OA} = [\omega^2/(2c^2)]\,\mathrm{Re}\{\Gamma^a\}L$ (compare with formula (3.48)) is the angle of polarization azimuth rotation due to conventional optical activity in a single pass through a gyrotropic crystal of length L. Here we have used the procedure for the treatment of symmetric and antisymmetric nonlocality tensor $\Gamma^{(1)}_{ijn}$ components outlined in Refs. [8] to [10]. The antisymmetric components should be treated as *i*-tensors, and the symmetric components as time nonreversal *c*-tensors: *i*-tensors do not change their sign with reversal of the propagation direction; *c*-tensors do.

Substituting $\Theta_{\rightarrow}$ into the expression for $\Theta_{\leftarrow}$, as an initial condition, gives a formula for the change $\delta\Theta$ in the polarization azimuth of a linearly polarized light beam which propagates in the crystal and is then reflected back through it:

$$\delta\Theta = \Theta_{\leftarrow} - \Theta_0 = -2\cosh(\xi_{OA})\cos(2\Theta_0 + \Theta_{OA})\,\mathrm{Re}\left\{\frac{v_s}{v_a}\sin(v_a z)\right\} \qquad \text{(A4.8)}$$

where $\xi_{OA} = \mathrm{Im}(v_a L)$ is the ellipticity due to conventional circular dichroism. For $\xi_{OA} \ll 1$,

$$\delta\Theta = -2\frac{\Theta_{TNOA}}{\Theta_{OA}}\cos\left[2\left(\Theta + \frac{\Theta_{OA}}{2}\right)\right]\sin(\Theta_{OA}) \qquad \text{(A4.9)}$$

Here $\Theta_{TNOA} = [\omega^2/(2c^2)]\,\mathrm{Re}\{\Gamma^s\}L$ is a figure of merit for dissipative nonreciprocal rotation, which is can only be seen in the spectral range of noticeable absorption.

Nonreciprocal rotation is a characteristic feature of nonlocal birefringence and has been seen experimentally in a double-pass configuration in a $Bi_{12}SiO_{20}$ crystal [11]. On the first pass through the crystal, polarization plane rotation is due to both conventional optical activity ($\Gamma^{(1,a)}_{ijn} = -\Gamma^{(1,a)}_{jin}$) and nonreciprocal rotation ($\Gamma^{(1,s)}_{ijn} = \Gamma^{(1,s)}_{jin}$) (see Fig. A4.2). When the light beam is reflected exactly

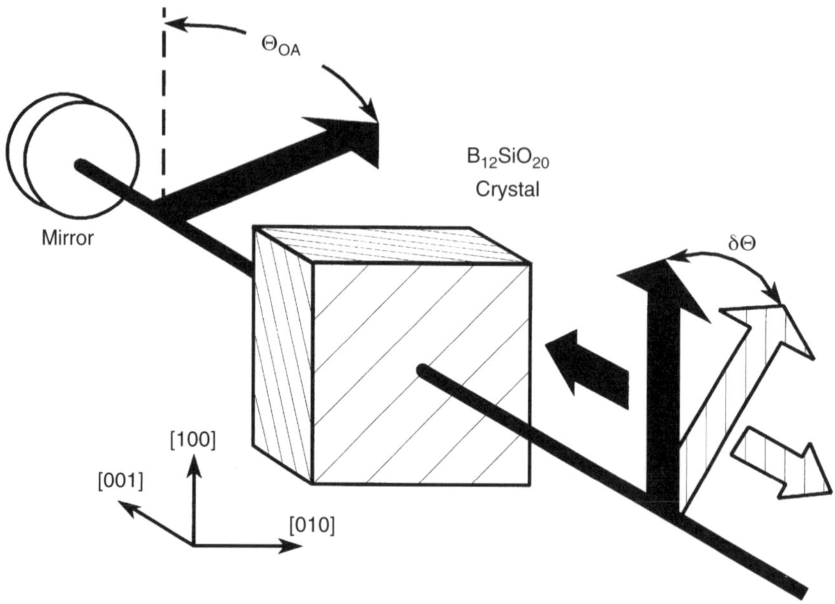

Fig. A4.2 Schematic of the experiment for measurement of nonreciprocal polarization rotation (Reproduced by permission of the Optical Society of America from Ref. [11])

back though the sample, the conventional optical activity is compensated for but the nonreciprocal rotation accumulates. Using argon ion and HeNe lasers, the effect was measured at 514.5, 501.7, 496.5, 476.5, 457.9 and 632.8 nm as a function of the crystal orientation with respect to the incident light polarization azimuth (rotating the crystal around the [001] direction) (see Fig. A4.3).

The characteristic oscillatory orientational dependence of the polarization plane rotation, predicted by formula (A4.9), is the most distinctive feature of the nonreciprocal effect observed. This formula also predicts a $\cos[2(\Theta + \Theta_{OA}/2)]$ dependence for the nonreciprocal rotation; i.e. measurements of this dependence on Θ are expected to have an offset $\eta = \Theta_{OA}/2$ with respect to a $-\cos(2\Theta)$ function (the dotted line in Fig. A4.3). The inset in Fig. A4.3 shows the experimentally measured dependence of η on Θ_{OA}. It correlates very well with prediction $\eta = \Theta_{OA}/2$.

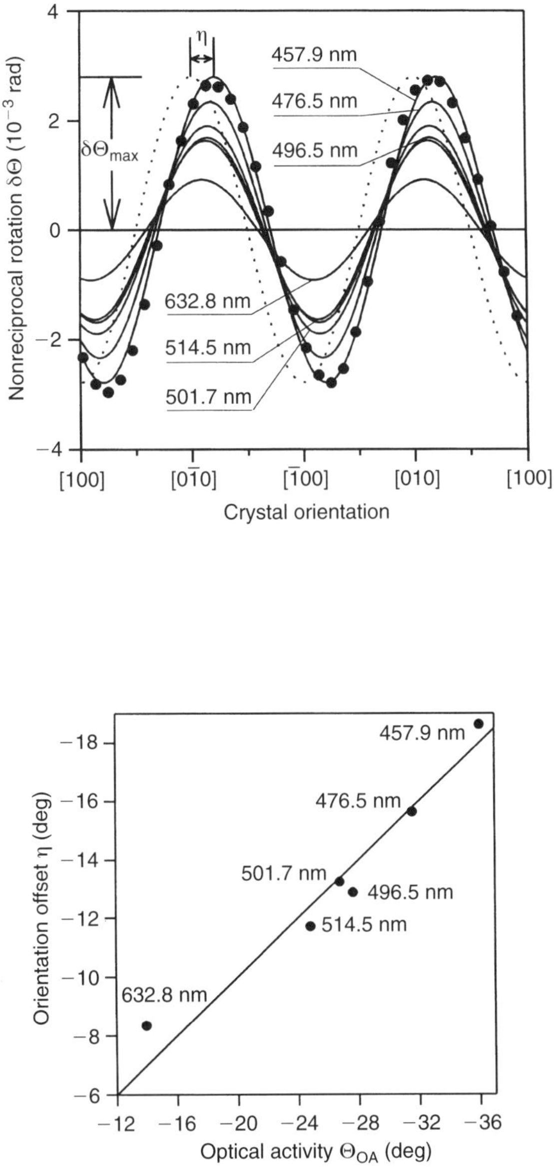

Fig. A4.3 The dependence, in $Bi_{12}SiO_{20}$, of the nonreciprocal component of the polarization azimuth rotation $\delta\Theta$ on the polarization azimuth Θ_0 of the incident light wave with respect to the crystal axes at different wavelengths (upper graph). The dependence of the orientation offset η on the single-pass optical activity Θ_{OA} at different wavelengths in $Bi_{12}SiO_{20}$ (lower graph) (Reproduced by permission of Optical Society of America from Ref. [11])

REFERENCES

1. L.D. Barron, *Molecular Light Scattering and Optical Activity*, Cambridge University Press, London and New York, 1982, pp. 127–8.
2. R.M. Hornreich, *J. Appl. Phys.*, **39**, 432 (1968).
3. R.M. Hornreich and S. Shtrikman. *Phys. Rev.*, **171**, 1065 (1968).
4. L.D. Landau and E.M. Lifshitz, *Electrodynamics of Continuous Media*, Pergamon Press, Oxford, 1984, pp. 362–5. A detailed justification of antisymmetry of the nonlocality tensor associated with a microscopically reversible interaction may be found in A.R. Bungay, Yu. P. Svirko and N.I. Zheludev, *Phys. Rev. B*, **47**, 16141 (1993).
5. Yu. P. Svirko and N.I. Zheludev, *Opt. Lett.*, **20**, 1809 (1995).
6. A.R. Bungay, S.V. Popov, Yu.P. Svirko and N.I. Zheludev, *Chem. Phys. Lett.*, **217**, 249 (1994).
7. A. Malinowski, Yu. P. Svirko and N.I. Zheludev, *J. Opt. Soc. Am. B*, **13**, 1641 (1996).
8. R.R. Birss and R.G. Shrubsall, *Phil. Mag.*, **15**, 687 (1967).
9. G.A. Smolenskii, R.V. Pisarev and I.G. Sini, *Soviet Phys.-Usp.*, **18**, 410 (1976).
10. V.A. Kizel and B.I. Burkov, *Gyrotropy of Crystals*, Nauka, Moscow, 1980, pp. 19–22.
11. P.J. Bennett, S. Dhanjal, Yu. P. Svirko and N.I. Zheludev, *Opt. Lett.*, **21**, 1955 (1996).
12. Z.G. Pan, S.J. Jiang, M. Dagenais, R.A. Morgan, K. Kojima, M.T. Asom, R.E. Leibenguth, G.D. Guth and M.W. Focht, *Appl. Phys. Lett.*, **63**, 2999 (1993).
13. S.J. Jiang, Z.Q. Pan, M. Dagenais, R.A. Morgan and K. Kojima, *Appl. Phys. Lett.*, **63**, 3545, 63 (1993).
14. A.K.J. Vandoorn, M.P. Vanexter and J.P. Woerdman, *Appl. Phys. Lett.*, **8**, 1041 (1996).

Appendix 5 Symmetry Operations and Reflective Optical Activity

In this appendix we consider the symmetry of reflective optical activity, i.e. the symmetrical properties of the change in the polarization of light reflected normally from the boundary between two linear isotropic gyrotropic media.

We start this analysis with the constitutive equation of inhomogeneous medium (6.1):

$$P_i(\mathbf{r}, \omega) = \frac{1}{4\pi}\{[\epsilon_{ij}(\mathbf{r}, \omega) - \delta_{ij}]E_j(\mathbf{r}, \omega) + \nabla_n[\Gamma^{(1)}_{ijn}(\mathbf{r}, \omega)E_j(\mathbf{r}, \omega)]\} \tag{A5.1}$$

Equation (A5.1) was derived assuming only the casualty principle and weak nonlocality of the optical response. As shown in Section 6.1 (see also Refs. [1] and [2]), the boundary conditions may be written as:

$$[\mathbf{z} \times (\mathbf{E}^{\alpha} - \mathbf{E}^{\beta})] = 0 \tag{A5.2}$$

$$[\mathbf{z} \times (\mathbf{B}^{\alpha} - \mathbf{B}^{\beta})]_i = \frac{1}{c}\frac{\partial}{\partial t}(\Gamma^{(1)\beta}_{ijz} E^{\beta}_j - \Gamma^{(1)\alpha}_{ijz} E^{\alpha}_j) \tag{A5.3}$$

(compare with equations (6.11) and (6.12) where we dropped nonlinearity). Here $\mathbf{z}$ is a unit vector normal to the interface and the indices α and β label the gyrotropic media on either sides of the boundary.

Following the procedure described in Ref. [1] or, alternatively, using formulae (6.21) to (6.31) we can derive the complex reflection amplitude coefficients $\mathbf{R}_{\pm}$ for circularly polarized waves. For two isotropic gyrotropic media, when $\epsilon^{\alpha,\beta}_{ij} = (n_{\alpha,\beta})^2\delta_{ij}$ and $\Gamma^{(1)\alpha,\beta}_{ijz} = (2c/\omega)v_{\alpha,\beta}e_{ijz}$, they are

$$R_{\pm} = \frac{(n_{\alpha} - n_{\beta}) \pm 2(v_{\alpha} - v_{\beta})(1 + n_{\beta}/n_{\alpha})^{-1}}{n_{\alpha} + n_{\beta}} \tag{A5.4}$$

The difference between the complex reflection coefficients for the right- and left-handed circular polarizations leads, in particular, to polarization azimuth rotation $\Delta\Theta^{\mathrm{R}}$ for a light wave reflected from the boundary. Another effect related to this difference is the nonzero ellipticity of reflected waves even for strictly linearly polarized incident waves.

Let us now analyse formula (A5.4) from the symmetrical point of view. The effect of optical activity itself, i.e. circular birefringence and dichroism in *transmitted* light is a simultaneously **P** and **T** invariant phenomenon. Indeed, spatial inversion (**P**) reverses the direction of the wave propagation and both the handedness of the medium and the electromagnetic wave (see Fig. A5.1). For example, the refraction and absorption coefficients for right polarized waves in left-hand rotating quartz are exactly equal to those for left polarized waves in right-hand rotating crystals. Thus, optical activity conserves parity. Time inversion (**T**) affects only the direction of propagation, leaving the wave and medium handedness unaffected: it does not matter from which direction a wave is launched into an optically active isotropic medium; the refraction and absorption coefficients will be the same and the polarization azimuth rotation will have the

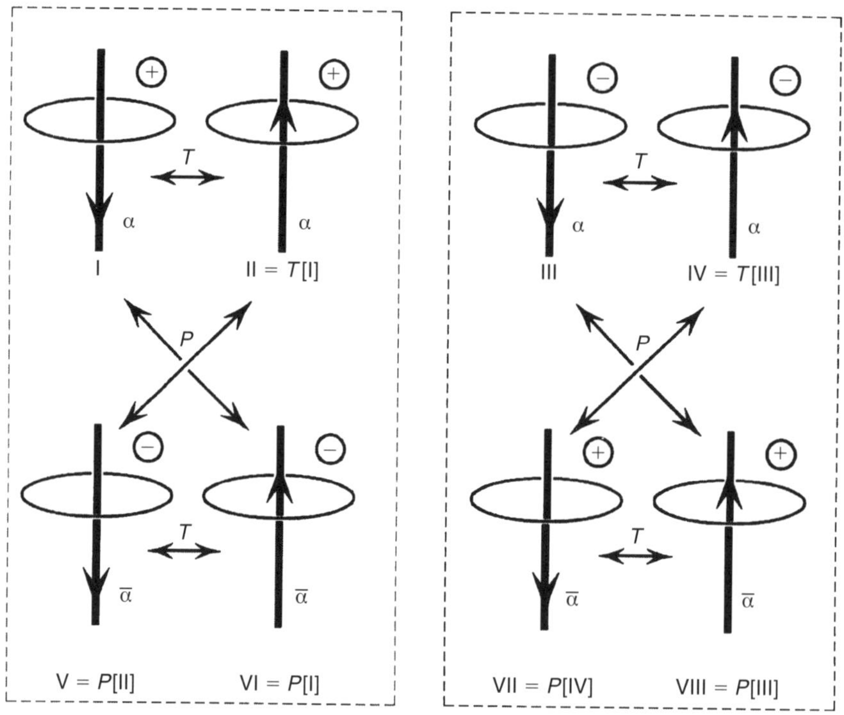

Fig. A5.1 Bulk optical activity (OA) in transmission obeys time-reversibility (**T**) and parity (**P**) simultaneously. Time reversal (**T**) converts cases I into II, and III into IV. The parity operator (**P**) reverses the handedness of media ($P(\alpha) = \overline{\alpha}$) (a left rotating medium becomes right rotating and vice versa). It converts cases I into VI, and II into V, III into VIII, and IV into VII. There are only two independent groups of propagation parameters such as the refractive indices and absorption coefficients. They are the same for cases I, II, V and VI and for cases III, IV, VII and VIII (Reproduced by permission of the Royal Society of Chemistry from Ref. [2])

same sense. This is known as reciprocity of optical activity. Appendix 4 gives a more detailed wave picture of this reciprocity.

The existence of a boundary between two media inevitably creates a different situation: the boundary itself implies the existence of a polar vector which is directed from one medium to the other (see Fig. A5.2). Consequently, we do not expect the reflectivity to conserve parity. Indeed, under **P**, i.e. space inversion, one must make the following substitutions: $n_\alpha \rightleftarrows n_\beta$, $v_\alpha \rightleftarrows -v_\beta$ and $\pm \rightleftarrows \mp$ (see Ref. [2]). In this case formula (A5.4) becomes

$$R_{\pm}(P \Rightarrow -P) = -\frac{(n_\alpha - n_\beta) \pm 2(v_\alpha - v_\beta)(1 + n_\alpha/n_\beta)^{-1}}{n_\alpha + n_\beta} \tag{A5.5}$$

which does not coincide with equation (A5.4) even by modulus. The difference

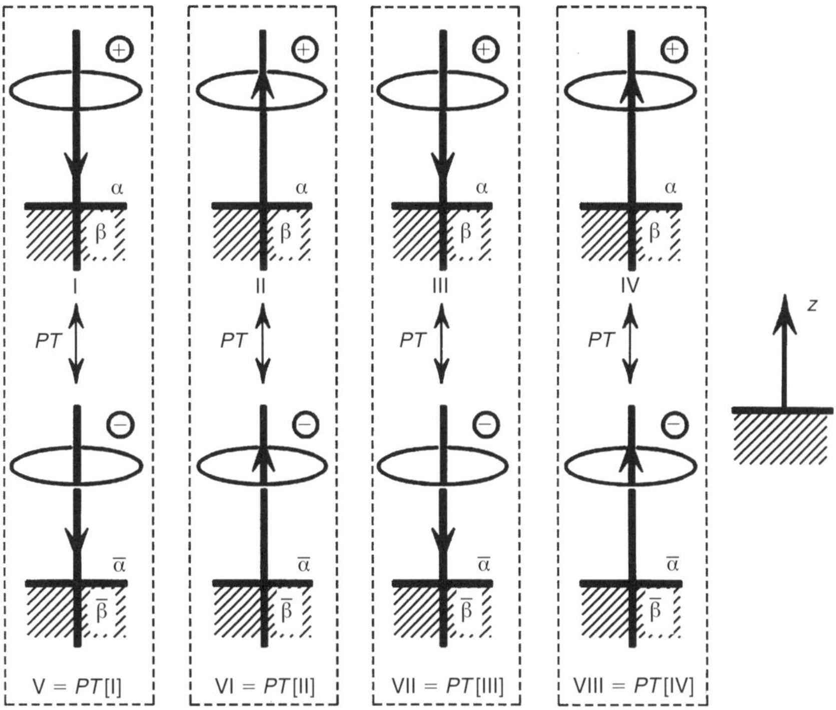

Fig. A5.2 In reflective optical activity the presence of the interface between media α and β critically affects the symmetry of the situation. Reversibility and parity, which are appropriate for bulk optical activity, fail because of the importance of the relative directions of the wave and the boundary normal vectors. All four reflection coefficients in cases I, II, III and IV are different. However, reflective optical activity obeys **PT**-invariance. Since the **PT** operator converts I into V, II into VI, III into VII and IV into VIII, the corresponding pairs of reflective coefficients are equal (Reproduced by permission of the Royal Society of Chemistry from Ref. [2])

between moduluses of reflection coefficients in formulae (A5.5) and (A5.4) results from nonlocality. It disappears for media with local optical response where $v_\alpha = v_\beta = 0$. For example, the reflection coefficient for a right-hand polarized wave approaching the surface of left-hand rotating quartz from a vacuum is not the same as that for a left-hand polarized wave approaching the boundary of a right-hand rotating modification of the same crystal from inside the crystal (compare cases I and V on Fig. A5.2).

More exciting and less evident is the fact that the reflective process is **T**-nonreversible. Indeed, because time inversion ($\mathbf{T} \Rightarrow -\mathbf{T}$) affects only the direction of propagation, leaving the wave and medium handedness unaffected, to obtain the reflection coefficient in the time-reversed scenario one must undertake the following substitutions in equation (A5.4): $n_\alpha \rightleftarrows n_\beta$; $v_\alpha \rightleftarrows v_\beta$:

$$R_\pm(T \Rightarrow -T) = -\frac{(n_\alpha - n_\beta) \pm 2(v_\alpha - v_\beta)(1 + n_\alpha/n_\beta)^{-1}}{n_\alpha + n_\beta} \qquad \text{(A5.6)}$$

This is similar to equation (A5.5) but evidently does not coincide with equation (A5.4) even by modulus. Thus, reflection from an optically active medium does not conserve time reversibility. For example, the reflection coefficient for a right-hand polarized wave approaching the surface of left-hand rotating quartz from a vacuum is not the same as that for the same wave approaching the boundary of the same crystal from the inside (compare cases I and II on Fig. A5.2). Importantly, the difference between the complex amplitude reflection coefficients for counter-propagating waves does not necessarily imply that the boundary of an optically active material has unidirectional energy-valve properties. These depend on the definition of energy flow density and the photometric procedures and definitions inside optically active media. For the time being such definitions is a disputable and unresolved problem in optical electrodynamics.

Nevertheless, as is the case for most optical phenomena, reflective optical activity is a **PT**-invariant effect [2]. Indeed, simultaneous application of **P** and **T** inversions converts the handedness of the light wave ($\pm \rightleftarrows \mp$) and the media on both sides of the boundary ($v_{\alpha,\beta} \rightleftarrows -v_{\alpha,\beta}$) whilst keeping the direction of propagation the same. Formula (A5.4) for the reflective coefficients is not affected:

$$R_\pm(P \Rightarrow -P;\ T \Rightarrow -T) \equiv R_\pm \qquad \text{(A5.7)}$$

Thus, reflective optical activity provides a very clear example of a situation where a lack of space inversion symmetry leads to a time-nonreversible optical effect while more solid **PT** symmetry holds. Similar examples of time nonreversible carrier transport in noncentrosymmetric crystals are given in Ref. [3]. We feel it is appropriate to quote at this stage L.D. Barron [4] who commented on the fundamental importance of the nonreversibility of reflective optical activity:

> I am fascinated by the non-reversality of reflective optical activity as discussed in this paper (paper [2] in the reference list where the analysis outlined above was first published). It appears to be a manifestation in a new context of something I first

pointed out a few year ago in the context of absolute asymmetric synthesis [5]; namely that in the presence of an influence which breaks P and T separately but which is PT invariant overall, a breakdown of microscopic reversibility can occur if the interacting particles are chiral. Symmetry is recovered in the form of a new principle of enantiomeric microscopic reversibility based on the equality of the scattering amplitude for the forward process with that for the time-reversed process involving the P-enantiomeric particles. The fundamental physics of the non-reversibility identified here in reflective optical activity appears to be identical although the context is more complicated. It is also intriguing how the symmetry, kinetic and thermodynamic aspects of the analysis parallel those associated with the breakdown of microscopic reversibility routinely observed in the decays of the neutral K-meson (e.g. Ref. [6] and references therein), the only difference being that the enantiomorphism in the elementary particle case is with respect to CP, rather than P in the molecular/crystal optics case.

REFERENCES

1. A.R. Bungay, Yu.P. Svirko and N.I. Zheludev, *Phys. Rev. B*, **47**, 16141 (1993).
2. Yu.P. Svirko and N.I. Zheludev, *Faraday Discussions*, **99**, 359 (1994).
3. V.I. Belincher and B.I. Sturman, *Soviet Phys. Usp.*, **23**, 1299 (1980).
4. L.D. Barron, Speech at 99th Faraday Discussion on vibrational optical activity, *Faraday Discussions*, **99**, 389 (1994).
5. L.D. Barron, *Chem. Phys. Lett.*, **135**, 1 (1987).
6. L.D. Barron, *Science*, **266**, 1491 (1994).

Index

Index compiled by G. Jones